Lecture Notes in Bioinformatics

T0237923

Subseries of Lecture Notes in Computer Science

Francesco Masulli   Roberto Tagliaferri
Gennady M. Verkhivker (Eds.)

Computational Intelligence Methods for Bioinformatics and Biostatistics

5th International Meeting, CIBB 2008
Vietri sul Mare, Italy, October 3-4, 2008
Revised Selected Papers

 Springer

Series Editors

Sorin Istrail, Brown University, Providence, RI, USA
Pavel Pevzner, University of California, San Diego, CA, USA
Michael Waterman, University of Southern California, Los Angeles, CA, USA

Volume Editors

Francesco Masulli
Università di Genova
Dipartimento di Informatica e Scienze dell'Informazione (DISI)
Via Dodecaneso 35, 16146 Genova, Italy
E-mail: masulli@disi.unige.it

Roberto Tagliaferri
Università di Salerno
Dipartimento di Matematica ed Informatica (DMI)
Via Ponte don Melillo, 84084 Fisciano (SA), Italy
E-mail: rtagliaferri@unisa.it

Gennady M. Verkhivker
The University of Kansas
School of Pharmacy, Department of Pharmaceutical Chemistry
2095 Constant Avenue, Lawrence, KS 66047, USA
E-mail: verk@ku.edu

Library of Congress Control Number: 2009929551

CR Subject Classification (1998): H.2.8, F.2.1, I.2, G.2.2, J.3, E.1

LNCS Sublibrary: SL 8 – Bioinformatics

ISSN   0302-9743
ISBN-10   3-642-02503-X Springer Berlin Heidelberg New York
ISBN-13   978-3-642-02503-7 Springer Berlin Heidelberg New York

springer.com

© Springer-Verlag Berlin Heidelberg 2009
Printed in Germany

Typesetting: Camera-ready by author, data conversion by Scientific Publishing Services, Chennai, India
Printed on acid-free paper   SPIN: 12697106   06/3180   5 4 3 2 1 0

Preface

This volume contains a selection of the best contributions delivered at the 5th International Meeting on Computational Intelligence Methods for Bioinformatics and Biostatistics (CIBB 2008) held at IIASS "E. R. Caianiello", Vietri sul Mare, Salerno, Italy during October 3–4, 2008.

The CIBB meeting series is organized by the Special Interest Group on Bioinformatics of the International Neural Network Society (INNS) to provide a forum open to researchers from different disciplines to present and discuss problems concerning computational techniques in bioinformatics, systems biology and medical informatics with a particular focus on neural networks, machine learning, fuzzy logic, and evolutionary computational methods. Previous CIBB meetings were held with an increasing number of participants within the format of a special session of larger conferences, namely, WIRN 2004 in Perugia, WILF 2005 in Crema, FLINS 2006 in Genoa and WILF 2007 in Camogli. Given the great success of the special session at WILF 2007 that included 26 papers after a strong selection, the 2008 edition of CIBB was organized, at last, as an autonomous conference, governed by its own Steering Committee.

CIBB 2008 attracted 69 paper submissions from all over the world. A rigorous peer-review selection process was applied to ultimately select the papers included in the program of the conference. This volume collects the best contributions presented at the conference. Moreover, the volume also includes three presentations from keynote speakers.

The success of this conference is to be credited to the contribution of many people. In the first place, we would like to thank the organizers of the special sessions for their strong effort in attracting so many good papers that extend and deepen the main topics of CIBB. Moreover, special thanks are due to the Program Committee members and reviewers for their commitment to the task of providing high-quality reviews. Last, but not least, we would like to thank the keynote speakers Mario Lauria (Systems Biology Lab, TIGEM, Telethon Institute of Genetics and Medicine, Naples, Italy), Nicolas Le Novere (Computational Neurobiology, EMBL-EBI, Wellcome-Trust Genome Campus, Hinxton, UK), and Giorgio Valentini (DSI, Dipartimento di Scienze dell'Informazione, Università degli Studi di Milano, Milan, Italy).

While we are concluding this editorial effort, a special thought goes to Tonina Starita who pioneered the applications of computational intelligence to the biomedical field and left us on August 2008. Our tender thanks go to her.

October 2008

Francesco Masulli
Roberto Tagliaferri
Gennady M. Verkhivker

Preface

Organization

The 5th CIBB meeting was a joint operation of the Special Interest Group on Bioinformatics of INNS with the International Society for Computational Biology (ISCB), and the collaboration of the Gruppo Nazionale Calcolo Scientifico (GNCS), the Italian Neural Networks Society (SIREN), the Department of Computer and Information Sciences of the University of Genova, Italy (DISI), and the Department of Mathematics and Computer Science of the University Salerno, Italy (DMI).

Conference Chairs

Francesco Masulli	University of Genoa, Italy & Temple University, USA
Roberto Tagliaferri	University of Salerno, Italy
Gennady M. Verkhivker	University of Kansas, USA & UCSD, USA

Program Committee

Klaus-Peter Adlassnig	Medical University of Vienna, Austria
Joaquin Dopazo	C.I. Príncipe Felipe, Valencia, Spain
Antonio Giordano	University of Siena, Italy & Temple University, USA
Emmanuel Ifeachor	University of Plymouth, UK
Samuel Kaski	Helsinki University of Technology, Finland
Natalio Krasnogor	University of Nottingham, UK
Giancarlo Mauri	University of Milano Bicocca, Italy
Oleg Okun	University of Oulu, Finland
Giulio Pavesi	University of Milan, Italy
David Alejandro Pelta	University of Granada, Spain
Graziano Pesole	University of Bari, Italy
Leif E. Peterson	Baylor College of Medicine, Houston, USA
Giancarlo Raiconi	University of Salerno, Italy
Volker Roth	ETH Zurich, Switzerland
Giuseppe Russo	Temple University, Philadelphia, USA
Giorgio Valentini	University of Milan, Italy
Giorgio Valle	University of Padua, Italy
Jean-Philippe Vert	Ecole des Mines de Paris, France

CIBB Steering Committee

Pierre Baldi	University of California, Irvine, USA
Alexandru Floares	Oncological Institute Cluj-Napoca, Romania

Jon Garibaldi	University of Nottingham, UK
Francesco Masulli	University of Genova, Italy & Temple University, USA
Roberto Tagliaferri	University of Salerno, Italy

Special Session Organizers

J. DeLeo, A. Floares	*Intelligent Systems for Medical Decisions Support*
L. Milanesi, R. Rizzo, A. Urso	*Computational Intelligence for Biological Data Visualization*
R. Battiti, M. Brunato, A. Passerini	*Machine Learning and Intelligent Optimization in Bioinformatics*

Referees

(in addition to previous committees)

A. Avogadri	V. Fabry-Asztalos	F. Pappalardo
J. Bacardit	B. Fischer	A. Passerini
M. Biba	H. Fröhlich	G. Pavesi
F. Camastra	S. Kaski	D. Pelta
R. Casadio	D. Liberati	C. Pizzuti
M. Cannataro	A. Maratea	V. Plagianakos
M. Ceccarelli	P.L. Martelli	S. Rovetta
J. Chen	A. Micheli	D. Saccà
L. Coelho	S. Motta	F.M. Schleif
M. Cristani	O. Okun	A. Zell
D. di Bernardo	L. Palopoli	

Local Scientific Committee

Ida Bifulco	NEuRoNe Lab, DMI, University of Salerno, Italy
Loredana Murino	NEuRoNe Lab, DMI, University of Salerno, Italy
Francesco Napolitano	NEuRoNe Lab, DMI, University of Salerno, Italy

Financing Institutions

GNCS, Gruppo Nazionale Calcolo Scientifico, Italy
DMI, University of Salerno, Italy

Table of Contents

Special Session: ISMDS - Intelligent Systems for Medical Decisions Support

Special Session: Computational Intelligence for Biological Data Visualization

Coarse-Grained Modeling of the HIV–1 Protease Binding Mechanisms: I. Targeting Structural Flexibility of the Protease Flaps and Implications for Drug Design

Gennady M. Verkhivker[1,2]

[1] Department of Pharmaceutical Chemistry, School of Pharmacy
and Center for Bioinformatics, The University of Kansas
2030 Becker Drive, Lawrence, KS 66047 USA
verk@ku.edu
[2] Department of Pharmacology, University of California San Diego,
9500 Gilman Drive, La Jolla CA 92093-0636 USA
gverkhiv@ucsd.edu

Abstract. We propose a coarse–grained model to study binding mechanism of the HIV–1 protease inhibitors using long equilibrium simulations with an ensemble of the HIV–1 protease crystal structures. A microscopic analysis suggests a binding mechanism, in which the HIV–1 protease drugs may exploit the dynamic equilibrium between thermodynamically stable, high affinity complexes with the closed form of the HIV–1 protease and meta–stable intermediate complexes with the alternative structural forms of the protease. We have found that formation of the hydrophobic interaction clusters with the conserved flap residues may stabilize semi–open and open forms of the enzyme and lead to weakly bound, transient inhibitor complexes. The results suggest that inhibitors may function through a multi-mechanistic effect of stabilizing structurally different conformational states of the protease, highlighting the molecular basis of the flap residues in developing drug resistance.

Keywords: Protein flexibility, Monte Carlo simulations, protease flaps, binding mechanism, drug design.

1 Introduction

Human immunodeficiency virus type 1 protease (HIV–1 PR), a member of the aspartic protease family of enzymes, is essential for life cycle of the HIV–1 virus, which causes acquired immunodeficiency syndrome (AIDS) Current therapeutic agents that can inhibit HIV–1 PR include saquinavir, ritonavir, indinavir, nelfinavir, amprenavir, lopnavir, atazanavir, and tipranavir, with several others under clinical investigation [1]. The clinical effectiveness of the HIV–1 PR drugs is often hindered by the emergence of resistant variants, resulting from point mutations in various regions of HIV–1 PR [2]. The mutations of L10, M46, I54,

F. Masulli, R. Tagliaferri, and G.M. Verkhivker (Eds.): CIBB 2008, LNBI 5488, pp. 1–12, 2009.

V82, I84, and L90 residues can confer resistance to almost all protease drugs, and are located in different regions of protease, including the active site (V82 and I84), the flap region (M46 and I54), and the dimerization interface (L10 and L90) [3]. The V82F/I84V double mutation is the key residue mutation of the HIV–1 PR drug resistance because it significantly lowers binding affinity of all protease inhibitors [4]. The thermodynamic characterization of the HIV–1 PR inhibitors has shown that the binding affinity of the first generation HIV–1 PR drugs is entropically driven and, except for ritonavir, is characterized by an unfavorable binding enthalpy [5]. A second-generation of HIV–1 PR inhibitors have significantly higher binding affinities and their binding affinity is determined by the favorable enthalpy and entropy changes [6]. Distinct conformational states of HIV–1 PR, which have been discovered from the crystal structures of HIV–1 PR have enhanced the current understanding of the HIV–1 PR dynamics and function [7]. Crystal structures of the HIV–1 PR complexes have shown that the protease flaps can control access to the active site by assuming a spectrum of distinct conformational states, ranging from closed, when the active site is occupied by a ligand, to semi–open, observed in the free HIV–1 PR. Theoretical studies have enabled a molecular level analysis of the HIV–1 PR flexibility [8]. Multiscale simulations using unconstrained all–atom molecular dynamics (MD) simulations of HIV–1 PR have demonstrated a possibility for a reversible flap opening and evidence for multiple transitions between the closed, semi–open and open HIV–1 PR conformations [9,10]. A curled conformational state of the HIV–1 PR flap tips has been discovered in long MD simulations, suggesting an asymmetrical flap closing mechanism which may be triggered by the assembly of a hydrophobic intermediate cluster formed by the Ile50 residues of both monomers [11,12]. A similar mechanism of the HIV–1 PR flaps opening induced by substrate binding was also observed in computer simulations [13,14]. The recent classification and analysis of the apo HIV–1 PR crystal structures has shown that the unliganded enzyme may favor semi–open, curled and open flap forms [15]. The mechanism of flap opening could be altered in the common V82F/I84V mutant [16] and the amprenavir resistant E35D mutant [17], resulting in an increased flexibility of the flaps, thereby affecting the conformational equilibrium between the closed and semi–open conformations of the protease. An alternative mechanism has arisen from simulations of amprenavir and amprenavir–based second generation inhibitors TMC126 and TMC114 [18]. In these long MD simulations, the wild-type and the V82F/I84V protease mutant have exhibited similar dynamic characteristics of the flap opening. This study suggested that the effect of the mutations may not dramatically affect the equilibrium between the semi–open and closed conformations and that the inability of the inhibitors accommodate to the distorted binding site of the V82F/I84V may be primarily responsible for drug resistance. MD simulations of the drug-resistant M46I/G51D double mutant [19] have demonstrated that the mutated residues may be involved in flap curling and thereby contribute to drug resistance. The M46I/G51D double mutation can delay flap opening and stabilize an ensemble of semi–open flap conformations, thus enhancing steric hindrance during the inhibitor entrance to the active site. To accurately describe the thermodynamics and kinetics of the HIV–1 PR mechanism, statistically significant conformational transitions between structurally different forms of the protease require a considerably longer simulation time scale, which is unrealistic to expect from all-atom MD. The simplified coarse–grained energy model of the HIV–1 PR dynamics has allowed to perform long time scale simulations and evaluate the equilibrium thermodynamics of the flap gating mechanism [20,21,22]. The existing body of structural and thermodynamic data have suggested that accumulation of the non-active site mutants may affect drug binding to HIV–1 PR by altering a dynamic equilibrium between conformational states of HIV–1 PR with the closed, semi–open and open flap conformations, ultimately causing deleterious changes in the stability of the HIV–1 PR binding site and decreased affinity with the inhibitors. In the present study, we suggest that the ensemble of the HIV–1 PR crystal structures provides a coarse–grained, yet an adequate model of major conformational states of HIV–1 PR, which reflect a multitude of functionally relevant flap motions. Parallel simulated tempering dynamics with an ensemble of the HIV–1 PR crystal structures facilitates conformational transitions between structurally different conformational states and provides an opportunistic solution to achieve an adequate characterization of the conformational sampling and binding energetics. The results of our study suggest that inhibitors may function through a multi–mechanistic effect of stabilizing structurally different conformational states of the protease by targeting structural flexibility of the flap residues.

2 Materials and Methods

We have employed a conformational ensemble of the apo HIV–1 PR structures [15,23] which includes semi–open conformations (pdb entries 1hhp, 3hvp); curled flaps conformations (pdb entries 3phv wild-type, 2g69 P53L mutant, 2hb4 2.15 Å wild type, 2hb2 2.3 Å 6–fold mutant); and the open flaps conformations (pdb entries 2pc0 1.4 Å resolution wild-type, 1rpi 9-fold mutant, 1tw7 10–fold mutant) ((Figure 1). We have used HIV–1 PR crystal structures complexes with ritonavir (pdb entries 2B60, 1RL8), indinavir (pdb entries 1HSG, 2B7Z), saquinavir (pdb entry 1HXB) and nelfinavir(pdb entry 1OHR). Theses HIV–1 PR conformations in complexes with the clinically approved HIV–1 inhibitors (Figure 1D) represent a closed form of the enzyme. We assume that the ensemble of the HIV–1 PR crystal structures can thereby represent a coarse–grained model of the flap-opening mechanism. The protonation state of active Asp25 and Asp25' may assume three possible states (diprotonated, monoprotonated, and deprotonated) depending on the inhibitor bound. In this study, the protonation state of Asp25/Asp25' was adjusted to monoprotonated with only one of the two active site aspartates is protonated while the other remains in the carboxylated form. We have equilibrated these HIV–1 PR structures at 300K using MD simulations performed with the GROMACS package and OPLS/AMBER force field [24,25]. The following steps were carried out: 1) a steepest descent energy minimization; 2) equilibration of water for 300 ps at 300K keeping the heavy atoms of the protein constrained; 3) a 300 ps dynamics at 300K at constant volume to thermalize the system.

The molecular recognition energetic model and simulated tempering sampling protocol used in equilibrium simulations have been documented in details in our previous studies (see for example, [26,27]). We have employed the AMBER force field [28] combined with an implicit solvation model [29]. The dispersion–repulsion and electrostatic terms have been modified to include a soft core component that was originally developed in free energy simulations to remove the singularity in the potentials and improve numerical stability of the simulations [30]. Equilibrium simulations with an ensemble of the HIV–1 conformational states are carried out using parallel simulated tempering dynamics [31] with 300 replicas of the ligand-protein system attributed respectively to 300 different temperature levels that are uniformly distributed in the range between 3300K and 300K. Monte Carlo moves are performed simultaneously and independently for each replica at the corresponding temperature level. This process of swapping configurations is repeated 100 times after each simulation cycle for all replicas. The inhibitor conformations and orientations are sampled in a parallelepiped that encompasses the superimposed HIV–1 PR structures with a large 20.0 Å cushion added to every side of the box surrounding the binding interface. The protein structure of each complex is held fixed in its minimized and equilibrated conformation, while rigid body degrees of freedom and the inhibitor rotatable angles are treated as independent variables. We have equilibrated the system for 3,000 cycles and collected data during 3,000 cycles with each simulation cycle consisting of 1,000,000 Monte Carlo steps at each temperature level.

Binding free energies are calculated using the molecular mechanics (MM) AMBER force field [28] and the solvation energy term based on continuum generalized Born and solvent accessible surface area (GB/SA) solvation model [32]. The binding free energy of the ligand–protein complex is computed as :

$$G_{bind} = G_{complex} - G_{protein} - G_{ligand} \qquad (1)$$

$$G_{molecule} = G_{solvation} + E_{MM} - TS_{solute} \qquad (2)$$

where $G_{molecule}$ is the solvation free energy, E_{MM} is the molecular mechanical energy of the molecule summing up the electrostatic E_{es} interactions, van der Waals contributions E_{vdw}, and the internal strain energy E_{int}. Because of significant variances in computing solute entropy using the MM/GBSA approach, this term was not included in the total binding free energy value. Solvation free energy $G_{solvation}$ is divided into nonpolar G_{nonpol} and polar G_{pol} components. Binding free energy evaluations are carried out using MD trajectories of the HIV–1 PR complexes. The structures for the uncomplexed protease and the inhibitor are generated by separating the protein and inhibitor coordinates, followed by an additional minimization of the unbound protein and inhibitor.

3   Results and Discussion

In our model, conformational landscape of HIV–1 PR is described by crystal structures with the closed, semi–open and open forms of the flaps, which are assumed to exist in the dynamic equilibrium, important in regulating molecular recognition of the enzyme with the inhibitors. By considering only a finite number of representative, structurally different conformational states of HIV–1 PR, this coarse-grained approximation allows to explore both longer time simulation scales and atomic resolution details of the inhibitor–protease interactions. The molecular basis of the HIV–1 PR binding mechanisms with ritonavir, indinavir, saquinavir, and nelfinavir (Figure 1D) has been investigated using Monte Carlo simulations with the ensemble of the HIV–1 PR crystal structures. Time–dependent history of the important microscopic parameters describing dynamics of the inhibitor–protease binding is monitored at a range of simulated temperatures. We report the root mean square deviation (RMSD) values of the sampled inhibitor conformations from the respective crystal structures (Figure 2) and the evolution history of the conformational transitions between different structural forms of HIV–1 PR at the room temperature (Figure 3). This analysis provides atomic level details of the inhibitor mobility and favorable binding modes of the inhibitors associated with the respective conformational states of HIV–1 PR. In agreement with the experiment, the thermodynamically most stable inhibitor conformations are formed in the complexes with the closed HIV–1 PR structures ((Figure 3). The predicted dominant inhibitor states conform with RMSD = 1.0 Å to the crystallographic binding mode of the inhibitors. The detected stability of the crystallographic conformations is determined by a network of interactions formed by the HIV–1 PR inhibitors in the active site through hydrogen bonds with the carboxyl groups of the essential Asp-25/Asp-25' catalytic residues and Asp-29/Asp-29' residues, accompanied by additional hydrogen bonding to the backbone amide nitrogen of Ile-50 and Ile-50 via intervening water molecules.

In accordance with the experiment, the crystal structures of the HIV–1 PR inhibitors dominate the equilibrium population of by forming high affinity complexes with the respective co-crystal HIV–1 PR conformations (Figure 4A,B). Frequent and reversible excursions have been detected between long-lived complexes of the HIV–1 PR inhibitors with the closed form of the enzyme and meta-stable intermediates formed by the inhibitors with the semi–open and open HIV–1 structures. We argue that detected in simulations equilibrium conformational transitions between alternative structural forms of HIV–1 PR may be relevant for the inhibitor–protein thermodynamics. We have found that long equilibrium simulations can produce frequent and reversible excursions between crystal structure complexes of the HIV–1 PR inhibitors and meta-stable intermediates formed with the semi–open and open enzyme forms (Figure 4B). The emergence of meta–stable and weakly bound intermediate complexes with the semi–open and open forms of HIV–1 PR contributes appreciably to the equilibrium population of the inhibitor complexes. The total binding free energies of the HIV–1 PR inhibitors are computed using the obtained equilibrium distribution of the inhibitor–protease states. This approach provides an effective estimate of the binding affinities for a panel of studied HIV–1 PR drugs, which is in a good agreement with the experimental activities (Figure 4C). Hence, a microscopic

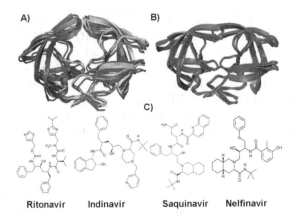

Fig. 1. (A) Superposition of the apo HIV–1 PR structures including semi–open conformations in red (pdb entries 1hhp, 3hvp); curled flaps conformations in yellow (pdb entries 3phv, 2g69, 2hb4, 2hb2 2); and the open conformations in blue (pdb entries 2pc0, 1rpi, 1tw7). (B) Superposition of the HIV–1 PR crystal structures complexes in the closed form with ritonavir, indinavir, saquinavir, and nelfinavir. (C) Chemical structures of Ritonavir, Indinavir, Saquinavir, and Nelfinavir used in this study.

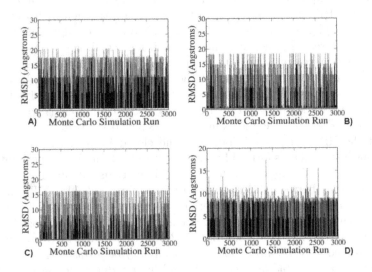

Fig. 2. Evolution of the RMSD values between simulated at T=300K conformations of ritonavir (A), indinavir (B), saquinavir (C) and nelfinavir (D) and the respective crystal structures of the inhibitor complexes

model of the inhibitor binding using a combination of a coarse-grained simulation model and a subsequent binding free energy refinement can reproduce the thermodynamic stability and binding affinity of the HIV–1 PR drugs.

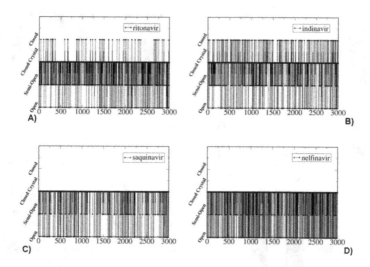

Fig. 3. Evolution of the conformational transitions between closed, semi–open and open forms of the HIV–1 PR during equilibrium simulations at T=300K with ritonavir (A), indinavir (B), saquinavir (C) and nelfinavir (D)

3.1   Targeting Structural Flexibility of the HIV–1 PR Flaps

We have observed that asymmetric targeting of the HIV–1 PR flap residues may lead to structurally similar intermediate complexes formed with the semi–open and open HIV–1 PR forms for all studied inhibitors (Figure 4D). In these complexes, hydrogen bonds are asymmetrically formed with the Ile50, Gly52 and Phe53 flap residues. Moreover, in the intermediate inhibitor complex with the open flaps a favorable hydrophobic cluster of interactions may be formed by the P1 Phe group of ritonavir stacking against the hydrophobic residues Pro79 and Pro81 on the inner surface of the HIV–1 PR loop, and further consolidated by the interactions with Ile54 and Ile47 of the flaps. The cumulative effect of the stabilizing flap–flap interactions and interactions, which may be formed by the inhibitors with the conserved flap residues, may contribute to the formation of the weakly bound intermediate complexes. The asymmetric targeting of the HIV–1 PR flaps and formation of the hydrophobic interaction clusters with the conserved flap residues may also facilitate curling of the flap tips motions, observed in MD simulations [11,12,13,14], which are arguably necessary for an efficient inhibitor entry into the active site. Consequently, equilibrium conformational transitions seen between these intermediate associations and thermodynamically stable crystal structure complexes may have a functional role, which may become more apparent on longer biologically relevant timescales. These results are also consistent with the recent MD studies, in which dynamics of the HIV–1 PR substrate binding was shown to induce opening and closing of the flaps through assembly of a hydrophobic cluster, consisting of Ile50 and Phe53 residues [11,12]. While the HIV–1 PR drugs were initially designed to stabilize

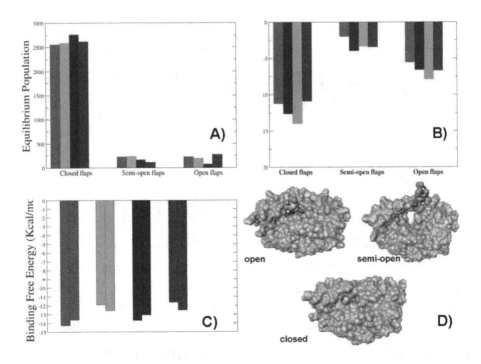

Fig. 4. (A) The equilibrium distribution of the conformational states of the HIV–1 PR obtained from equilibrium simulations with the inhibitors at T=300K. Ritonavir (shown in red), Indinavir (shown in green), Saquinavir (shown in blue), and Nelfinavir (shown in maroon). (B) The computed binding free energies of the HIV–1 PR inhibitors in complexes with the closed HIV–1 PR conformations, semi–open and open forms of the HIV–1 PR. Ritonavir (red), Indinavir (green), Saquinavir (blue), and Nelfinavir (maroon) (C) The total computed (left bar) and experimental (right bar) binding free energies of the HIV–1 PR inhibitors : Ritonavir (red), Indinavir (green), Saquinavir (blue), and Nelfinavir (maroon). (D) Connolly surface representation of the open, semi–open and closed conformational states of the HIV–1 PR (shown in light blue) bound to ritonavir (shown in CPK model and default color).

the closed form, it is possible that high affinity, active site inhibitors may express their function through a multi–mechanistic effect, in which they can stabilize several conformational forms of the protease (open, semi–open and closed forms). The ability of certain non–active site mutants to alter the dynamic equilibrium between conformational states of HIV–1 PR has been shown to result in excessive stabilization of the semi–open conformation and deleterious structural changes of the active site. The results of our study indicate that asymmetric targeting of the conserved flap residues and formation of weakly bound, transient inhibitor complexes with the semi–open and open forms of HIV–1 PR may play a role for capturing incoming inhibitors and assist in the flap opening regulation. We may speculate that equilibrium transitions between weakly bound transient complexes may facilitate flaps transitions from a semi–open form, which is dominant for the unliganded protease, to a more open conformation and thereby promote the inhibitor entry into the active site. At the subsequent stage of the binding reaction closing of the flaps, induced by the favorable accommodation of the inhibitor in the active site, would result in the thermodynamically stable, high affinity complex.

3.2 Diversity of Binding Mechanisms and Implications for Inhibitor Design

An example of the inhibitor binding to multiple binding sites of HIV–1 PR was experimentally observed in recent structural studies by Weber and coworkers [33,34,35,36]. Ultra–high resolution crystal structures of the TMC114 inhibitor (darunavir) [33,34] complexes with the HIV–1 PR variants containing the drug-resistant mutation V32I and the M46L mutant have revealed binding two distinct sites, one in the active-site cavity and the second on the surface of one of the flexible flaps of the protease dimer [35]. The existence of the second binding site has suggested a dual mechanism for the high effectiveness of TMC114 against drug-resistant forms of HIV–1 PR and avenues for a potential design of new inhibitors. The crystal structure of the unliganded HIV–1 PR F53L mutant [36] has suggested a distinct mechanism for drug resistance, attributed to the additional mobility of the flaps and the loss of attractive interactions with Phe53. Similar to our findings, these structural studies have also suggested that forming interactions with the conserved flap residues Ile50 and Phe53 may be important for regulation the dynamics of flap opening events and play a functional role in binding approaching substrates during. While some inhibitors have revealed binding to surface sites in the flap region of HIV–1 PR [37,38], the functional relevance of such complexes for the inhibition of the protease is not entirely clear. Indeed, another HIV–1 PR structure with a peptide inhibitor bound to a similar second site was also reported [39]. In contrast to TMC114, this peptide inhibitor forms only water-mediated contacts with the flap residues, while direct hydrogen bonds are observed with other symmetry-related PR molecules. The diversity of the HIV–1 PR inhibition scenarios include the conventional active site binding and alternative inhibitory mechanisms, based on blocking the assembly of the HIV–1 PR homodimer and disrupting the dimeric interface. Currently clinically approved antiviral drugs against HIV–1 PR are the active site inhibitors, binding to the closed conformation of HIV–1 PR. The results of our study suggest an alternative discovery strategy in which HIV–1 PR inhibitors may be designed to modulate dynamics on the conformational landscape of HIV–1 PR and target flaps flexibility in the semi–open and curled conformations. While the HIV–1 PR drugs were initially designed to stabilize the closed form, it is possible that high affinity, active site inhibitors may express their function through a multi-mechanistic effect, in which they can stabilize several conformational forms of the protease.

References

1. de Clercq, E.: New anti-HIV agents and targets. Med. Res. Rev. 22, 531–565 (2002)
2. Barbaro, G., Scozzafava, A., Mastrolorenzo, A., Supuran, C.T.: Highly active antiretroviral therapy: Current state of the art, new agents and their pharmacological interactions useful for improving therapeutic outcome. Curr. Pharm. Des. 11, 1805–1843 (2005)
3. D'Aquila, R.T., Schapiro, J.M., Brun-Vezinet, F., Clotet, B., Conway, B., Demeter, L.M., Grant, R.M., Johnson, V.A., Kuritzkes, D.R., Loveday, C., Shafer, R.W., Richman, D.D.: Drug Resistance Mutations in HIV–1. Top HIV Med. 10, 21–25 (2002)
4. Ohtaka, H., Schon, A., Freire, E.: Multidrug resistance to HIV–1 protease inhibition requires cooperative coupling between distal mutations. Biochemistry 42, 13659–13666 (2003)
5. Velazquez-Campoy, A., Todd, M.J., Freire, E.: HIV–1 protease inhibitors: enthalpic versus entropic optimization of the binding affinity. Biochemistry 39, 2201–2207 (2000)
6. Ohtaka, H., Freire, E.: Adaptive inhibitors of the HIV–1 protease. Prog. Biophys. Mol. Biol. 88, 193–208 (2005)
7. Wlodawer, A.: Rational approach to AIDS drug design through structural biology. Annu. Rev. Med. 53, 595–614 (2002)
8. Hornak, V., Simmerling, C.: Targeting structural flexibility in HIV–1 protease inhibitor binding. Drug Discov. Today 12, 132–138 (2007)
9. Hornak, V., Okur, A., Rizzo, R.C., Simmerling, C.: HIV–1 protease flaps spontaneously close to the correct structure in simulations following manual placement of an inhibitor into the open state. J. Am. Chem. Soc. 128, 2812–2813 (2006)
10. Hornak, V., Okur, A., Rizzo, R.C., Simmerling, C.: HIV–1 protease flaps spontaneously open and reclose in molecular dynamics simulations. Proc. Natl. Acad. Sci. U. S. A. 103, 915–920 (2006)
11. Scott, W.R., Schiffer, C.A.: Curling of flap tips in HIV–1 protease as a mechanism for substrate entry and tolerance of drug resistance. Structure 8, 1259–1265 (2000)
12. Kurt, N., Scott, W.R., Schiffer, C.A., Haliloglu, T.: Cooperative fluctuations of unliganded and substrate-bound HIV–1 protease: a structure-based analysis on a variety of conformations from crystallography and molecular dynamics simulations. Proteins 51, 409–422 (2003)
13. Toth, G., Borics, A.: Closing of the flaps of HIV–1 protease induced by substrate binding: a model of a flap closing mechanism in retroviral aspartic proteases. Biochemistry 45, 6606–66014 (2006)
14. Toth, G., Borics, A.: Flap opening mechanism of HIV–1 protease. J. Mol. Graph. Model. 24, 465–474 (2006)
15. Heaslet, H., Rosenfel, R., Giffin, M., Lin, Y.C., Tam, K., Torbett, B.E., Elder, J.H., McRee, D.E., Stout, C.D.: Conformational flexibility in the flap domains of ligand-free HIV protease. Acta Crystallogr. D. Biol. Crystallogr. 63, 866–875 (2007)
16. Perryman, A.L., Lin, J.H., McCammon, J.A.: HIV–1 protease molecular dynamics of a wild-type and of the V82F/I84V mutant: possible contributions to drug resistance and a potential new target site for drugs. Protein Sci. 13, 1108–1123 (2004)
17. Meiselbach, H., Horn, A.H., Harrer, T., Sticht, H.: Insights into amprenavir resistance in E35D HIV–1 protease mutation from molecular dynamics and binding free-energy calculations. J. Mol. Model. 13, 297–304 (2007)

18. Hou, T., Yu, R.: Molecular dynamics and free energy studies on the wild-type and double mutant HIV–1 protease complexed with amprenavir and two amprenavir-related inhibitors: mechanism for binding and drug resistance. J. Med. Chem. 50, 1177–1188 (2007)

19. Lauria, A., Ippolito, M., Almerico, A.M.: Molecular dynamics studies on HIV–1 protease: a comparison of the flap motions between wild type protease and the M46I/G51D double mutant. J. Mol. Model. 13, 1151–1156 (2007)

20. Tozzini, V., Trylska, J., Chang, C.E., McCammon, J.A.: Flap opening dynamics in HIV–1 protease explored with a coarse-grained model. J. Struct. Biol. 157, 606–615 (2007)

21. Chang, C.E., Shen, T., Trylska, J., Tozzini, V., McCammon, J.A.: Gated binding of ligands to HIV–1 protease: Brownian dynamics simulations in a coarse-grained model. Biophys. J. 90, 3880–3885 (2006)

22. Chang, C.E., Trylska, J., Tozzini, V., McCammon, J.A.: Binding pathways of ligands to HIV–1 protease: coarse-grained and atomistic simulations. Chem. Biol. Drug. Des. 69, 5–13 (2007)

23. Berman, H.M., Westbrook, J., Feng, Z., Gilliland, G., Bhat, T.N., Weissig, H., Shindyalov, I.N., Bourne, P.E.: The Protein Data Bank. Nucleic Acids R. 28, 235–242 (2000)

24. Spoel, D.V.D., Lindahl, E., Hess, B., Groenhof, G., Mark, A.E., Berendsen, H.J.C.: GROMACS: fast, flexible, and free. J. Comput. Chem. 26, 1701–1718 (2005)

25. Christen, M., Hunenberger, P.H., Bakowies, D., Baron, R., Burgi, R., Geerke, D.P., Heinz, T.N., Kastenholz, M.A., Krautler, V., Oostenbrink, C., Peter, C., Trzesniak, D., van Gunsteren, W.F.: The GROMOS software for biomolecular simulation: GROMOS 2005. J. Comput. Chem. 26, 1719–1751 (2005)

26. Verkhivker, G.M.: Computational proteomics of biomolecular interactions in the sequence and structure space of the tyrosine kinome: deciphering the molecular basis of the kinase inhibitors selectivity. Proteins 66, 912–929 (2007)

27. Verkhivker, G.M.: In silico profiling of tyrosine kinases binding specificity and drug resistance using Monte Carlo simulations with the ensembles of protein kinase crystal structures. Biopolymers 85, 333–348 (2007)

28. Cornell, W.D., Cieplak, P., Bayly, C.L., Gould, I.R., Merz, K.M., Ferguson, D.M., Spellmeyer, D.C., Fox, T., Caldwell, J.W., Kollman, P.A.: A second generation force field for simulation of proteins, nucleic acids, and organic molecules. J. Amer. Chem. Soc. 117, 5179–5197 (1995)

29. Stouten, P.F.W., Frömmel, C., Nakamura, H., Sander, C.: An effective solvation term based on atomic occupancies for use in protein simulations. Mol. Simul. 10, 97–120 (1993)

30. Beutler, T.C., Mark, A.E., van Schaik, R.C., Gerber, P.R., van Gunsteren, W.: Avoiding singularities and numerical instabilities in free energy calculations based on molecular simulations. Chem. Phys. Lett. 222, 529–539 (1994)

31. Sugita, Y., Okamoto, Y.: Replica-exchange molecular dynamics method for protein folding. Chem. Phys. Lett. 314, 141–151 (1999)

32. Kollman, P.A., Massova, I., Reyes, C., Kuhn, B., Huo, S., Chong, L., Lee, M., Lee, T., Duan, Y., Wang, W., Donini, O., Cieplak, P., Srinivasan, J., Case, D.A., Cheatham, T.E.: Calculating structures and free energies of complex molecules: combining molecular mechanics and continuum models. Acc. Chem. Res. 33, 889–897 (2000)

33. Tie, Y., Boross, P.I., Wang, Y.-F., Gaddis, L., Hussain, A.K., Leshchenko, S., Ghosh, A.K., Louis, J.M., Harrison, R.W., Weber, I.T.: High resolution crystal structures of HIV–1 protease with a potent non-peptide inhibitor (UIC-94017) active against multi-drug-resistant clinical strains. J. Mol. Biol. 338, 341–352 (2004)
34. Kovalevsky, A.Y., Tie, Y., Liu, F., Boross, P., Wang, Y.-F., Leshchenko, S., Ghosh, A.K., Harrison, R.W., Weber, I.T.: Effectiveness of nonpeptidic clinical inhibitor TMC114 to highly drug resistant mutations D30N, I50V, L90M of HIV–1 protease. J. Med. Chem. 49, 1379–1387 (2006)
35. Kovalevsky, A.Y., Liu, F., Leshchenko, S., Ghosh, A.K., Louis, J.M., Harrison, R.W., Weber, I.T.: Ultra-high resolution crystal structure of HIV–1 protease mutant reveals two binding sites for clinical inhibitor TMC114. J. Mol. Biol. 363, 161–173 (2006)
36. Liu, F., Kovalevsky, A.Y., Louis, J.M., Boross, P.I., Wang, Y.F., Harrison, R.W., Weber, I.T.: Mechanism of drug resistance revealed by the crystal structure of the unliganded HIV–1 protease with F53L mutation. J. Mol. Biol. 358, 1191–1199 (2006)
37. Sperka, T., Pitlik, J., Bagossi, P., Tozser, J.: Beta-lactam compounds as apparently uncompetitive inhibitors of HIV–1 protease. Bioorg. Med. Chem. Lett. 15, 3086–3090 (2005)
38. Judd, D.A., Nettles, J.H., Nevins, N., Snyder, J.P., Liotta, D.C., Tang, J., Ermolieff, J., Schinazi, R.F., Hill, C.L.: Polyoxometalate HIV–1 protease inhibitors: a new mode of protease inhibition. J. Am. Chem. Soc. 123, 886–897 (2001)
39. Brynda, J., Rezacova, P., Fabry, M., Horejsi, M., Stouracova, R., Soucek, M., Hradilek, M., Konvalinka, J., Sedlacek, J.: Inhibitor binding at the protein interface in crystals of a HIV–1 protease complex. Acta Crystallog. sect. D 60, 1943–1948 (2004)

Coarse-Grained Modeling of the HIV−1 Protease Binding Mechanisms: II. Folding Inhibition

Gennady M. Verkhivker[1,2]

[1] Department of Pharmaceutical Chemistry, School of Pharmacy
and Center for Bioinformatics, The University of Kansas
2030 Becker Drive, Lawrence, KS 66047 USA
verk@ku.edu
[2] Department of Pharmacology, University of California San Diego,
9500 Gilman Drive, La Jolla CA 92093-0636 USA
gverkhiv@ucsd.edu

Abstract. Evolutionary and structurally conserved fragments 24-34 and 83–93 from each of the HIV-1 protease (HIV-1 PR) monomers constitute the critical components of the HIV−1 PR folding nucleus. It has been recently discovered that the peptide with the amino acid sequence NIIGRNLLTQI identical to the corresponding segment 83-93 of the HIV−1 PR monomer, can inhibit folding of HIV-1 PR. We have previously shown that this peptide can form stable complexes with the folded HIV−1 PR monomer by targeting the conserved segment 24–34 of the folding nucleus (folding inhibition) and by interacting with the antiparallel termini β–sheet region (dimerization inhibition). In this follow-up study, we propose a generalized, coarse–grained model of the folding inhibition based simulations with an ensemble of both folded and partially unfolded HIV-1 PR conformational states. Using a dynamic equilibrium between low–energy complexes formed with the folded and partially unfolded HIV-1 PR monomers, the NIIGRNLLTQI peptide may effectively intervene with the HIV−1 PR folding and dimerization. The performed microscopic analysis reconciles the experimental and computational results and rationalizes the molecular basis of folding inhibition.

Keywords: HIV-1 protease, folding inhibitors, protein conformational ensembles, molecular docking, Monte Carlo simulations, drug design.

1 Introduction

Human immunodeficiency virus type 1 protease (HIV−1 PR), which exists as a homodimeric enzyme, plays an important role in processing the viral polypeptide precursors, critical for the life-cycle of HIV, and presents an important target for the design of specific antiviral agents [1]. Structural studies [2,3,4,5,6,7] and computer simulations [8,9,10,11,12,13] have provided important insights in understanding HIV-1 PR flexibility and dynamics associated with the binding of HIV−1 PR substrates and inhibitors. Crystal structures of the HIV-1 PR

F. Masulli, R. Tagliaferri, and G.M. Verkhivker (Eds.): CIBB 2008, LNBI 5488, pp. 13–24, 2009.
© Springer-Verlag Berlin Heidelberg 2009

complexes [2,3,4] have revealed that the protease flaps can exhibit a considerable mobility ranging from a "closed" form, when the active site is occupied by a ligand, to a "semi–open", typically observed in the free HIV–1 PR, and a "wide-open" structure, which may permit the substrate and inhibitor access to the active site [4]. The solution NMR experiments of the free HIV–1 PR have shown that the ensemble of the HIV–1 PR unbound structures can include "open flap" conformations as an outcome of a rare event in the slow conformational exchange [5,6,7]. The formation of the HIV–1 PR homodimer is necessary for enzymatic activity because each subunit contributes one of the two catalytic aspartic acid residues that form the active site. The two major areas that constitute the dimer interface are the active site region 24–29, that encompasses the triplet Asp-25/Thr-26/Gly-27 (DTG) forming the "fireman's grip" hydrogen bond network, and the four-stranded anti-parallel β–sheet, that is formed by interdigitation of the C–terminal and N–terminal residues of HIV–1 PR. These interfaces are assembled from the evolutionary and structurally conserved segments of HIV–1 PR, corresponding to the active site residues 24-34 and a stretch of amino acids 83-93 encompassing a single α–helix (Figure 1A). It has been shown that these conserved regions may constitute the HIV–1 PR stabilization core as evident from a strong protection pattern obtained for the folding HIV–1 PR segments 24–34, 74-78 and 83-93 [14]. The NMR studies of the early folding hierarchy in HIV–1 PR, which monitored the protein response to different degrees of denaturation with guanidine hydrochloride, have suggested that even under the strongest denaturing conditions hydrophobic clusters and native secondary structural elements can be transiently formed [15,16]. While the stabilization core of HIV–1PR includes residues from the active site, hinge region and dimerization domain, which may be contiguous or in close spatial proximity, the formation of the native interface between conserved protease segments 24–34 and 83-93 is critical for assembly of the HIV–1 PR folding nucleus. Intrinsic structural and folding preferences in the HIV–1 PR precursor have important regulatory roles in the autoprocessing reaction and generation of the mature enzyme as HIV–1 PR activation is tightly coupled to folding [17,18,19]. The low dimer stability of the HIV–1 PR precursor relative to the mature protease, which is essential to allow initial recruitment of the polyproteins, may be arguably an apparent effect of the equilibrium between the partially folded and the folded, enzymatically active HIV–1 PR dimer. A systematic understanding of the HIV–1 PR folding and dimerization requires structural analysis of the HIV–1 PR monomer in its precursor and mature forms. A comparison of the solution NMR structures of the HIV–1 PR monomers with the subunit of the uninhibited HIV–1 PR dimer has demonstrated that, with the exception of the terminal regions (residues 1-10 and 91-95) that are disordered, the tertiary folds of the HIV–1 PR monomer and a single subunit of the HIV–1 PR dimer are essentially identical [20,21,22,23]. The diversity of the HIV–1 PR inhibition scenarios may include the conventional active site binding and alternative inhibitory mechanisms, based on blocking the assembly of the HIV-1 PR homodimer and disrupting the dimeric interface [24]. Design of HIV– PR dimerization inhibitors is typically focused on disrupting four-stranded β–sheet and targeting flexible N– and C–termini that constitute most of the dimer interface. It was recently proposed that the peptide LDTGADDTVLE (p–S2) and the peptide NIIGRNLLTQI (p–S8), with the sequences identical to that of the conserved fragments 24-34 (S2) and 83-93 (S8) from the protease folding nucleus, may act as unconventional therapeutic agents intervening with the HIV–1 PR folding and dimerization [25,26,27,28]. The results of biochemical assays have revealed that the peptide p–S8, with a sequence NIIGRNLLTQI identical to that of the HIV–1 PR folding fragment 83-93 (S8) (Figure 1A), exhibits biological activity and can inhibit the HIV–1 PR folding and formation of the active dimer in the micromolar range (Ki = 2.58 μM) [25,26]. Structural analysis of the folding inhibition mechanism has been performed by means of circular dichroism (CD) spectroscopy, which has indicated that the p–S8 peptide can inhibit catalytic activity of the enzyme by binding to the conserved segment 24–34 and intervening with the assembly of the HIV–1 PR folding nucleus [27]. The changes in the CD spectrum of HIV–1 PR upon incubation with the p–S8 peptide, measured under the same conditions as the ones used in the activity assays, have demonstrated a loss of the β–sheet content in the protease from 30% in the native folded conformation to 14%. Consequently, the biochemical and CD spectrum experiments indicates that the inhibition mechanism of the p–S8 peptide may proceed through partial destabilization of the native HIV–1 PR structure. Theoretical models have suggested that the detected inhibitory activity of the folding inhibitor may be controlled by the stable local elementary structure (LES) of the peptide which enables specific binding with the complementary segment 24–34 of the folding HIV–1 PR nucleus, thereby effectively intervening with the HIV–1 PR folding and dimerization [25,26,27]. We have recently used a hierarchical modeling strategy to dissect the molecular and energetic basis of the HIV–1 PR folding inhibition at atomic resolution [28]. This approach included coarse-grained molecular docking of the flexible p–S8 peptide with the ensembles of folded HIV–1 PR dimers and monomers and all-atom MD simulations of the flexible peptide–HIV–1 PR monomer complexes using the predicted folding and dimerization binding modes of the p–S8 peptide. Using this hierarchical simulation approach, we demonstrated that the folding inhibitor may exhibit both folding and dimerization modes of inhibition by targeting conservative elements of the HIV–1 PR folding core and dimerization interface [28]. In the present work, we broaden the scope of the initial theoretical framework and propose a generalized model of folding inhibition using simulations with the ensembles of folded and partially unfolded HIV–1 PR conformations. We assume that the ensemble of the HIV–1 PR folded and partially folded conformational states provides a coarse–grained, yet a reasonable model of major conformational states of HIV–1 PR formed during the folding process. The results of our studies suggest that a mechanism of the folding inhibition may be largely governed by the minimally frustrated inhibitor interactions with the conserved segments of the HIV–1 PR folding nucleus.

2   Materials and Methods

The conformational ensemble of the HIV–1 monomeric conformations is obtained from the crystal structures of the unliganded HIV–1 PR and HIV–1 PR complexes with the inhibitors, followed by subsequent minimization and thermal equilibration procedures. Three partially unfolded conformations of the HIV monomer (Figure 1B,C,D) have been generated starting from the native HIV–1 PR conformation (pdb code 1BVG), by gradually pulling apart 24–34 and 83–93 segments of the enzyme. Starting from the thus generated folded and partially unfolded HIV–1 PR monomers, we have subsequently thermalized these structures at 300K by molecular dynamics simulations performed with the GROMACS package and employing OPLS force field within a cubic box of 6.7 nm around the protein, with 9428 water molecules and 2 chloride ions [30]. More precisely, we have carried out the following steps: 1) a steepest descent energy minimization; 2) equilibration of water for 100 ps at 100K keeping constrained the heavy atoms of the protein; 3) the same at 200K with spring constant 500; 4) the same at 300K with a spring constant 250; 5) a 100 ps dynamics at 300K of the solution, at constant volume; 6) a 100 ps dynamics at 300K and constant pressure, in order to recover a realistic density; 7) a 4 ns dynamics at 300K at constant volume to thermalize the system. The molecular recognition energetic model and simulated tempering sampling protocol used in equilibrium simulations have been documented in details in our previous studies [31,32]. We summarize here the major components of the simulation model. We have employed the AMBER force field [33] combined with an implicit solvation model [34]. Equilibrium simulations with an ensemble of the HIV–1 conformational states are carried out using parallel simulated tempering dynamics [35] with 300 replicas of the ligand-protein system attributed respectively to 300 different temperature levels that are uniformly distributed in the range between 3300K and 300K. The peptide p–S8 conformations and orientations are sampled in a parallelepiped that encompasses the superimposed HIV–1 PR structures with a large 20.0 Å cushion added to every side of the box surrounding the binding interface. Monte Carlo moves are performed simultaneously and independently for each replica at the corresponding temperature level. After each simulation cycle, that is completed for all replicas, exchange of configurations for every pair of adjacent replicas at neighboring temperatures is introduced. The m-th and n–th replicas, described by a common Hamiltonian $H(X_1, ..., X_m, ..., X_n, ..., X_N)$, are associated with the inverse temperatures β_m and β_n, and the corresponding conformations X_m and X_n. The exchange of conformations between adjacent replicas m and n is accepted or rejected according to Metropolis criterion. We have adapted an efficient simulated tempering protocol, where the highest temperature is defined as the 'melting' temperature of the system. In this procedure, by starting at an arbitrary temperature, 100–500 Monte Carlo moves are attempted and the acceptance ratio is determined. If this parameter is less than 80 %, the temperature is doubled. When the acceptance ratio exceeds this threshold, s sufficient number of moves to completely 'melt' the system is applied and this temperature gets chosen as the highest temperature level.

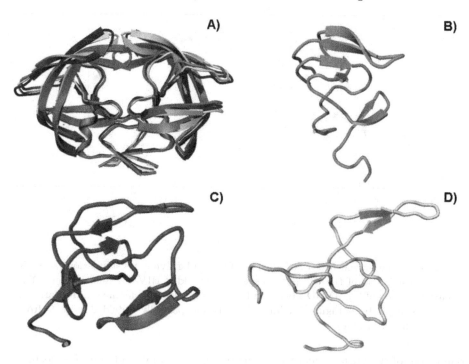

Fig. 1. (A) Superposition of the HIV–1 PR crystal structure with the closed conformation of the flaps, a representative of the "closed" enzyme form (pdb entry 1BVG, in
green); a "semi–open" form of the unliganded HIV–1 PR (pdb entry 1HHP, in yellow)
and a the "wide-open" form of the HIV–1 PR mutant dimer (pdb entry 1TW7, in blue).
The HIV–1 PR conformations are obtained using crystal structures of the HIV–1 free
dimer and complexes with the inhibitors, with subsequent thermal equilibration and
minimization. The superposition of the HIV–1 PR crystal structures into a common
reference frame is based on the similarity of these structurally conserved regions of
HIV–1 PR. (B-D) A collection of computer–generated partially unfolded structures of
the HIV–1 PR monomer. These partially unfolded HIV–1 PR monomeric states were
formed by gradually breaking contacts between segments 83–93 and 24–34 of the folding nucleus, followed by subsequent refinement with the energy minimization and 4 ns
of molecular dynamics simulations at 300K.

3   Results and Discussion

Conformational landscape of HIV–1 PR, which is employed in the present study,
includes not only a diverse ensemble of the crystal and NMR structures of HIV–
1 PR dimers and monomers, but also a collection of computer–generated partially unfolded HIV–1 PR monomers (Figure 1). Since the HIV–1 PR active
dimer is likely a result of the assembly of the folded HIV–1 PR monomers, we
have initially constructed a number of partially unfolded HIV–1 PR monomeric
states by gradually breaking contacts between segments 83–93 and 24–34 in
a single subunit of the native HIV–1 PR dimer (pdb entry 1BVG). Subsequent

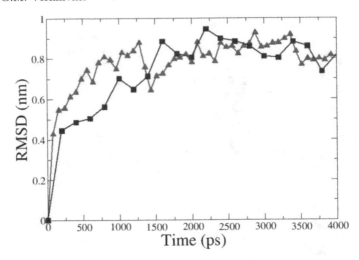

Fig. 2. Time–dependent history of the 4 ns of molecular dynamics simulations at 300K produced using two of the computer–generated unfolded HIV–1 PR monomers. The evolution history of the RMSD's between the C_α atoms in the reference to the partially unfolded state B from Figure 1 (shown in red) and partially unfolded state C from Figure 1 (shown in blue).

refinement of the initial structures by energy minimization and 4 ns of molecular dynamics simulations at 300K have produced three structurally different, meta–stable intermediates, which may illustrate a possible hierarchy of the HIV–1 PR monomer folding (Figure 2). While the ensemble of the HIV–1 PR denat–urated and partially unfolded states is highly dynamic and heterogeneous, the elements of the residual native structure and local structural preferences are re–tained in this otherwise heterogeneous dynamic model [15,16]. In the partially unfolded HIV–1 PR monomers (Figure 1B,C,D), there are clusters of native in–teractions involving segments 24–34 and 83–93 of the HIV–1 PR folding nucleus. In addition, some native secondary structural elements are also retained in these partially unfolded HIV–1 PR conformations. Hence, it may be reasonable to as–sume that these computer–generated, partially unfolded HIV–1 PR monomers may partly reflect conformational heterogeneity of the unfolded HIV–1 PR en–semble. The molecular mechanism of folding inhibition for the p–S8 peptide is analyzed using the results of three independent simulations with the HIV–1 PR conformational ensembles which include protease dimers, monomers and partially unfolded monomers. For each of the three independent simulations, we have selected a different ensemble of structurally different, partially unfolded monomeric states (Figure 3). The equilibrium distribution of the HIV–1 PR con–formational states forming low–energy complexes with the inhibitor is primarily determined by contributions of partially unfolded and folded HIV–1 monomers. Furthermore, the results of simulations are relatively insensitive to the specific structural characteristics of the partially unfolded HIV–1 PR monomers and reveal a similar equilibrium distribution of the HIV–1 PR states favorably inter–acting with the inhibitor (Figure 3).

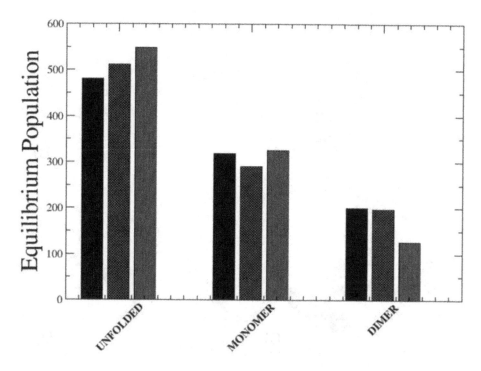

Fig. 3. The equilibrium distribution of the HIV–1 PR conformational states forming low–energy complexes with the p–S8 peptide from three independent simulations with the HIV–1 PR conformational ensembles which include HIV–1 PR dimers, monomers and partially unfolded monomers. We have selected a different ensemble of the partially unfolded conformations, obtained by collecting samples from three independent 4 ns molecular dynamics runs at 300K, which have produced three structurally different, partially unfolded monomer intermediates. The results are insensitive to the specific structural characteristics of the three different unfolded HIV–1 conformational ensembles.

While computer–generated partially unfolded monomers may only illustrate the highly heterogeneous unfolded ensemble of HIV–1 PR, structural diversity of these meta–stable intermediate states and a different degree of unfolding provide means to investigate possible binding scenarios of the inhibitor intervention with the HIV–1 PR folding. The important and somewhat unexpected result of simulations with conformational ensembles of HIV–1 PR is the dominant contribution of the partially unfolded states in the equilibrium distribution of low–energy complexes. Strikingly, despite considerable structural variations between three ensembles of partially unstructured HIV–1 PR monomers, the folding inhibitor and the active analogs tend to target the same active site residues from the conserved segment 24–34 of the HIV–1 PR folding nucleus (Figure 4). Moreover, during binding with the partially unfolded HIV–1 PR monomers, the p–S8 peptide folds into the structure of the 83–93 segment, similar to the peptide folded structure in complexes with the folded HIV–1 PR monomers [28].

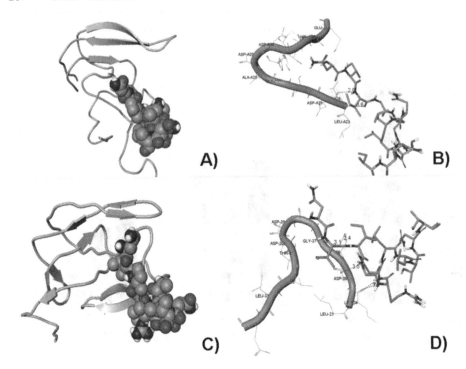

Fig. 4. Structural analysis of the p–S8 peptide binding with the ensemble of partially unfolded monomers obtained from first simulation (upper panel, A,B) and second simulation (lower panel, C,D) (A,C) folded conformation of a flexible p–S8 peptide (in CPK) interacting with the partially unfolded HIV–1 PR monomer (shown in light blue ribbons). (B,D) A close-up of the binding interface formed by the folded p–S8 peptide (shown in all–atom representation) with the segment 24–34 of the HIV–1 PR partially unfolded conformation.

Structural analysis of the inhibitor interactions formed with the partially unfolded ensemble I (Figure 4A,B) has shown that the inhibitor can locate the 24–34 fragment of the HIV–1 PR folding nucleus to form hydrogen bonding between (a) carbonyl oxygen of Ile–85 and NH of Leu–24 and (b) NH group of Ile–85 and carbonyl oxygen of Leu–24. The concomitant formation of the folded peptide structure allows to form a low–affinity complex with the partially unstructured HIV–1 PR monomer. The extensive network of specific interactions formed by the folded p–S8 peptide with the partially unfolded ensemble II (Figure 4C,D) can allow the inhibitor to directly interfere with the assembly of the HIV–1 PR folding nucleus. The hydrogen bonding is formed between (a) carbonyl oxygen of Ile–84 from the peptide and NH of and backbone NH of Gly–27; (b) carbonyl oxygen of Ile–84 and the side-chain hydroxyl group of Thr–26; (c) backbone NH of Arg–87 and carbonyl oxygen of Leu–24; and (d) backbone NH of Asn–88 and carbonyl oxygen of Leu–24. Similarly, the inhibitor interactions with the partially unfolded ensemble III are also characterized by a network of hydrogen bonds formed with Leu–24, Asp–25 and Gly–27 protease residues.

Interestingly, a network of specific interactions formed by the folded p–S8 peptide with the HIV–1 PR monomers does not usually include contribution of highly structured residues from a α–helix region. In contrast, the α–helix region of the folded peptide provides a considerable degree of structural mimicry with the second monomer at the binding interface. This observation can be further analyzed in the context of coupling between folding and binding where the native topology may often govern the choice of the binding mechanism for systems where two highly flexible protein chains form a homodimer. According to the fly–casting mechanism of molecular recognition [36], even in the presence of stable folded monomers, a flexible region of one of the binding partners may be utilized to bind weakly and speed–up the reaction, followed by a folding transition to the final complex. Consequently, folding of the p–S8 peptide may be partly facilitated by the binding requirements to interact with the 24–34 segment using a less structured peptide motif and ultimately form a functionally relevant complex with the HIV–1 PR monomer.

4   Conclusions

The ultimate experiments probing the folding inhibition mechanism require a combination of biochemical studies and structural determination of the peptide–protease complexes. X-ray crystallography techniques are not readily applicable for studies of the inhibitor binding with the HIV–1 PR monomer, since the necessary high concentrations of the enzyme would tend to shift the equilibrium toward the active dimer. NMR studies of the inhibitor intervention with the assembly of the HIV–1 PR monomers are quite challenging, but may in principle assess the molecular basis of folding inhibition. Considering experimental challenges in dissecting the molecular basis of folding inhibition at atomic resolution, the presented computational analysis provides a useful insight into possible binding scenarios and suggests a feasible mechanism, which is consistent with the available experimental data. The performed simulations provide the atomic–level insight into the binding mechanism of the folding inhibitors, which may act through direct intervention with the assembly of the HIV–1 PR folding nucleus and subsequent destabilization of the active HIV–1 PR dimer. We propose that folding inhibitors may effectively intervene with the HIV–1 PR folding and dimerization using a dynamic equilibrium between low–energy complexes formed with the folded and partially unfolded HIV–1 PR monomers. Despite diversity of the binding scenarios, the prevailing mechanism of folding inhibition may be largely determined by the unfrustrated inhibitor interactions with the conserved segment 24–34 of the HIV–1 PR folding nucleus. Theoretical and experimental studies of the folding HIV–1 PR inhibitors may ultimately facilitate understanding of the viral life cycle and develop novel therapeutic strategies to combat drug resistance.

References

1. Kohl, N.E., Emini, E.A., Schleif, W.A., Davis, L.J., Heimbach, J.C., Dixon, R.A., Scolnick, E.M., Sigal, I.S.: Active human immunodeficiency virus protease is required for viral infectivity. Proc. Natl. Acad. Sci. U. S. A. 85, 4686–4690 (1988)
2. Wlodawer, A., Vondrasek, J.: Inhibitors of HIV–1 protease: A major success of structure-assisted drug design. Annu. Rev. Biophys. Biomol. Struct. 27, 249–284 (1998)
3. Vondrasek, J., Wlodawer, A.: HIVdb: a database of the structures of human immunodeficiency virus protease. Proteins 49, 429–431 (2002)
4. Martin, P., Vickrey, J.F., Proteasa, G., Jimenez, Y.L., Wawrzak, Z., Winters, M.A., Merigan, T.C., Kovari, L.C.: "Wide-open" 1.3 A structure of a multidrug-resistant HIV–1 protease as a drug target. Structure 13, 1887–1895 (2005)
5. Ishima, R., Freedberg, D.I., Wang, Y.X., Loui, J.M., Torchia, D.A.: Flap opening and dimer-interface flexibility in the free and inhibitor bound HIV protease and their implications for function. Structure 7, 1047–1055 (1999)
6. Freedberg, D.I., Ishima, R., Jacob, J., Wang, Y.X., Kustanovich, I., Louis, J.M., Torchia, D.A.: Rapid structural fluctuations of the free HIV protease flaps in solution: relationship to crystal structures and comparison with predictions of dynamics calculations. Protein Sci. 11, 221–232 (2002)
7. Katoh, E., Louis, J.M., Yamazaki, T., Gronenborn, A.M., Torchia, D.A., Ishima, R.: A solution NMR study of the binding kinetics and the internal dynamics of an HIV–1 protease-substrate complex. Protein Sci. 12, 1376–1385 (2003)
8. Scott, W.R., Schiffer, C.A.: Curling of flap tips in HIV–1 protease as a mechanism for substrate entry and tolerance of drug resistance. Structure 8, 1259–1265 (2000)
9. Kurt, N., Scott, W.R., Schiffer, C.A., Haliloglu, T.: Cooperative fluctuations of unliganded and substrate-bound HIV–1 protease: a structure-based analysis on a variety of conformations from crystallography and molecular dynamics simulations. Proteins 51, 409–422 (2003)
10. Perryman, A.L., Lin, J.H., McCammon, J.A.: HIV–1 protease molecular dynamics of a wild-type and of the V82F/I84V mutant: possible contributions to drug resistance and a potential new target site for drugs. Protein Sci. 13, 1108–1123 (2004)
11. Perryman, A.L., Lin, J.H., McCammon, J.A.: Restrained molecular dynamics simulations of HIV–1 protease: the first step in validating a new target for drug design. Biopolymers 82, 272–284 (2007)
12. Hornak, V., Okur, A., Rizzo, R.C., Simmerling, C.: HIV–1 protease flaps spontaneously close to the correct structure in simulations following manual placement of an inhibitor into the open state. J. Am. Chem. Soc. 128, 2812–2813 (2006)
13. Hornak, V., Okur, A., Rizzo, R.C., Simmerling, C.: HIV–1 protease flaps spontaneously open and reclose in molecular dynamics simulations. Proc. Natl. Acad. Sci. U S A 103, 915–920 (2006)
14. Wallqvist, A., Smythers, G., Covell, G.: A cooperative folding unit in HIV–1 protease. Implications for protein stability and occurrence of drug-induced mutations. Protein Eng. 11, 999–1005 (1998)
15. Bhavesh, N.S., Sinha, R., Mohan, P.M., Hosur, R.V.: NMR elucidation of early folding hierarchy in HIV–1 protease. J. Biol. Chem. 278, 19980–19985 (2003)
16. Chatterjee, A., Hosur, R.V.: Following autolysis in proteases by NMR: insights into multiple unfolding pathways and mutational plasticities. Biophys. Chem. 123, 1–10 (2006)

17. Louis, J.M., Wondrak, E.M., Kimmel, A.R., Wingfield, P.T., Nashed, N.T.: Proteolytic processing of HIV–1 protease precursor, kinetics and mechanism. J. Biol. Chem. 274, 23437–23442 (1999)

18. Louis, J.M., Clore, G.M., Gronenborn, A.M.: Autoprocessing of HIV–1 protease is tightly coupled to protein folding. Nat. Struct. Biol. 6, 868–875 (1999)

19. Chatterjee, A., Mridula, P., Mishra, R.K., Mittal, R., Hosur: Folding regulates autoprocessing of HIV–1 protease precursor. J. Biol. Chem. 280, 11369–11378 (2005)

20. Ishima, R., Ghirlando, R., Tozser, J., Gronenborn, A.M., Torchia, D.A., Louis, J.M.: Folded monomer of HIV–1 protease. J. Biol. Chem. 276, 49110–49116 (2001)

21. Louis, J.M., Ishima, R., Nesheiwat, I., Pannell, L.K., Lynch, S.M., Torchia, D.A., Gronenborn, A.M.: Revisiting monomeric HIV–1 protease. Characterization and redesign for improved properties. J. Biol. Chem. 278, 6085–6092 (2003)

22. Ishima, R., Torchia, D.A., Lynch, S.M., Gronenborn, A.M., Louis, J.M.: Solution structure of the mature HIV–1 protease monomer: insight into the tertiary fold and stability of a precursor. J. Biol. Chem. 278, 43311–43319 (2003)

23. Ishima, R., Torchia, D.A., Louis, J.M.: Mutational and structural studies aimed at characterizing the monomer of HIV–1 protease and its precursor. J. Biol. Chem. 282, 17190–17199 (2007)

24. Bannwarth, L., Reboud-Ravaux, M.: An alternative strategy for inhibiting multidrug-resistant mutants of the dimeric HIV-1 protease by targeting the subunit interface. Biochem. Soc. Trans. 35, 551–554 (2007)

25. Broglia, R.A., Provasi, D., Vasile, F., Ottolina, G., Longhi, R., Tiana, G.: A folding inhibitor of the HIV-1 protease. Proteins 62, 928–933 (2005)

26. Broglia, R.A., Tiana, G., Sutto, L., Provasi, D., Simona, F.: Design of HIV-1-PR inhibitors that do not create resistance: blocking the folding of single monomers. Protein Sci. 14, 2668–2681 (2005)

27. Bonomi, M.F., Gervasio, L., Tiana, G., Provasi, D., Broglia, R.A., Parrinello, M.: Insight into the folding inhibition of the HIV-1 protease by a small peptide. Biophys. J. 93, 2813–2821 (2007)

28. Verkhivker, G., Tiana, G., Camilloni, C., Provasi, D., Broglia, R.A.: Atomistic simulations of the HIV-1 protease folding inhibition. Biophys. J. 95, 550–562 (2008)

29. Berman, H.M., Westbrook, J., Feng, Z., Gilliland, G., Bhat, T.N., Weissig, H., Shindyalov, I.N., Bourne, P.E.: The Protein Data Bank. Nucleic Acids R. 28, 235–242 (2000)

30. Christen, M., Hunenberger, P.H., Bakowies, D., Baron, R., Burgi, R., Geerke, D.P., Heinz, T.N., Kastenholz, M.A., Krautler, V., Oostenbrink, C., Peter, C., Trzesniak, D., van Gunsteren, W.F.: The GROMOS software for biomolecular simulation: GROMOS 2005. J. Comput. Chem. 26, 1719–1751 (2005)

31. Verkhivker, G.M.: Computational proteomics of biomolecular interactions in the sequence and structure space of the tyrosine kinome: deciphering the molecular basis of the kinase inhibitors selectivity. Proteins 66, 912–929 (2007)

32. Verkhivker, G.M.: In silico profiling of tyrosine kinases binding specificity and drug resistance using Monte Carlo simulations with the ensembles of protein kinase crystal structures. Biopolymers 85, 333–348 (2007)

33. Cornell, W.D., Cieplak, P., Bayly, C.L., Gould, I.R., Merz, K.M., Ferguson, D.M., Spellmeyer, D.C., Fox, T., Caldwell, J.W., Kollman, P.A.: A second generation force field for simulation of proteins, nucleic acids, and organic molecules. J. Amer. Chem. Soc. 117, 5179–5197 (1995)

24   G.M. Verkhivker

34. Stouten, P.F.W., Frömmel, C., Nakamura, H., Sander, C.: An effective solvation term based on atomic occupancies for use in protein simulations. Mol. Simul. 10, 97–120 (1993)
35. Sugita, Y., Okamoto, Y.: Replica-exchange molecular dynamics method for protein folding. Chem. Phys. Lett. 314, 141–151 (1999)
36. Shoemaker, B.A., Portman, J.J., Wolynes, P.G.: Speeding molecular recognition by using the folding funnel: the fly-casting mechanism. Proc. Natl. Acad. Sci. U S A 97, 8868–8873 (2000)

Unsupervised Stability-Based Ensembles to Discover Reliable Structures in Complex Bio-molecular Data

Alberto Bertoni and Giorgio Valentini

DSI, Dipartimento di Scienze dell' Informazione,
Università degli Studi di Milano,
Via Comelico 39, 20135 Milano, Italia
{bertoni,valentini}@dsi.unimi.it

Abstract. The assessment of the reliability of clusters discovered in bio-molecular data is a central issue in several bioinformatics problems. Several methods based on the concept of stability have been proposed to estimate the reliability of each individual cluster as well as the "optimal" number of clusters. In this conceptual framework a clustering ensemble is obtained through bootstrapping techniques, noise injection into the data or random projections into lower dimensional subspaces. A measure of the reliability of a given clustering is obtained through specific stability/reliability scores based on the similarity of the clusterings composing the ensemble. Classical stability-based methods do not provide an assessment of the statistical significance of the clustering solutions and are not able to directly detect multiple structures (e.g. hierarchical structures) simultaneously present in the data. Statistical approaches based on the chi-square distribution and on the Bernstein inequality, show that stability-based methods can be successfully applied to the statistical assessment of the reliability of clusters, and to discover multiple structures underlying complex bio-molecular data. In this paper we provide an overview of stability based methods, focusing on stability indices and statistical tests that we recently proposed in the context of the analysis of gene expression data.

1 Introduction

Clustering of complex of bio-molecular data represents one of the main problems in bioinformatics [1]. Classes of co-expressed genes, classes of functionally related proteins, or subgroups of patients with malignancies differentiated at bio-molecular level can be discovered through clustering algorithms, and several other tasks related to the analysis of bio-molecular data require the development and application of unsupervised clustering techniques [2, 3, 4]. From a general standpoint the discovered clusters depend on the clustering algorithm, the initial condition, the parameters of the algorithm, the distance or correlation measure applied to the data and other clustering and data-dependent factors [5].

Moreover the bioinformatics domain raises specific and challenging problems that characterize clustering applications in bio-molecular biology and medicine.

F. Masulli, R. Tagliaferri, and G.M. Verkhivker (Eds.): CIBB 2008, LNBI 5488, pp. 25–43, 2009.

In particular the integration of multiple data sources [6], the very high dimensionality [7, 8], and the visualization of the data [9, 10], as well as interactive data analysis in clustering genomic data [11, 12] represent relevant topics in the unsupervised analysis of bio-molecular data.

Another relevant problem is the assessment of the reliability of the discovered clusters, as well as the proper selection of the "natural" number of clusters underlying bio-molecular data [13, 14]. Indeed in many cases we have no sufficient biological knowledge to "a priori" evaluate both the number of clusters (e.g. the number of biologically distinct tumor classes), as well as the validity of the discovered clusters (e.g. the reliability of new discovered tumor classes). Note that this is an intrinsically "ill-posed" problem, since in unsupervised learning we lack an external objective criterion, that is we have not an equivalent of a priori known class label as in supervised learning, and hence the evaluation of the validity/reliability of the discovered classes becomes elusive and difficult.

Most of the works focused on the estimate of the number of clusters in gene expression data [15, 16, 17, 18, 19], while the problem of stability of each individual cluster has been less investigated. Nevertheless, the stability and reliability of the obtained clusters is crucial to assess the confidence and the significance of a bio-medical discovery [20, 21].

Considerings the complexity and the characteristics of the data used in bioinformatics applications (e.g. the low cardinality and very high dimensionality of DNA microarray data), classical parametric methods in many cases may fail to discover structures in the data. This is the main reason why non parametric methods, based on the concept of the stability, have been recently introduced in the context of significant bioinformatics problems.

In particular, several methods based on the concept of stability have been proposed to estimate the "optimal" number of clusters in complex bio-molecular data [22, 23, 24, 17, 25, 26]. In this conceptual framework multiple clusterings are obtained by introducing perturbations into the original data, and a clustering is considered reliable if it is approximately maintained across multiple perturbations. Several perturbation techniques have been proposed, ranging from bootstrap techniques [19, 16, 23], to random projections to lower dimensional subspaces [21, 27] to noise injection procedures [20].

Another major problem related to stability-based methods is to estimate the statistical significance of the structures discovered by clustering algorithms. To face this problem we proposed a χ^2-based statistical test [26] and a test based on the classical *Bernstein inequality* [28, 29]. These statistical tests may be applied to any stability method based on the distribution of similarity measures between pairs of clusterings. We experimentally showed that by this approach we may discover multiple structures simultaneously present in the data (e.g. hierarchical structures), associating a *p-value* to the clusterings selected by a given stability-based method for model order selection [30, 31].

In this paper we introduce the main concepts behind stability based methods, focusing on the work developed in [27, 26, 29]. More precisely, in the next section an overview of the main characteristics of stability-based methods is given.

Then in Sect. 3 a stability index, proposed in [26] to assess the reliability of a clustering solution, is described. Sect. 4 introduces two statistical tests to assess the significance of overall clustering solutions, while Sect. 5 provides an introduction to stability indices proposed in [27] to estimate the reliability of each individual cluster inside a given clustering. Then the main drawbacks and limitations of the proposed approaches, as well as new research lines are discussed and the conclusions end the paper. In the appendix 7, we briefly describe the main characteristics of two R packages implementing the stability indices and statistical tests described in the previous sections.

2   An Overview of Stability Based Methods

A major requirement for clustering algorithms is the reproducibility of their solutions on other data sets drawn from the same source. In this context, several methods based on the concept of stability have been proposed to estimate the "optimal" number of clusters in clustered data [23, 24]: multiple clusterings are obtained by introducing perturbations into the original data, and a clustering is considered reliable if it is approximately maintained across multiple perturbations.

A general stability-based algorithmic scheme for assessing the reliability of clustering solutions may be summarized in the following way:

1. For a fixed number k of clusters, randomly perturb the data many times according to a given perturbation procedure.
2. Apply a given clustering algorithm to the perturbed data
3. Apply a given clustering similarity measure to multiple pairs of k-clusterings obtained according to steps 1 and 2.
4. Use appropriate similarity indices (stability scores) to assess the stability of a given clustering.
5. Repeat steps 1 to 4 for multiple values of k and select the most stable clustering(s) as the most reliable.

Several approaches have been proposed to implement the first step: a random "perturbation" of the data may be obtained through bootstrap samples drawn from the available data [19, 23], or random noise injection into the data [20] or random subspace [21] or random projections into lower dimensional subspaces [27].

The application of a given algorithm (step 2) represents a choice based on "a priori" knowledge or assumptions about the characteristics of the data. To estimate the similarity between clusterings (step 3), classical measures, such as the Rand Index [32], or the Jaccard or the Fowlkes and Mallows coefficients [5] or their equivalent dot-product representations [16] may be applied. More precisely, for a given clustering algorithm $\mathcal{C}$ applied to a data set X, we may obtain the following clustering:

$$\mathcal{C}(X, k) = < A_1, A_2, \ldots, A_k >, \quad \cup_{i=1}^{k} A_i = X \qquad (1)$$

For each clustering $C = \mathcal{C}(X, k)$ we may obtain a *pairwise similarity matrix M* with $n \times n$ elements, where n is the cardinality of X:

$$
M_{i,j} = \begin{cases} 1, & \text{if } \exists r \in \{1, \ldots, k\}, \ x_i \in A_r \text{ and } x_j \in A_r, \ i \neq j \\ 0, & \text{otherwise} \end{cases} \tag{2}
$$

Given two clusterings $C^{(1)}$ and $C^{(2)}$ obtained from the same data set X, we may compute the corresponding similarity matrices $M^{(1)}$ and $M^{(2)}$. Then we count the number of entries $M_{i,j}$ for which $M^{(1)}$ and $M^{(2)}$ have corresponding values equal to 1 (that is the number of entries N_{11} for which the clusterings agree about the membership of a pair of examples to the same cluster). Equivalently we may compute N_{10}, that is the number of entries for which a given pair of examples belongs to the same cluster in $C^{(1)}$, but does not belong to the same cluster in $C^{(2)}$. N_{01} and N_{00} can be computed in the same way. From this quantities we may compute the classical similarity measures between clusterings:
the *Matching* coefficient:

$$
M(C^{(1)}, C^{(2)}) = \frac{N_{00} + N_{11}}{N_{00} + N_{11} + N_{10} + N_{01}} \tag{3}
$$

the *Jaccard* coefficient:

$$
M(C^{(1)}, C^{(2)}) = \frac{N_{11}}{N_{11} + N_{10} + N_{01}} \tag{4}
$$

and the *Fowlkes and Mallows* coefficient:

$$
M(C^{(1)}, C^{(2)}) = \frac{N_{11}}{\sqrt{(N_{01} + N_{11})(N_{10} + +N_{11})}} \tag{5}
$$

Several stability indices for model order selection have been proposed in the literature (see, e.g. [20, 21, 23, 24]): very schematically they can be divided into indices that use statistics of the similarity measures [21, 27] or their overall empirical distribution [16, 26].

The last step, that is the selection of the most stable/reliable clustering, given a set of similarity measures and the related stability indices, has been usually approached by choosing the best scored clustering (according to the chosen stability index). A major problem in this last step is represented by the estimate of the statistical significance of the discovered solutions.

3   A Stability Index Based on the Distribution of the Similarity Measures

In [26] we extended the approach proposed by *Ben-Hur, Ellisseeff and Guyon* [16], by providing a quantitative estimate of a *stability score* based on the overall distribution of the similarities between pairs of clusterings.

Let be $\mathcal{C}$ a clustering algorithm, $\rho(D)$ a given random perturbation procedure applied to a data set D and *sim* a suitable similarity measure between two clusterings (e.g. the Fowlkes and Mallows similarity [33]). For instance ρ may be a random projection from a high dimensional to a low dimensional subspace [34], or a bootstrap procedure to sample a random subset of data from the original data set D [16].

We define S_k ($0 \leq S_k \leq 1$) as the random variable given by the similarity between two k-clusterings obtained by applying a clustering algorithm $\mathcal{C}$ to pairs D_1 and D_2 of random independently perturbed data. The intuitive idea is that if S_k is concentrated close to 1, the corresponding clustering is stable with respect to a given controlled perturbation and hence it is reliable.

As an example, consider a a 1000-dimensional synthetic multivariate gaussian data set with relatively low cardinality (60 examples), characterized by a two-level hierarchical structure, highlighted by the projection of the data into the two main principal components (Fig. 1): indeed a two-level structure, with respectively 2 and 6 clusters is self-evident in the data.

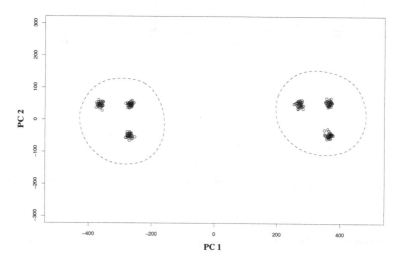

Fig. 1. A two-level hierarchical structure with 2 and 6 clusters is revealed by principal components analysis (data projected into the two components with highest variance)

We can estimate S_k, for the number of clusters k varying e.g. from 2 to 9. This can be performed by using 100 pairs of *Bernoulli* projections [26], with a distortion bounded to 20 % with respect to the original data, yielding to random projections from 1000 to 479-dimensional subspaces, and using PAM (Partitioning Around Medoids) as clustering algorithm [35]. The distribution of the similarity values is depicted in Fig. 2: the histograms of the similarity measures for $k = 2$ and $k = 6$ clusters are tightly concentrated near 1, showing that these clusterings are very stable, while for other values of k the similarity measures are spread across multiple values.

These results suggest that we could try to exploit the *cumulative distribution* of the similarities between pairs of clusterings to compute a reliable *stability score*

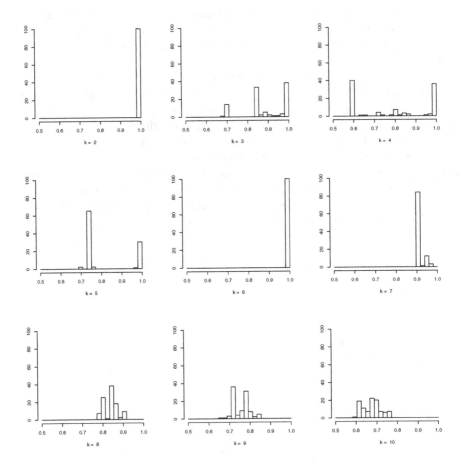

Fig. 2. Histograms of the similarity measure distributions for different numbers of clusters

for a given k-clustering. Indeed if the similarities are spread across multiple values, the clustering is unstable, while if they are cumulated close to 1 the clustering is stable. In the following we derive a more formal derivation of a stability score based on the cumulative distribution of the similarity measure between clusterings.

Let be $f_k(s)$ the density function of the random variable S_k, and

$$F_k(\bar{s}) = \int_{-\infty}^{\bar{s}} f_k(s)ds \qquad (6)$$

its cumulative distribution function.

We define $g(k)$ as the integral of the cumulative distribution function:

$$g(k) = \int_0^1 F_k(s)ds \tag{7}$$

Intuitively $g(k)$ represents the "concentration" of the similarity values close to 1; that is, if $g(k) \simeq 0$ then the distribution of the values of S_k is concentrated near 1, or, in other words, the k-clustering is stable. On the other hand, if $g(k) \simeq 1$ then the clusterings are totally unstable, while if the distribution is close to the uniform distribution, we have $g(k) \simeq 1/2$.

We may directly estimate eq. 7 by numerical integration, or we may more easily obtain $g(k)$ from the estimate of the expectation $E[S_k]$:

$$E[S_k] = \int_0^1 sf_k(s)ds = \int_0^1 sF'_k(s)ds$$
$$= sF_k(s)\, |_0^1 - \int_0^1 F_k(s)ds = 1 - \int_0^1 F_k(s)ds \tag{8}$$

Hence from eq. 8 we may easily compute $g(k)$:

$$g(k) = \int_0^1 F_k(s)ds = 1 - E[S_k] \tag{9}$$

Eq. 9 shows that we have a very stable and reliable clustering ($E[S_k]$ close to 1), if and only if $g(k)$ is close to 0.

In practice we can compute the empirical means ξ_k of the similarity values, while varying the number of clusters k from 2 to H and then we can perform a sorting of the obtained values:

$$(\xi_2, \xi_3, \ldots, \xi_H) \overset{sort}{\to} (\xi_{p(1)}, \xi_{p(2)}, \ldots, \xi_{p(H-1)}) \tag{10}$$

where p is the permutation index such that $\xi_{p(1)} \geq \xi_{p(2)} \geq \cdots \geq \xi_{p(H-1)}$. Roughly speaking, this ordering represents the "most reliable" $p(1)$-clustering down to the least reliable $p(H-1)$-clustering and ξ_k provides a *stability score* of the obtained k-clustering.

4   Statistical Tests to Assess the Significance of Overall Clustering Solutions

In [26] we proposed a χ^2 test to assess the significance of clustering solutions and to discover multiple structures underlying gene expression data. Moreover, in [28, 36], we proposed a distribution-free approach that does not assume any "a priori" distribution of the similarity measures, and that does not require any user-defined additional parameter, using the classical Bernstein inequality [37].

4.1 A χ^2-Based Test to Discover Multiple Structures in Bio-molecular Data

Consider a set of k-clusterings $k \in \mathcal{K}$, where $\mathcal{K}$ is a set of numbers of clusters. By estimating the expectations $E[S_k]$ or equivalently by computing eq. 7 through numerical integration, we obtain a set of values $\mathcal{G} = \{g_k | k \in \mathcal{K}\}$. We can sort G obtaining $\hat{\mathcal{G}}$ with values $\hat{g}_i$ in ascending order. For each k-clustering we consider two groups of pairwise clustering similarities values separated by a threshold t^o (a reasonable threshold could be $t^o = 0.9$.). Thus we may obtain: $P(S_k > t^o) = 1 - F_k(s = t^o)$, where $F_k(s = t^o)$ is computed according to eq. 6. If n represents the number of trials for estimating the value of S_k then $x_k = P(S_k > t^o)n$ is the number of times for which the similarity values are larger than t^o. The x_k may be interpreted as the successes from $|\mathcal{K}|$ binomial populations with parameters θ_k. If the number of trials n is sufficiently large, and setting X_k as a random variable that counts how many times $S_k > t^o$, we have that the following random variable, for sufficiently large values of n is distributed according to a normal distribution:

$$\frac{X_k - n\theta_k}{\sqrt{n\theta_k(1 - \theta_k)}} \sim N(0, 1) \tag{11}$$

A sum of i.i.d. squared normal variables is distributed according to a χ^2 distribution:

$$\sum_{k \in \mathcal{K}} \frac{(X_k - n\theta_k)^2}{n\theta_k(1 - \theta_k)} \sim \chi^2 \tag{12}$$

Considering the null hypothesis H_0: all the θ_k are equal to θ, where the unknown θ is estimated through its pooled estimate $\hat{\theta} = \frac{\sum_{k \in \mathcal{K}} x_k}{|\mathcal{K}| \cdot n}$, then the null hypothesis may be evaluated against the alternative hypothesis that the θ_k are not all equal using the statistic

$$Y = \sum_{k \in \mathcal{K}} \frac{(x_k - n\hat{\theta})^2}{n\hat{\theta}(1 - \hat{\theta})} \sim \chi^2_{|\mathcal{K}|-1} \tag{13}$$

If $Y \geq \chi^2_{\alpha, |\mathcal{K}|-1}$ we may reject the null hypothesis at α significance level, that is we may conclude that with probability $1 - \alpha$ the considered proportions are different, and hence that at least one k-clustering significantly differ from the others. Using the above test we start considering all the k-clustering. If a significant difference is registered according to the statistical test we exclude the last clustering (according to the sorting of $\mathcal{G}$). This is repeated until no significant difference is detected (or until only 1 clustering is left out): the set of the remaining (top sorted) k-clusterings represents the set of the estimate stable number of clusters discovered (at α significance level).

It is worth noting that the above χ^2-based procedure can be also applied to automatically find the optimal number of clusters independently of the applied perturbation method.

Anyway, note that with the previous χ^2-based statistical test we implicitly assume that some probability distributions are normal. Moreover test results depend on the choice of user-defined parameters (the threshold t^o). Using the classical Bernstein inequality [37] we may apply a partially "distribution independent" approach to assess the significance of the discovered clustering.

4.2   A *Bernstein Inequality*-Based Test to Discover Multiple Structures in Bio-molecular Data

We briefly recall the Bernstein inequality, because this inequality is used to build-up our proposed hypothesis testing procedure, without introducing any user defined parameter.

Bernstein inequality. If $Y_1, Y_2, \ldots, Y_n$ are independent random variables s.t. $0 \leq Y_i \leq 1$, with $\mu = E[Y_i], \sigma^2 = Var[Y_i], \bar{Y} = \sum Y_i/n$ then

$$Prob\{\bar{Y} - \mu \geq \Delta\} \leq e^{\frac{-n\Delta^2}{2\sigma^2 + 2/3\Delta}} \qquad (14)$$

Consider the following random variables:

$$P_i = S_{p(1)} - S_{p(i)} \quad \text{and} \quad X_i = \xi_{p(1)} - \xi_{p(i)} \qquad (15)$$

We start considering the first and last ranked clustering $p(1)$ and $p(H)$. In this case the null hypothesis becomes: $E[S_{p(1)}] \leq E[S_{p(H)}]$, that is: $E[S_{p(1)}] - E[S_{p(H)}] = E[P_H] \leq 0$. The distribution of the random variable X_H (eq. 15) is in general unknown; anyway note that in the Bernstein inequality no assumption is made about the distribution of the random variables Y_i (eq. 14). Hence, fixing a parameter $\Delta \geq 0$, considering true the null hypothesis $E[P_H] \leq 0$, and using Bernstein inequality, we have:

$$Prob\{X_H \geq \Delta\} \leq Prob\{X_H - E[P_H] \geq \Delta\} \leq e^{\frac{-n\Delta^2}{2\sigma^2 + 2/3\Delta}} \qquad (16)$$

Considering an instance (a measured value) $\hat{X}_H$ of the random variable X_H, if we let $\Delta = \hat{X}_H$ we obtain the following probability of type I error:

$$P_{err}\{X_H \geq \hat{X}_H\} \leq e^{\frac{-n\hat{X}_H^2}{2\sigma_H^2 + 2/3\hat{X}_H}}$$

with $\sigma_H^2 = \sigma_{p(1)}^2 + \sigma_{p(H)}^2$.

If $P_{err}\{X_H \geq \hat{X}_H\} < \alpha$, we reject the null hypothesis: a significant difference between the two clusterings is detected at α significance level and we continue by testing the $p(H-1)$ clustering. More in general if the null hypothesis has been rejected for the $p(H-r+1)$ clustering, $1 \leq r \leq H-2$ then we consider the $p(H-r)$ clustering, and by union bound we can estimate the type I error:

$$P_{err}(H-r) = Prob\{ \bigvee_{H-r \leq i \leq H} X_i \geq \hat{X}_i \} \leq \sum_{i=H-r}^{H} Prob\{X_i \geq \hat{X}_i\} \leq \sum_{i=H-r}^{H} e^{\frac{-n\hat{X}_i^2}{2\sigma_i^2 + 2/3\hat{X}_i}} \qquad (17)$$

As in the previous case, if $P_{err}(H-r) < \alpha$ we reject the null hypothesis: a significant difference is detected between the reliability of the $p(1)$ and $p(H-r)$ clustering and we iteratively continue the procedure estimating $P_{err}(H-r-1)$.

This procedure stops if either of these cases succeeds:

I) The null hypothesis is rejected till to $r = H - 2$, that is $\forall r, 1 \leq r \leq H - 2$, $P_{err}(H - r) < \alpha$: all the possible hypotheses have been rejected and the only reliable clustering at α-significance level is the top ranked one, that is the $p(1)$ clustering.

II) The null hypothesis cannot be rejected for $r < H - 2$, that is, $\exists r, 1 \leq r \leq H - 2$, $P_{err}(H - r) \geq \alpha$: in this case the clusterings that are significantly less reliable than the top ranked $p(1)$ clustering are the $p(r + 1), p(r + 2), \ldots, p(H)$ clusterings.

Note that in this second case we cannot state that there is no significant difference between the first r top-ranked clusterings, since the upper bound provided by the Bernstein inequality is not guaranteed to be tight. To answer to this question, we may apply the χ^2-based hypothesis testing proposed in [26] to the remaining top ranked clusterings to establish which of them are significant at α level, but in this case we need to assume that the similarity measures between pairs of clusterings are distributed according to a normal distribution.

For applications of the χ^2-based and the *Bernstein inequality*-based to the analysis of bio-molecular data see e.g. [26, 28, 29]. The experimental results show that Bernstein test is more sensitive to multiple structures underlying the data, but at the same time more susceptible to false positives with respect to the χ^2 test.

5 Stability Indices for the Assessment of the Reliability of Individual Clusters

In this section we provide an overview of the approach proposed in [27] to assess the validity of each individual cluster, using random projections to lower dimensional subspaces as perturbation methods.

5.1 Perturbations through Randomized Embedding

Dimensionality reduction may be obtained by mapping points from a high to a low-dimensional space, approximately preserving some characteristics, i.e. the distances between points. In this context randomized embeddings with low distortion represent a key concept. Randomized embeddings have been successfully applied both to combinatorial optimization and data compression [38].

A *randomized embedding* between L_2 normed metric spaces with distortion $1 + \epsilon$, with $\epsilon > 0$ and failure probability P is a distribution probability over mappings $\mu : \mathbb{R}^d \to \mathbb{R}^{d'}$, such that for every pair $p, q \in \mathbb{R}^d$, the following property holds with probability $1 - P$:

$$\frac{1}{1 + \epsilon} \leq \frac{||\mu(p) - \mu(q)||_2}{||p - q||_2} \leq 1 + \epsilon \qquad (18)$$

The main result on randomized embedding is due to Johnson and Lindenstrauss [39], who proved the existence of a randomized embedding $\mu : \mathbb{R}^d \to \mathbb{R}^{d'}$

with distortion $1 + \epsilon$ and failure probability $e^{\Omega(-d'\epsilon^2)}$, for every $0 < \epsilon < 1/2$. As a consequence, for a fixed data set $S \subset \mathbb{R}^d$, with $|S| = n$, by union bound, for all $p, q \in S$, it holds:

$$Prob\left(\frac{1}{1+\epsilon} \leq \frac{||\mu(p) - \mu(q)||_2}{||p - q||_2} \leq 1 + \epsilon\right) \geq 1 - n^2 e^{\Omega(-d'\epsilon^2)} \qquad (19)$$

Hence, by choosing d' such that $n^2 e^{\Omega(-d'\epsilon^2)} < 1/2$, it is proved the following: *Johnson-Lindenstrauss (JL) lemma*: Given a set S with $|S| = n$ there exists a $1 + \epsilon$-distortion embedding into $\mathbb{R}^{d'}$ with $d' = c \, \log n/\epsilon^2$, where c is a suitable constant.

The embedding exhibited in [39] consists in random projections from $\mathbb{R}^d$ into $\mathbb{R}^{d'}$, represented by matrices $d' \times d$ with random orthonormal vectors. Similar results may be obtained by using simpler embeddings, represented through random $d' \times d$ matrices $P = 1/\sqrt{d'}(r_{ij})$, where r_{ij} are random variables such that:

$$E[r_{ij}] = 0, \qquad Var[r_{ij}] = 1$$

For sake of simplicity, we call random projections even this kind of embeddings. In particular in [34] matrices are proposed such that their entries are uniformly chosen in $\{-1, 1\}$, or in $\{-\sqrt{3}, 0, \sqrt{3}\}$, by choosing 0 with probability 2/3 and $-\sqrt{3}$ or $\sqrt{3}$ with probability 1/6. In this case the *JL lemma* holds with $c \simeq 4$.

Consider now a data set represented by a $d \times n$ matrix X whose columns represent n d-dimensional observations. Suppose that $d' = 4 \, \log n/\epsilon^2 << d$; the *JL lemma* guarantees the existence of a $d' \times d$ matrix P such that the columns of the "compressed" data set $X^P = PX$ have approximately the same distance (up to a distortion $1 + \epsilon$) of the corresponding columns in X. Moreover there is a randomized algorithm that, having in input X, outputs X^P in time $\mathcal{O}(dd'n)$ with high confidence.

5.2 Stability Measures for Individual Clusters

The *JL* lemma shows that we may generate relatively low-distorted random projected data. Our aim is to exploit random projections to estimate stability of clusters, because random projections do not induce relevant distortions (as long as we provide a projection into a sufficiently high-dimensional subspace).

Given a finite set $X \subset \mathbb{R}^d$, we denote (with abuse of notation) with X the metric space $< X, f >$, where $f(x, y) = ||x - y||_2$, $x, y \in \mathbb{R}^d$. In the following of this section we consider a fixed random projection $\mu : \mathbb{R}^d \to \mathbb{R}^{d'}$ that verifies the *JL* lemma, and we propose a stability index for clustering by using a pairwise similarity matrix between the projected examples.

Let $\mathcal{C}$ be a clustering algorithm, that, having in input X, outputs a set of k clusters:

$$\mathcal{C}(X) =< A_1, A_2, \ldots, A_k >, \; A_j \subset X, 1 \leq j \leq k \qquad (20)$$

Then we compute a "similarity" matrix M, with indices in X, using the following algorithm:

1. Generate t independent projections $\mu_i : \mathbb{R}^d \to \mathbb{R}^{d'}$, $1 \le i \le t$, such that $d' = 4 \frac{\log|X| + \log t}{\epsilon^2}$
2. Apply $\mathcal{C}$ to the new projected data $\mu_i(X)$, obtaining a set of clusterings, for $1 \le i \le t$:

$$\mathcal{C}(\mu_i(X)) = < B_1^i, \ldots, B_k^i >, \; B_j^i \subset X_i, 1 \le j \le k \qquad (21)$$

where B_j^i is the jth cluster of the ith clustering.
3. Set the elements M_{xy} of the similarity matrix:

$$M_{xy} = \frac{1}{t} \sum_{j=1}^{k} \sum_{i=1}^{t} \chi_{B_j^i}(\mu_i(x)) \cdot \chi_{B_j^i}(\mu_i(y)) \qquad (22)$$

where $\chi_{B_j^i}$ is the characteristic function for the cluster B_j^i.

Since the elements M_{xy} measure the occurrences of the examples $\mu_i(x), \mu_i(y) \in \mu_i(X)$ in the same clusters B_j^i for $1 \le i \le t$, then M represents the "tendency" of the projections to belong to the same cluster. It is easy to see that $0 \le M_{xy} \le 1$, for each $x, y \in X$.

Using the similarity matrix M (eq. 22) we propose the following *stability index* s for a cluster A_i [27]:

$$s(A_i) = \frac{1}{|A_i|(|A_i| - 1)} \sum_{(x,y) \in A_i \times A_i, x \neq y} M_{xy} \qquad (23)$$

The index $s(A_i)$ estimates the stability of a cluster A_i in the original non projected space, by measuring how much the projections of the pairs $(x, y) \in A_i \times A_i$ occur together in the same cluster in the projected subspaces. The stability index has values between 0 and 1: values near 1 denote stable clusters, while lower values indicate less reliable clusters. The above stability index is very similar to that proposed by [23]. The main difference of our approach consists in the way the similarity matrix is computed: we applied randomized projections into lower dimensional subspaces, while [23] applied bootstrap techniques.

An overall measure of the stability of the clustering in the original space may be obtained averaging between the stability indices:

$$S(k) = \frac{1}{k} \sum_{i=1}^{k} s(A_i) \qquad (24)$$

In this case also we have that $0 \le S(k) \le 1$, where k is the number of clusters.

Experimental applications of the stability indices (eq. 23 and 24) to the discovery of bio-molecular subclasses of malignancies are described in [40, 41, 27].

6 Drawbacks of Stability-Based Methods and New Research Lines

Despite their successful application to several real-world problems, and in particular in bioinformatics, the theory underlying stability-based methods is not well-understood and several problems remain open from a theoretical standpoint [42]. Moreover, using clustering stability in a high sample setting can be problematic. In particular it has been shown the bounding the difference between the finite sample stability and the "true stability" can exist only if one makes strong assumptions on the underlying distribution [43].

Moreover stability-based method may converge to a suboptimal solution owing to the shape of the data manifold and not to the real structure of the data [13], especially if the distribution of the data obey to a some rule of symmetry.

A problem that cannot be directly addressed by stability-based methods is the detection of "no structure" in the data. However, we may obtain an indirect evidence of "no structure" if the stability scores are always very low and comparable for a large set of numbers of clusters, or if the statistical tests consider equally reliable all or a large part of the possible clusterings.

Another problem relies on the characteristics of the perturbations that may induce bias into the stability indices used to estimate the reliability of the discovered clusters. In particular if the intensity of the perturbation is too high, significant distortions can be introduced, and the structure of the data cannot be preserved. For instance we showed that random subspace perturbations can induce significant distortions into real gene expression data. We showed also that random projections obeying the Johnson-Lindenstrauss lemma may induce bounded distortions, thus providing a theoretically-founded way to perturb the data approximately preserving their underlying structure [27, 26]. Unfortunately similar results are not available when we introduce perturbations through resampling techniques or noise injection into the data.

Apart from these theoretical problems, that need to be considered in future research work, we would like to cite at least two other problems that to our opinion are relevant in the bioinformatics context.

The first one is related to problems characterized by a very high number of possible clusters and clusterings. These problems naturally come from genomics and proteomics: consider, e.g. the unsupervised search for functional classes of genes or proteins. In this context the number of possible clusters is too high for classical stability based methods (consider e.g. Gene Ontology taxonomy that includes thousands of possible functional classes [44]), and the the iterative approach is too computationally expensive. For relatively moderate sized problems parallel computation could be a solution, but from a more general standpoint the problem is too complex and requires the development of new specific algorithms. A possible solution could be the reduction of possible candidate clusters, making some assumption about the characteristics of the data. To this end approaches based on hierarchical clustering ensembles and non parametric tests have been recently proposed [45, 46].

A related problem of paramount importance in medicine is the detection of stable clusters considering at the same time both patients and the genes involved in subclass separation. To this end a new approach based on stability indices for biclustering algorithms has been recently proposed [47].

A second problem is related to data integration. Indeed different sources of biomolecular data are available for unsupervised analysis and for the analysis of the reliability of clustering results. Even if this topic has been investigated in supervised analysis of bio-molecular data [48, 49], largely less efforts have been devoted to the unsupervised analysis and in particular to the integration of multiple sources of data in the context of stability based methods. However, the integration of multiple sources of data to assess the validity of clustering solutions should in principle significantly improve the reliability of stability-based methods.

7   Conclusions

We presented an overview of stability based methods to estimate the reliability of clusterings discovered in bio-molecular data. These methods, if jointly used with statistical tests specifically designed to discover multiple structures, can be successfully applied to assess the statistical significance and to discover multiple structures in complex bio-molecular data. Summarizing, stability based methods can be applied for:

1. Assessment of the reliability of a given clustering solution
2. Assessment of the reliability of a each cluster inside a clustering
3. Assessment of the reliability of each example to a given cluster
4. Clustering model order selection: selection of the "natural" number of clusters.
5. Assessment of the statistical significance of a given clustering solution
6. Discovery of multiple structures underlying the data

In this introduction we focused on methods, without discussing in detail applications to real bio-molecular data. Anyway, bioinformatics applications of stability based methods can be found in most of the papers cited in this paper (see e.g. [21, 16, 20, 40]). Several problems not discussed in the paper remain open, ranging from the applicability of stability-based methods to problems characterized by very high number of examples and clusters (e.g.: discovery of functional classes of proteins), to their theoretical foundations [42].

References

[1] Dopazo, J.: Functional interpretation of microarray experiments. OMICS 3 (2006)
[2] Gasch, P., Eisen, M.: Exploring the conditional regulation of yeast gene expression through fuzzy k-means clustering. Genome Biology 3 (2002)
[3] Dyrskjøt, L., Thykjaer, T., Kruhøffer, M., Jensen, J., Marcussen, N., Hamilton-Dutoit, S., Wolf, H., Ørntoft, T.: Identifying distinct classes of bladder carcinoma using microarrays. Nature Genetics 33, 90–96 (2003)
[4] Kaplan, N., Friedlich, M., Fromer, M., Linial, M.: A functional hierarchical organization of the protein sequence space. BMC Bioinformatics 5 (2004)
[5] Jain, A., Murty, M., Flynn, P.: Data Clustering: a Review. ACM Computing Surveys 31, 264–323 (1999)

[6] Kasturi, J., Acharya, R.: Clustering of diverse genomic data using information fusions. Bioinformatics 21, 423–429 (2005)

[7] Avogadri, R., Valentini, G.: Fuzzy ensemble clustering based on random projections for dna microarray data analysis. Artificial Intelligence in Medicine (2008), doi:10.1016/j.artmed.2008.07.014

[8] Swift, S., Tucker, A., Liu, X.: An analysis of scalable methods for clustering high-dimensional gene expression data. Annals of Mathematics, Computing and Teleinformatics 1, 80–89 (2004)

[9] Napolitano, F., Raiconi, G., Tagliaferri, R., Ciaramella, A., Staiano, A., Miele, G.: Clustering and visualization approaches for human cell cycle gene expression data analysis. Int. J. Approx. Reasoning 47, 70–84 (2008)

[10] Azuaje, F., Dopazo, J.: Data Analysis and Visualization in Genomics and Proteomics. Wiley, Chichester (2005)

[11] Giardine, B., Riemer, C., Hardison, R., Burhans, R., Elnitski, L., Shah, P., Zhang, Y., Blankenberg, D., Albert, I., Taylor, J., Miller, W., Kent, W., Nekrutenko, A.: Galaxy: a platform for interactive large-scale genome analysis. Genome Res. 15, 1451–1455 (2005)

[12] Ciaramella, A., Cocozza, S., Iorio, F., Miele, G., Napolitano, F., Pinelli, M., Raiconi, G., Tagliaferri, R.: Interactive data analysis and clustering of genomic data. Neural Networks 21, 368–378 (2008)

[13] Handl, J., Knowles, J., Kell, D.: Computational cluster validation in post-genomic data analysis. Bioinformatics 21, 3201–3215 (2005)

[14] Dudoit, S., Fridlyand, J.: Bagging to improve the accuracy of a clustering procedure. Bioinformatics 19, 1090–1099 (2003)

[15] Ben-Dor, A., Shamir, R., Yakhini, Z.: Clustering gene expression patterns. Journal of Computational Biology 6, 281–297 (1999)

[16] Ben-Hur, A., Ellisseeff, A., Guyon, I.: A stability based method for discovering structure in clustered data. In: Altman, R., Dunker, A., Hunter, L., Klein, T., Lauderdale, K. (eds.) Pacific Symposium on Biocomputing, Lihue, Hawaii, USA, vol. 7, pp. 6–17. World Scientific, Singapore (2002)

[17] Dudoit, S., Fridlyand, J.: A prediction-based resampling method for estimating the number of clusters in a dataset. Genome Biology 3, 1–21 (2002)

[18] Yeung, K., Haynor, D., Ruzzo, W.: Validating clustering for gene expression data. Bioinformatics 17, 309–318 (2001)

[19] Kerr, M., Curchill, G.: Bootstrapping cluster analysis: assessing the reliability of conclusions from microarray experiments. PNAS 98, 8961–8965 (2001)

[20] McShane, L., Radmacher, D., Freidlin, B., Yu, R., Li, M., Simon, R.: Method for assessing reproducibility of clustering patterns observed in analyses of microarray data. Bioinformatics 18, 1462–1469 (2002)

[21] Smolkin, M., Gosh, D.: Cluster stability scores for microarray data in cancer studies. BMC Bioinformatics 36 (2003)

[22] Bittner, M., Meltzer, P., Chen, Y., Jiang, Y., Seftor, E., Hendrix, M., Radmacher, M., Simon, R., Yakhini, Z., Ben-Dor, A., Sampas, N., Dougherty, E., Wang, E., Marincola, F., Gooden, C., Lueders, J., Glatfelter, A., Pollock, P., Carpten, J., Gillanders, E., Leja, D., Dietrich, K., Beaudry, C., Berens, M., Alberts, D., Sondak, V.: Molecular classification of malignant melanoma by gene expression profiling. Nature 406, 536–540 (2000)

[23] Monti, S., Tamayo, P., Mesirov, J., Golub, T.: Consensus Clustering: A Resampling-based Method for Class Discovery and Visualization of Gene Expression Microarray Data. Machine Learning 52, 91–118 (2003)

[24] Lange, T., Roth, V., Braun, M., Buhmann, J.: Stability-based validation of clustering solutions. Neural Computation 16, 1299–1323 (2004)

[25] Valentini, G.: Clusterv: a tool for assessing the reliability of clusters discovered in DNA microarray data. Bioinformatics 22, 369–370 (2006)

[26] Bertoni, A., Valentini, G.: Model order selection for bio-molecular data clustering. BMC Bioinformatics 8 (2007)

[27] Bertoni, A., Valentini, G.: Randomized maps for assessing the reliability of patients clusters in DNA microarray data analyses. Artificial Intelligence in Medicine 37, 85–109 (2006)

[28] Bertoni, A., Valentini, G.: Discovering Significant Structures in Clustered Bio-molecular Data Through the Bernstein Inequality. In: Apolloni, B., Howlett, R.J., Jain, L. (eds.) KES 2007, Part III. LNCS, vol. 4694, pp. 886–891. Springer, Heidelberg (2007)

[29] Bertoni, A., Valentini, G.: Discovering multi-level structures in bio-molecular data through the Bernstein inequality. BMC Bioinformatics 9 (2008)

[30] Valentini, G.: Mosclust: a software library for discovering significant structures in bio-molecular data. Bioinformatics 23, 387–389 (2007)

[31] Bertoni, A., Valentini, G.: Randomized embedding cluster ensembles for gene expression data analysis. In: SETIT 2007 - IEEE International Conf. on Sciences of Electronic, Technologies of Information and Telecommunications, Hammamet, Tunisia (2007)

[32] Rand, W.: Objective criteria for the evaluation of clustering methods. J. Am. Stat. Assoc. 66, 846–850 (1971)

[33] Jain, A., Dubes, R.: Algorithms for clustering data. Prentice Hall, Englewood Cliffs (1988)

[34] Achlioptas, D.: Database-friendly random projections. In: Buneman, P. (ed.) Proc. ACM Symp. on the Principles of Database Systems. Contemporary Mathematics, pp. 274–281. ACM Press, New York (2001)

[35] Kaufman, L., Rousseeuw, P.: Finding Groups in Data: An Introduction to Cluster Analysis. Wiley, New York (1990)

[36] Bertoni, A., Valentini, G.: Assessment of clusters reliability for high dimensional genomic data. In: BITS 2007, Bioinformatics Italian Society Meeting, Napoli Italy (2007)

[37] Hoeffding, W.: Probability inequalities for sums of independent random variables. J. Amer. Statist. Assoc. 58, 13–30 (1963)

[38] Indyk, P.: Algorithmic Applications of Low-Distortion Geometric Embeddings. In: Proceedings of the 42nd IEEE symposium on Foundations of Computer Science, Washington DC, USA, pp. 10–33. IEEE Computer Society, Los Alamitos (2001)

[39] Johnson, W., Lindenstrauss, J.: Extensions of Lipshitz mapping into Hilbert space. In: Conference in modern analysis and probability. Contemporary Mathematics, Amer. Math. Soc., vol. 26, pp. 189–206 (1984)

[40] Valentini, G., Ruffino, F.: Characterization of lung tumor subtypes through gene expression cluster validity assessment. RAIRO - Theoretical Informatics and Applications 40, 163–176 (2006)

[41] Bertoni, A., Valentini, G.: In: Random projections for assessing gene expression cluster stability. In: IJCNN 2005, The IEEE-INNS International Joint Conference on Neural Networks, Montreal (2005)

[42] Ben-David, S., von Luxburg, U., Pal, D.: A sober look at clustering stability. In: Lugosi, G., Simon, H.U. (eds.) COLT 2006. LNCS, vol. 4005, pp. 5–19. Springer, Heidelberg (2006)

[43] Ben-David, S., von Luxburg, U.: Relating clustering stability to properties of cluster boundaries. In: 21st Annual Conference on Learning Theory (COLT 2008). LNCS, pp. 379–390. Springer, Heidelberg (2008)

[44] Harris, M., et al.: The Gene Ontology (GO) database and informatics resource. Nucleic Acid Res. 32, D258–D261 (2004)

[45] Brehelin, L., Gascuel, O., Martin, O.: Using repeated measurements to validate hierarchical gene clusters. Bioinformatics 24, 682–688 (2008)

[46] Avogadri, R., Brioschi, M., Ruffino, F., Ferrazzi, F., Beghini, A., Valentini, G.: An algorithm to assess the reliability of hierarchical clusters in gene expression data. In: Lovrek, I., Howlett, R.J., Jain, L.C. (eds.) KES 2008, Part III. LNCS, vol. 5179, pp. 764–770. Springer, Heidelberg (2008)

[47] Filippone, M., Masulli, F., Rovetta, S.: Stability and Performances in Biclustering Algorithms. In: Masulli, F., Tagliaferri, R., Verhivker, G.M. (eds.) CIBB 2008. LNCS (LNBI), vol. 5488, pp. 91–101. Springer, Heidelberg (2009)

[48] Troyanskaya, O., et al.: A Bayesian framework for combining heterogeneous data sources for gene function prediction (in *saccharomices cerevisiae*). Proc. Natl. Acad. Sci. USA 100, 8348–8353 (2003)

[49] Guan, Y., Myers, C., Hess, D., Barutcuoglu, Z., Caudy, A., Troyanskaya, O.: Predicting gene function in a hierarchical context with an ensemble of classifiers. Genome Biology 9 (2008)

[50] Alizadeh, A., Eisen, M., Davis, R., Ma, C., Lossos, I., Rosenwald, A., Boldrick, J., Sabet, H., Tran, T., Yu, X., Powell, J., Yang, L., Marti, G., Moore, T., Hudson, J., Lu, L., Lewis, D., Tibshirani, R., Sherlock, G., Chan, W., Greiner, T., Weisenburger, D., Armitage, J., Warnke, R., Levy, R., Wilson, W., Grever, M., Byrd, J., Botstein, D., Brown, P., Staudt, L.: Distinct types of diffuse large B-cell lymphoma identified by gene expression profiling. Nature 403, 503–511 (2000)

[51] Lapointe, J., Li, C., Higgins, J., van de Rijn, M., Bair, E., Montgomery, K., Ferrari, M., Egevad, L., Rayford, W., Bergerheim, U., Ekman, P., DeMarzo, A., Tibshirani, R., Botstein, D., Brown, P., Brooks, J., Pollack, J.: Gene expression profiling identifies clinically relevant subtypes of prostate cancer. PNAS 101, 811–816 (2004)

Appendix: R Software Packages Implementing Stability Based Methods

Two main *R* packages, implementing stability based methods, are freely available on the web:

1. *Mosclust*: Model order selection for **clust**ering problems. It implements stability based methods to discover the number of clusters and multiple structures underlying bio-molecular data [30]
2. *Clusterv*: **Cluster** **v**alidation. It implements a set of functions to assess the reliability of individual clusters discovered by clustering algorithms [25]

Overview of the *Clusterv* R Package

The *clusterv* R package implements a set of functions to assess the reliability of clusters discovered by clustering algorithms [25] This library is tailored to the analysis of high dimensional data and in particular it is conceived for the analysis of the reliability of clusters discovered using DNA microarray data.

Indeed cluster analysis has been used for investigating structure in microarray data, such as the search of new tumor taxonomies [50],[3],[51]. It provides a way for validating groups of patients according to prior biological knowledge or to discover new "natural groups" inside the data. Anyway, clustering algorithms always find structure in the data, even when no structure is present instead. Hence we need methods for assessing the validity of the discovered clusters to test the existence of biologically meaningful clusters.

To assess the reliability of the discovered classes, *clusterv* provides a set of measures that estimate the stability of the clusters obtained by perturbing the original data set. This perturbation is achieved through random projections of the original high dimensional data to lower dimensional subspaces, approximately preserving the distances between examples, in order to avoid too large distortions of the data. These random projections are repeated many times and each time a new clustering is performed. The obtained multiple clusterings are then compared with the clustering for which we need to evaluate its reliability. Intuitively a cluster will be reliable if it will be maintained across multiple clusterings performed in the lower dimensional subspaces. The measures provided by *clusterv* are based on the evaluation of the stability of the clusters across multiple random projections. By these measures we can assess:

1. the reliability of single individual clusters inside a clustering
2. the reliability of the overall clustering (that is, an estimate of the "optimal" number of clusters)
3. the confidence by which example may be assigned to each cluster

The *clusterv* R source package is downloadable from the *clusterv* web-site: `http://homes.dsi.unimi.it/~valenti/SW/clusterv/`

Overview of the *Mosclust* R Package

The *mosclust* R package (that stands for **m**odel **o**rder **s**election for **clust**ering problems) implements a set of functions to discover significant structures in bio-molecular data [30]. One of the main problems in unsupervised clustering analysis is the assessment of the "natural" number of clusters. Several methods and software tools have been proposed to tackle this problem (see [13] for a recent review).

Recently, several methods based on the concept of stability have been proposed to estimate the "optimal" number of clusters in complex bio-molecular data [22, 23, 24, 17, 25]. In this conceptual framework multiple clusterings are obtained by introducing perturbations into the original data, and a clustering is considered reliable if it is approximately maintained across multiple perturbations.

Several perturbation techniques have been proposed, ranging form bootstrap techniques [19, 16, 23], to random projections to lower dimensional subspaces [21, 27] to noise injection procedures [20]. All these perturbation techniques are implemented in *mosclust*.

The library implements indices of stability/reliability of the clusterings based on the distribution of similarity measures between multiple instances of clusterings performed on multiple instances of data obtained through a given random perturbation of the original data.

These indices provides a "score" that can be used to compare the reliability of different clusterings. Moreover statistical tests based on χ^2 and on the classical Bernstein inequality [37] are implemented in order to assess the statistical significance of the discovered clustering solutions. By this approach we could also find multiple structures simultaneously present in the data. For instance, it is possible that data exhibit a hierarchical structure, with subclusters inside other clusters, and using the indices and the statistical tests implemented in *mosclust* we may detect them at a given significance level.

The *mosclust* R source package is downloadable from the *mosclust* web-site: `http://homes.dsi.unimi.it/~valenti/SW/mosclust/`

Comparative *In Silico* Evaluation of MYB Transcription Factors in Eucalyptus, Sugarcane and Rice Transcriptomes

Nina M. Soares-Cavalcanti, Ana C. Wanderley-Nogueira, Luis C. Belarmino,
Petra dos Santos Barros, and Ana M. Benko-Iseppon

Universidade Federal de Pernambuco, Center for Biological Sciences (CCB),
Genetics Department, Laboratory for Plant Genetics and Biotechnology,
Rua Prof. Moraes Rêgo s/no. Recife, PE, Brazil
Tel.: (81) 2126.8520; Fax: (81) 2126.8569
ana.benko.iseppon@pesquisador.cnpq.br

Abstract. Transcription factors are essential in plant growth and development regulation and during response activation to environmental
stimuli. This study examined 237,954 sugarcane and 110,000 eucalyptus
ESTs in search of genes related to MYB factors active during abiotic
stresses in arabidopsis and rice. Searches revealed relative abundance
and diversity of MYB candidate sequences in both transcriptomes. In
general, 71 and 42 MYB sequences have been selected in eucalyptus
and sugarcane, respectively. The generated phenogram, using selected
sequences and sequences from other organisms, revealed a sub-grouping
according to domain-repeats number; confirming the ancestral origin of
MYB genes. Expression pattern analysis revealed higher number of reads
in wood, flower and bark libraries in eucalyptus, and in tissues infected
by pathogens in sugarcane. Splicing pattern comparative analyses of arabidopsis and rice ORFs with eucalyptus and sugarcane ORFs revealed
that neither of them presented similarities regarding all exons and intron
junctions.

Keywords: Bioinformatics, Salinity Stress, Drought Stress, Saccharum,
Crop Evolution.

1   Introduction

Abiotic stresses, like drought and salinity, generally impose limitations to plant
growth and crop productivity, also limiting farming activities in many regions
throughout the world. Salinity limits vegetative and reproductive growth of
plants by inducing severe physiological dysfunctions and causing widespread
direct and indirect harmful effects, even at low salt concentrations. Once water
is fundamental to almost all aspects of plant physiology, its reduction imposes
strong and recurring selective pressure to plants [1].

Drought and salinity stresses are often interconnected, and cause cellular damage and secondary stresses, such as osmotic and oxidative stresses. Plants must

F. Masulli, R. Tagliaferri, and G.M. Verkhivker (Eds.): CIBB 2008, LNBI 5488, pp. 44–55, 2009.

be able to "sense" the environmental cues before being able to respond appropriately to the abiotic stress. Due to the complex nature of stress, multiple sensors, rather than a single sensor, are more likely to be responsible for the stress perception. After initial stress recognition, a signal transduction cascade is induced. Secondary messengers activate stress-responsive genes, like transcription factors (TF), generating the initial stress response [2]. Many TF families existed before the divergence of plants, animals and fungi; however, the size of TF families differs considerably among eukaryote kingdoms. For example, the MADS-box family has two members in *Drosophila*, two in *C. elegans* and four in *S. cerevisae*. However, it has more than 100 members in *Arabidopsis*, suggesting substantial amplification of this gene in the land plant lineage [3]. A typical plant TF contains a DNA-biding region, an oligomerization site, a transcription-regulation domain, and a nuclear localization signal [4].

The MYB superfamily is the largest TF gene family in plants, also present in all eukaryotes analyzed. The encoded proteins are crucial to control proliferation and differentiation in a number of cell types, and share the conserved MYB DNA-binding domain [5]. This domain generally comprises up to three imperfect repeats, each forming a helix-turn-helix (HTH) structure of about 35 amino acids, with three regularly spaced tryptophan residues, that form a tryptophan cluster in the three dimensional HTH structure [5]. Regarding MYB repeats, the proteins can be classified in three subfamilies R1/2, R2R3 and R1R2R3 [5], while in plants *MYB* genes bearing two repetitions are the most common [3]. The development of tolerant cultivars to abiotic stresses, as well as the development of technologies to over express TFs in transgenic plants may help in the tolerance and surveillance during drought and salinity stress, collaborating to an increase in the world crop production.

Little information is available about *MYB* diversity and abundance in plants with large genomes, as sugarcane or in woody angiosperms, as eucalyptus. The present work aimed to perform a data mining-based identification of *MYB* TF family in FOREST and SUCEST databases, as compared with data available for rice and arabidopsis sequences, comparing them with sequences deposited in public databases and literature data.

2 Methods

Complete full length cDNA sequences of *MYB* transcription factor genes from *A. thaliana* were obtained in fasta format from NCBI, based on Seki *et al.* (2002) [6] results. Each sequence was compared against the SUCEST, FORESTs and TGI (The Gene Index Project) databases using tBLASTn tool. Only sequences with e-value e^{-5} or less were considered. DNA sequences of eucalyptus, sugarcane and rice were translated using ORFfinder program (www.ncbi.nlm.nih.gov/gorf), and screened for similarity to entries in GenBank at NCBI (National Center for Biotechnology Information; http://www.ncbi.nlm.nih.gov/) using mainly BLASTx and BLASTp algorithms [7]. Moreover, the Blink option was used to obtain similar sequences, containing the searched domains, in other organisms. RPS_BLAST option was used to infer CD presence and integrity.

A multiple alignment was generated using the CLUSTALx program, aiming to evaluate the conservation among MYB sequences. For this alignment only sequences with complete domains were selected. The MEGA software (V. 4.0) allowed the dendrogram construction using Neighbor Joining method [8], gamma distance and Bootstrap test of inferred phylogeny (1,000 replications). The visualization of the trees was made using TreeView [9]. To generate an overall picture of the expression patterns of MYB TFs in sugarcane and eucalyptus, annotated reads were associated to respective library and used for a hierarchical clustering approach with normalized data using CLUSTER program [10]. Dendrograms including both axes (using the weighted pair-group for each gene class and library) were generated. On graphics (Figure 3), light gray means no expression and dark gray/black all degrees of expression.

The analyses of protein migratory behavior, using only the best matches of each seed sequence, were generated with the JvirGel program [11], using analytical modes for serial calculation of MW (molecular weight), pI (Isoelectric Point), pH-dependent charge curves and hydrophobicity profiles, a virtual 2D gel as JavaTM applet.

Aiming to infer about the putative mRNA splicing sites among both dicots (arabidopsis and eucalyptus) and monocots (rice and sugarcane), eucalyptus and sugarcane best hits were translated, having their splicing sites compared with ORFs from arabidopsis and rice, respectively.

3 Results

3.1 Eucalyptus, Sugarcane and Rice Orthologs

MYB candidates were identified in the studied databases, including 89 clusters in FOREST, 81 in SUCEST and 261 in TGI. Only for Ara_MYB_02 seed sequence no match could be uncovered in these databases (Table 1).

Considering the identified CDs in the three mentioned data banks, 149, 13 and 29 sequences presented the SANT and MYB domains combined as follows: 2xSANT, 2xMYB and 1xSANT + 1xMYB (where "2x" means "twice" and "1x" means "once"). Other 78 and 23 sequences presented also SANT and MYB complete CDs, nine with 3xSANT, two and 18 with SANT and MYB incomplete domains among other combinations including incomplete domains and 99 with significant homology but no CD. Moreover, one rice cluster presented two copies of SANT domain and one copy of the maize P_C domain (Table 1). In addition, the results showed that 35.2%, 63.4% and 1.4% of the eucalyptus clusters presented one, two and three MYB repetitions, respectively. In sugarcane and rice clusters with one MYB repetition represented by 35.7% and 32.4%, two MYB repetitions by 59.5% and 64.4% and three repetitions by 4.8% and 3.2%. In a general view 86, 79 and 259 clusters were obtained at FOREST, SUCEST and TGI projects, which participated to salinity+drought category (16, 19 and 59, respectively) and to drought category (70, 60 and 200, respectively). As a whole, results for *MYB* genes showed that the percentage of clusters in each stress category was similar considering all organisms. MYB analysis showed that most

Table 1. Data mining resources and results using arabidopsis seed sequences, matches of eucalyptus, sugarcane and rice with respective domains

Arabidopsis thaliana		Matches of Eucalyptus, Sugarcane and Rice										Domains Found					
Stress Type	Seed Sequence	Higher Match	e-value	Score	Fr.	Other up to E-05	Size nt	ORF	Dom 1	Start	End	Dom 2	Start	End	Dom 3	Start	End
		Euc_MYB_01	$4e^{-33}$	135	1	16	1713	409ic	MYB+bind	24	68	-	-	-	-	-	-
		Euc_MYB_02	$1e^{-27}$	117	3	16	789	195ic	SANT	137	180	-	-	-	-	-	-
Salinity and Drought	AY050961 Ara_MYB_01 type MYB-bind 2684 nt	Euc_MYB_03	$5e^{-08}$	52	1	16	1221	263ic	SANT	11	54	SANT	122	167	-	-	-
		Sug_MYB_01	$3e^{-31}$	129	1	19	3291	748cp	SANT	53	96	-	-	-	-	-	-
		Sug_MYB_02	$1e^{-21}$	97	3	19	670	191ic	MYB+bind	96	140	-	-	-	-	-	-
		Sug_MYB_03	$2e^{-07}$	50	-1	19	1483	300cp	SANT	19	66	SANT	127	172	-	-	-
		Ric_MYB_01	$2e^{-29}$	330	2	59	1007	246ic	MYB+bind	44	88	-	-	-	-	-	-
		Ric_MYB_02	$1.1e^{-28}$	330	3	59	2805	719cp	SANT	26	69	-	-	-	-	-	-
		Ric_MYB_03	$3.8e^{-07}$	126	1	59	934	182cp	SANT	26	75	SANT	114	159	-	-	-
Salinity, Drought & Cold	AV821864 Ara_MYB_02 no CD, 604 nt	no hits	-	-	-	-	-	-	-	-	-	-	-	-	-	-	-
		no hits	-	-	-	-	-	-	-	-	-	-	-	-	-	-	-
		no hits	-	-	-	-	-	-	-	-	-	-	-	-	-	-	-
		Euc_MYB_04	$8e^{-76}$	280	2	70	706	210ic	MYB+bind	14	61	MYB+bind	67	112	-	-	-
		Euc_MYB_05	$7e^{-61}$	230	1	70	995	219cp	SANT	16	61	SANT	69	112	-	-	-
		Euc_MYB_06	$1e^{-45}$	177	3	70	796	121cp	SANT	17	61	MYB+bind	67	112	-	-	-
		Euc_MYB_07	$3e^{-26}$	115	2	70	796	216ic	SANT	67	110	SANT	118	162	SANT	170	213
		Euc_MYB_08	$7e^{-48}$	187	1	70	779	216ic	SANT	24	67	-	-	-	-	-	-
		Euc_MYB_09	$1e^{-35}$	146	2	70	867	105cp	MYB	14	62	-	-	-	-	-	-
		Sug_MYB_04	$2e^{-61}$	232	2	60	687	226ic	SANT	38	83	SANT	91	134	-	-	-
Only Drought	AF386932 Ara_MYB_03 type 1148 nt	Sug_MYB_05	$3e^{-61}$	231	2	60	613	117ic	SANT	22	66	MYB+bind	72	115	-	-	-
		Sug_MYB_06	$1e^{-35}$	146	2	60	479	138ic	MYB+bind	24	71	MYB+bind	77	120	-	-	-
		Sug_MYB_07	$3e^{-29}$	125	1	60	600	109cp	SANT	16	61	-	-	-	-	-	-
		Sug_MYB_08	$3e^{-22}$	102	-2	60	1560	274cp	SANT	52	96	SANT	104	148	SANT	156	196
		Sug_MYB_09	$4e^{-21}$	99	1	60	449	145ic	MYB+bind	82	129	-	-	-	-	-	-
		Ric_MYB_04	$1.6e^{-66}$	679	2	200	1331	368cp	SANT	16	61	SANT	69	112	-	-	-
		Ric_MYB_05	$2.2e^{-49}$	518	3	200	1105	255cp	MYB	16	61	SANT	71	112	-	-	-
		Ric_MYB_06	$2.4e^{-42}$	452	1	200	996	331cp	SANT	16	61	SANT	71	112	-	-	-
		Ric_MYB_07	$2.8e^{-41}$	352	3	200	536	106ic	SANT	49	92	-	-	-	P_C	113	331
		Ric_MYB_08	$3e^{-39}$	422	1	200	1198	102cp	MYB	14	61	MYB	67	112	-	-	-
		Ric_MYB_09	$3.5e^{-26}$	300	1	200	873	290cp	MYB	94	137	-	-	-	-	-	-
		Ric_MYB_10	$6.6e^{-24}$	285	1	200	2572	587cp	SANT	70	113	SANT	121	165	SANT	173	216

Abbreviations: nt=nucleotides; fr.=frame; ORF=Open Reading Frame; Dom=Domain; Ara_=*Arabidopsis* seed sequence; Euc_=*Eucalyptus* cluster; Sug_=Sugarcane cluster; Ric_=Rice cluster; cp=complete; ic=incomplete; bind=DNA binding element; - =not found.

clusters (ca. 78%) belonged to drought category, against ca. 22% from salinity+drought category.

3.2 Isoelectric Point Analysis

The virtual electrophoresis evaluation of MYB proteins from eucalyptus, sugarcane and rice best matches revealed that all sequences presented isoelectric point from 4.17 to 10.69. Considering the molecular mass, values varied of 5.6 MW to 81.8 MW, with sequences presenting two groupings with few exceptions (Data not shown).

3.3 Phenetic Analysis

Multiple alignment of MYB sequences revealed a considerable degree of conservation among MYB family sequences considering fungi, algae, protozoan, plants and animals. The generated dendrogram using MYB proteins grouped sequences in 24 different branches, with a clear segregation of the organisms in subclades according to their higher taxonomic grouping (Figure 1).

However, in branches XIX and XX, the grouping of distant related species as micro algae (*Ostreococcus tauri*), a gymnosperm (*Sequoia sempervirens*) and a protozoan (*Dictyostelium discoideum*) together with yeast (*Saccharomyces cerevisae*) was remarkable. Also interesting was the grouping of members exclusively of the animal kingdom into the XVI branch, including insects, amphibian, fishes, birds and mammals (Figure 1). Angiosperms also grouped together, but analyzing internal grouping no distinction was revealed among monocots, dicots and

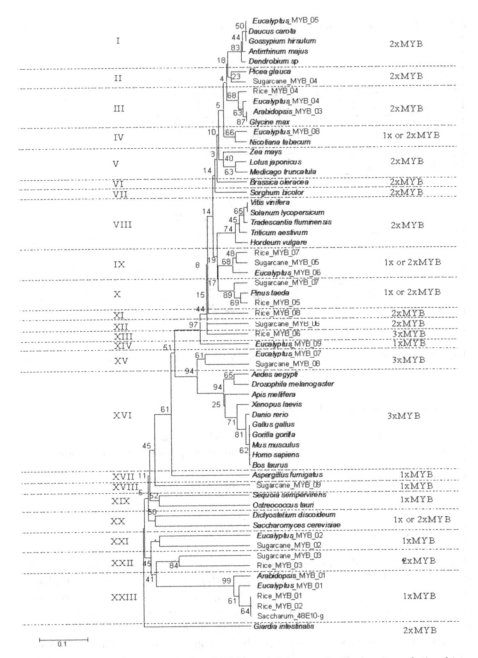

Fig. 1. Dendrogram generated after Neighbor Joining analysis, showing relationships among the MYB seed sequences of arabidopsis and orthologs of eucalyptus, sugarcane, rice and other organisms. Dotted lines delimit the main taxonomic units and roman numbers on the left of the dendrogram and between the dotted lines refer to the groupings. At the right of the groupings the number of MYB repetitions per group is given. The numbers on the left of the clusters are P values. Bar represents similarity coefficient.

also gymnosperms. Moreover, we could note that inferior organisms like the fungus *Aspergillus fumigatos* and the protozoan *Giardia intestinalis* did not grouped to any other organism, pertaining to individual branches. In most cases, grouping reflected the number of MYB CD repetitions and therefore reflected duplication and divergent evolution events.

3.4   Expression in Different Tissues

It was observed that most tissues presented expression of at least one MYB TF considering FOREST and SUCEST libraries (Figure 2A and B); with exception of a complete absence in the sugarcane NR library (that included all tissues) (Figure 2B). In a general view, MYB family was best represented in eucalyptus by reads from bark/wood tissues from 7-8 year-old trees of *E. grandis* followed by bud, flower and fruit tissues (Figure 2A, bottom). In sugarcane two tissues infected by bacteria presented higher expression (Figure 2B, bottom). Diverse transcripts obtained both in FOREST and SUCEST data bases were used to perform a hierarchical clustering analysis from ESTs permitting an evaluation of expression intensity considering co-expression in different libraries (black upper dendrogram) (Figure 2A and B).

Co-expression was observed regarding the two *MYB* genes in both eucalyptus and sugarcane, with higher expressed WD and AD libraries, respectively (for

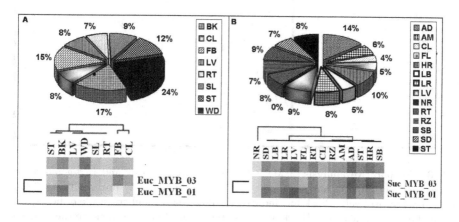

Fig. 2. Prevalence of MYB factors (upper) and *in silico* expression pattern (bottom) in FOREST (A) and SUCEST (B) libraries. Black indicates higher expression, gray lower expression, and light gray absence of expression in the corresponding tissue and cluster. Abbreviations: AD=tissues infected by *Gluconacetobacter diazotroficans*; AM=Apical meristem; CL=Callus; FL=Flower; HR=tissues infected with *Herbaspirillum rubrisubalbicans*; LB=Lateral bud; LR=Leaf roll; LV=Leaves; RT=Root; RZ=Stem-root transition; SB=Stalk bark; SD=Seeds; ST=Stem; NR=all libraries; BK=bark/duramen/pith and alburnum of *E. grandis*; LV=leaves; RT=roots; FB=flowers; CL=calli; ST=stem; SL=seedlings grown in the dark; WD=wood; Euc_MYB=Eucalyptus MYB candidates; Sug_MYB=sugarcane MYB candidates.

libraries abbreviations see legend of Figure 2). In *Eucalyptus* ST and BK, LV, WD, SL and RT and FB and CL libraries were co-expressed. Regarding sugarcane, co-expression was observed in SD, LB, LR, LV and FL, RT, CL, RZ, AM and AD and ST, HR AND SB libraries.

3.5 Splicing Analysis

Evaluation of *MYB* genes revealed the presence of three exons interrupted by two introns in MYB_03 (Figure 3A). ORF comparison revealed that neither eucalyptus nor sugarcane presented similarities regarding all exons. Moreover, in arabidopsis MYB_01 an alternative splicing was observed, being this sequence MYB_01.1 formed by four introns and four exons and MYB_01.2 formed by seven of each one (Figure 3B). Considering eucalyptus sequences, high similarities were obtained with the three first exons in the first case, and six exons in the second (Figure 3B). This same sequence analysis in rice revealed five exons and five introns with high similarities to sugarcane (Figure 3A and B).

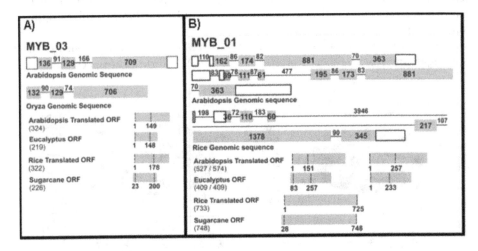

Fig. 3. Comparative analysis of MYB ORFs from arabidopsis and rice genomic sequences with eucalyptus and sugarcane clusters. White and gray boxes indicate UTRs and exons, respectively. Black line indicate introns. Black numbers inside of boxes correspond to the segment size. Black numbers above black line correspond to the segment size. Numbers below the organisms name correspond to ORFs size. Black numbers below the gray boxes (ORFs) correspond to alignment start and end positions. Black bars inside of these boxes delimit the alignment region. Abbreviations: ORF, Open Reading Frame.

4 Discussion

Modulation of transcriptional activity is fundamental for gene expression regulation associated with most biological phenomena and is largely mediated by proteins that interact directly or indirectly with specific DNA sequences (*cis* elements) in the promoter region of genes. In spite of their importance, such factors present often discrete expression levels enough to activate given genes, and their expression level may not be enough to be measured in microarray experiments. Despite of that, they have been identified in large databases as for *Eucalyptus* (with no previous analysis regarding such factors) and for sugarcane with a single evaluation of TF transcript abundance [12].

MYB family is represented by at least 71, 42 and 189 orthologous sequences in eucalyptus, sugarcane and rice genomes, respectively. Analysis of these sequences showed the presence of two categories of domains: SANT and MYB while comparisons among them showed that they were homologous domains. Thus, *Eucalyptus* and *S. officinarum* transcriptomes presented, at least, 25 and 15 MYBs type R1/R2 and one and two MYBs R1R2R3, respectively, whereas *O. sativa* genome contain 61 members with one repetition, 121 with two repetitions and seven with three repetitions. Of the seven rice candidates to MYB R1R2R3, one presented two SANT domains and one P_C domain, a similar feature found in maize sequences, a domain combination apparently absent in Magnoliophyta division that may have appeared after separation of Liliopsida and Magnoliopsida classes during divergent evolution.

The obtained results corroborate literature reports, showing high amount of MYB type R2R3, followed by MYB type R1/2 and in few cases R1R2R3. According to Jin and Martin (1999) [5], in plants the predominant family of MYB proteins presents two repeats, but three-repeat MYB together with a growing number of MYB proteins with a single MYB domain were recently identified in plants, as found also in the present work.

Yanhui *et al.* (2006) [13] identified 64 (33%) genes containing one repetition, 126 (65%) containing two repetitions and five (2%) containing three repetitions. A similar proportion to the one found by these authors was revealed in our results, in all transcriptomes studied. Evaluations of transcription profile by microarray using 7,000 genes in *Arabidopsis* revealed the expression of *MYB* genes are increased 5-fold or more in the first hours after abiotic stress induction [6], with similar results also found in a later experiment in rice [14]. These evaluations revealed that MYB is mainly active during initial salinity and drought stress perception. The identification of such multi function TFs, i.e., active in more than one stress type, are among the most important candidates for improvement of plant resistance to abiotic stress.

Considering our evaluation using virtual bidimensional electrophoresis migration for MYB sequences, two groups could be clearly identified, with only three sequences (two of rice and one of sugarcane) deviating especially with respect to their molecular weight, despite the similar pIs. The first group (with pI from 4.75 to 7.11) included mainly rice sequences, while the second group included most species of the three remaining species. Since the functional domains presented high degree of conservation, probably the divergent pattern of migration for some sequences reflects divergences of the extra-domain regions that are responsible for the acid and basic character of the proteins; these extra-domain variations are, also, responsible for the diversity of the sequences and for differences in the functionality of the genes.

Regarding generated dendrograms, including MYB proteins from diverse organisms, it was possible to discriminate groupings based on the number of MYB repetitions; the groups XIV, XVII to XIX, XXI and XXIII presented one repetition, the groups I to III, V to VIII, XI to XII, XXII and XXIV presented two repetitions and the groups XII and XV to XVI presented three repetitions. However, some clades containing sequences with one and two MYB repetitions side by side could be observed, as in groups IV, IX, X and XX. The groupings containing two repetition types were formed just by plants, as expected, since in plants the R2R3 type is predominant for MYB family [5], a type absent in *Saccharomyces cerevisiae* whilst only 1-3 copies *R2R3-MYB* genes per haploid genome are known in protists and animals [15]. Besides, the evolutionary pattern of the MYB factor seems to follow a sinarcheomorphic pattern, considering their presence in such different organisms. Regarding the multiple structural forms (number o MYB repetitions), we observed that the presence of three repetitions may be considered a synapomorphy of the animal kingdom, appearing as a monophyletic group. On the other hand, it was not possible to classify remaining types of CD combinations (R1/2 e R2R3), that formed merophyletic polyphyletic groups.

The fact that MYB proteins from animals have grouped separately seems to be related with the number of MYB repetitions. In animals MYB family presents as main characteristic the presence of three imperfect repetitions [16]. In addition, three-repeat MYBs have recently been identified in plants together with a growing number of MYB proteins with a single MYB domain [5]. Our results showed that MYB proteins with three repetitions were not closely related to animal proteins, revealing that these repetitions present specific motifs, in both plants and animals. Despite of that, the neighboring position of both plant sequences with R3 to the animal clade (XIII+XV and XVI, respectively), may indicate that the presence of the third domain is an ancestral feature in plants.

The junction in the same clade of organisms possessing one (R1/2) and two (R2R3) repetitions seem to be related with the functionality of these repetitions. According to Kamiya *et al.* (2002) [17] the DNA-binding specificity is determined primarily by R2 and R3, whereas R1 is not thought to be important for DNA binding specificity. This suggests that selective factors could have collaborated to the diversity of CD combinations in plants. All told, the generated dendrogram may reflect adaptation driven by environmental pressure over the analyzed taxa, rather than stochastic evolutionary patterns of MYB sequences, a gene family that appeared before splitting of the main kingdoms.

Since plants are sessile, transcriptional regulation is also important for adaptation to abiotic stresses, and for protection from biotic stress [18]. The tissues that first "sense" and respond to abiotic stress are roots and root-shoot transition region, for the reduction in water availability causing osmotic imbalance, and the leaves, for the alteration in transpiration and respiration rates [1]. In the FOREST database the *MYB* candidate genes were present in greater amount in WD (wood), BK (bark, duramen, pith and alburnum) and FB (bud, flower and fruit) libraries, as expected, since these tissues are closely related to plant development. The first one being directly involved in the sustentation, hindering the falling of the plant; the second directly related with plant growth and development, mainly with respect to plant support and cellular water balance; and the last in reproductive processes. In both, eucalyptus and sugarcane, the analysis of expression showed the existence of transcripts in almost all tissues, revealing the crucial importance of transcription factors for plants.

The results from SUCEST database, however, were surprising, with larger amount of reads from libraries generated from tissues infected by nitrogen fixing bacteria (*Herbaspirilum rubrisubalbicans* and by *Gluconacetobacter diazotroficans*). In these libraries the infection occurred during *in vitro* cultivation. So, the environmental stress can not be discarded as the inducing factor for higher expression, since the *in vitro* environment is largely known as an abiotic stress condition. On the other hand, higher amount of reads from leaves tissues were also found; a plant region closely related with sensing of abiotic stress and responsible for respiration and photosynthesis, processes involved directly with plant growth and development. An association of the here procured TFs with defense also against biotic stress can not be discarded. Tobacco plants presented increased resistance to osmotic stress but also to pathogen attack when a unique *ERF* gene (*Tsi*) was overexpressed [19], showing that responses to different stresses can be regulated by a single TF. Besides activation of *MYB* genes in bacteria infected tissues in sugarcane significant expression was also observed in SB (Stalk Bark) and LV (Leaves) - vascular tissues that need higher amounts of nutrients for plant development and tissues involved in plant respirations respectively - showing the participation of these genes both in defense against pathogen and in acquisition of tolerance to abiotic stresses.

Finally, the fact that MYB TFs were detected in almost all libraries in both FOREST and SUCEST databases, and that a single eucalyptus library was constructed under water deficit, supports the assumption that these genes are expressed not only under stress condition, but also in lower levels or in the absence of stress situations.

Regarding splicing comparative evaluation, TF sequences showed high conservation of splicing patterns. In fact, gene structure analyses of some TFs in different organisms indicated conservation, including the position of putative junction between introns and exons [20,21]. Studying rates and models of intron evolution in green plants Roy and Penny (2007) [22] compared 8,258 pairs of orthologous genes among *A. thaliana* and *O. sativa* and noted that 94.7% of intron positions were conserved comparing the two species and that only 5.3% of introns were species-specific.

The majority of arabidopsis (59%) and rice (53%) *MYB type-R2R3* genes presented conserved splicing patterns, comprising three exons and two introns; however, either one or both of the two introns can be absent. In addition, variable splicing patterns were detected in 22% of arabidopsis and 35% of rice *MYB type-R2R3* genes [23]. Our observations regarding eucalyptus and sugarcane ORFs revealed a high similarity with three exons present in MYB_03 from arabidopsis and also the corresponding rice paralog, leading to the recognition of at least two introns in MYB sequences of eucalyptus and sugarcane. Additionally, comparisons among ORFs aligned to MYB_01 in *Eucalyptus* and *S. officinarum* exposed at least three exon-exon junction sites revealing that eucalyptus may present two splicing forms for MYB_01 orthologs.

In general, *MYB* TF genes were found in all genomes studied, also considering non induced conditions. However the construction of additional libraries submitted to stress conditions could provide further candidates and give a best coverage of existing TFs in sugarcane and eucalyptus. The recognized sequences and especially their putative splicing sites bring valuable information for future *in vitro* and *in vivo* assays, especially regarding a better understanding of the functional role of their structural diversity.

The identified *MYB* orthologs from sugarcane and eucalyptus are also valuable for comparative studies of evolution among higher plants, including understanding of processes in woody and herbaceous plants. Furthermore, the obtained results are useful for generation of molecular markers in genetic mapping, germplasm screening for breeding purposes and comparative mapping. Since transcription factors are expressed in low levels, the knowledge of the *MYB* gene pool diversity is meaningful for the development of bioinformatic tools and automated assays related to differential expression.

Acknowledgments

The authors acknowledge the FOREST and SUGARCANE consortium for permission to access and mine their databases. To CNPq (Conselho Nacional de Desenvolvimento Científico e Tecnológico) we thank for fellowships and financial support.

References

1. Munns, R.: Comparative Physiology of Salt and Water Stress. Plant Cell Environ. 28, 239–250 (2002)
2. Wang, W., Vinocur, B., Altman, A.: Plant Responses to Drought, Salinity and Extreme Temperatures: Towards Genetic Engineering for Stress Tolerance. Planta 218, 1–14 (2003)
3. Riechmann, J.L., Herad, J., Martin, G., Reuber, L., Jiang, C.-Z., Keddie, J., Adam, L., Pineda, O., Ratcliffe, O.J., Samaha, R.R., Creelman, R., Pilgrim, M., Broun, P., Zhang, J.Z., Ghandehari, D., Sherman, B.K., Yu, G.: *Arabidopsis* Transcription Factors: Genome-Wide Comparative Analysis among Eukaryotes. Science 290, 2105–2110 (2000)
4. Liu, L., White, M.J., MacRae, T.H.: Transcription Factors and their Genes in Higher Plants: Functional Domains Evolution and Regulation. Eur. J. Biochem. 262, 247–257 (1999)
5. Jin, H., Martin, C.: Multifunctionality and Diversity within the Plant *MYB*-Gene Family. Plant Mol. Biol. 41, 577–585 (1999)

6. Seki, M., Narusaka, M., Ishida, J., Nanjo, T., Fujita, M., Oono, Y., Kamiya, A., Nakajima, M., Enju, A., Sakurai, T., Satou, M., Akiyama, K., Taji, T., Yamaguchi-Shinozaki, K., Carninci, P., Kawai, J., Hayashizaki, Y., Shinozaki, K.: Monitoring the Expression Profiles of 7,000 *Arabidopsis* Genes under Drought Cold and High-Salinity Stresses Using a Full-Length cDNA Microarray. Plant J. 31, 279–292 (2002)
7. Altschul, S.F., Gish, W., Miller, W., Myers, E.W., Lipman, D.J.: Basic Local Alignment Search Tool. J. Mol. Biol. 251, 403–410 (1990)
8. Sneath, P.H.A., Sokal, R.R.: Numerical Taxonomy. Freeman, San Francisco (1973)
9. Page, R.D.: Treeview Program Version 161. Comput. Appl. Biosci. 12, 357–358 (1996)
10. Eisen, M.B., Spellman, P.T., Brown, P.O., Botstein, D.: Cluster Analysis and Display of Genomic-Wide Expression Pattern. Proc. Natl. Acad. Sci. USA 95, 14863–14868 (1998)
11. Hiller, K., Schobert, M., Hundertmark, C., Jahn, D., Munch, R.: JVirGel: Calculation of Virtual Two-Dimensional Protein Gels. Nucleic Acids Res. 31, 3862–3865 (2003)
12. Papini-Terzi, F.S., Rocha, F.R., Vêncio, R.Z.N., Oliveira, K.C., Felix, J.M., Vicentini, R., Rocha, C.S., Simões, A.C.Q., Ulian, E.C., Mauro, S.M.Z., Silva, A.M., Pereira, C.A.B., Menossi, M., Souza, G.M.: Transcription Profiling of Signal Transduction-Related Genes in Sugarcane Tissues. DNA Res. 12, 27–38 (2005)
13. Yanhui, C., Xiaoyuan, Y., Kun, H., Meihua, L., Jigang, L., Zhaofeng, G., Zhiqiang, L.: The MYB Transcription Factor Superfamily of *Arabidopsis*: Expression Analysis and Phylogenetic Comparison with the Rice MYB Family. Plant Mol. Biol. 60, 107–124 (2006)
14. Rabbani, M.A., Maruyama, K., Abe, H., Khan, M.A., Katsura, K., Ito, Y., Yoshiwara, K., Seki, M., Shinozaki, K., Yamaguchi-Shinozaki, K.: Monitoring Expression Profiles of Rice Genes under Cold, Drought, and High-Salinity Stresses and Abscisic Acid Application Using cDNA Microarray and RNA Gel-Blot Analyses. Plant Physiol. 133, 1755–1767 (2003)
15. Thompson, M.A., Ramsay, R.G.: Myb: an Old Oncoprotein with New Roles. Bioessays 17, 341–350 (1995)
16. Lipsick, J.S.: One Billion Years of MYB. Oncogene 13, 223–235 (1996)
17. Kamiya, T., Kawabe, A., Miyashita, N.T.: Nucleotide Polymorphism at *Atmyb2* Locus of Wild Plant *Arabidopsis thaliana*. Genet. Res. 80, 89–98 (2002)
18. Shikata, M., Matsuda, Y., Ando, K., Nishii, A., Takemura, M., Yokota, A., Kohchi, T.: Characterization of *Arabidopsis* ZIM a Member of Novel Plant-Specific GATA Factor Gene Family. J. Exp. Bot. 55, 631–639 (2004)
19. Park, J.M., Park, C.J., Lee, S.B., Ham, B.K., Shin, R., Paek, K.H.: Overexpression of the Tobacco *Tsi1* Gene Encoding an EREBP/AP2-Type Transcription Factor Enhances Resistance Against Pathogen Attack and Osmotic Stress in Tobaccoy. Plant Cell 13, 1035–1046 (2001)
20. Wu, K.-L., Guo, Z.-J., Wang, H.-H., Li, J.: The WRKY Family of Transcription Factors in Rice and Arabidopsis and their Origins. DNA Res. 12, 9–26 (2005)
21. Reyes, J.C., Muro-Pastor, M.I., Florêncio, F.J.: The GATA Family of Transcription Factors in Arabidopsis and Rice. Plant Physiol. 134, 1718–1732 (2004)
22. Roy, S.W., Penny, D.: Patterns of Introns Loss and Gain in Plants: Intron Loss-Dominated Evolution and Genome-Wide Comparison of *O. sativa* and *A. thaliana*. Mol. Biol. Evol. 24, 171–181 (2007)
23. Jiang, C., Gu, X., Peterson, T.: Identification of Conserved Gene Structure and Carboxy-Terminal Motifs in the MYB Gene Family of *Arabidopsis* and *Oryza sativa* L. ssp. *indica*. Genome Biol. 5, 1–11 (2004)

Building Maps
of Drugs Mode-of-Action
from Gene Expression Data

Francesco Iorio[1,2], Roberto Tagliaferri[2], and Diego di Bernardo[1,3]

[1] Systems and Synthetic Biology Lab, TeleThon Institute of Genetics and Medicine
(TIGEM), Via Pietro Castellino, Naples, Italy
[2] Neural and Robotic Networks Lab (NeuRoNe Lab), Dept. of Mathematics and
Computer Science, University of Salerno, Via Ponte don Melillo, Fisciano (SA), Italy
[3] Dept. of Computer Science and Systems, University Federico II of Naples, P.le
Tecchio, Naples, Italy
{iorio,dibernardo}@tigem.it, rtagliaferri@unisa.it
http://dibernardo.tigem.it,
http://neuronelab.dmi.unisa.it

Abstract. We developed a data-mining and visualization approach to
analyze the mode-of-action (MOA) of a set of drugs. Starting from wide-
genome expression data following perturbations with different compounds
in a reference data-set, our method realizes an euclidean embedding providing a map of MOAs in which drugs sharing the therapeutic application
or a subset of molecular targets lies in close positions. First we build a low-
dimensional, visualizable space combining a *rank-aggregation* method and
a recent tool for the analysis of the enrichment of a set of genes in ranked
lists (based on the *Kolmogorov-Smirnov* statistic). This space is obtained
using prior knowledge about the data-set composition but with no assumptions about the similarities between different drugs. Then we assess that,
despite the complexity and the variety of the experimental conditions, our
aim is reached with good performance without across-condition normalization procedures.

Keywords: Drug mode-of-action, Connectivity Map, Gene Set Enrichment Analysis.

1 Introduction

Identifying pathways mediating a drug mode of action is a key challenge in
biomedicine. We demonstrated that using gene expression profiles in yeast, it is
possible to detect the mode of action of a drug candidate [1].

In the process of *Drug-Design* a crucial role is played by the *lead-identification*
step in which two main problems are typically tackled:

– We have a therapeutic molecule acting on a set of proteins and, among
  them, a particular protein of interest and we are searching for a more specific
  molecule with less side effects;

F. Masulli, R. Tagliaferri, and G.M. Verkhivker (Eds.): CIBB 2008, LNBI 5488, pp. 56–65, 2009.

– We have a new potential therapeutic molecule but we don't know its mode-of-action (MOA), i.e. the set of proteins the molecule interacts with, and we want to analyze its function.

In both cases, bioinformatic tools enabling the analysis of the new molecule MOA and the comparison between its MOA and those of known drugs are very useful.

In this work we focus on a data-mining approach whose aim is the realization of a map of MOAs in which drugs with similar effects are placed in close position. Secondly, we want this map is as *continuous* as possible. In other words, fixed two drugs, if we move from the first to the second one we will observe drugs whose MOA gradually switches from the MOA of the first drug to the MOA of the second one. We realized this map choosing as starting point a public data-set of wide-genome expression data upon which a non-parametric pattern matching procedure (based on the Kolmogorov-Smirnov statistic) was applied in combination with a rank aggregation method. A novel similarity metric between wide-genome expression profiles was developed and dimensionality reduction techniques were used. In the following sections a description of the data-set is provided, each step of our method is explained and results are shown.

2   Data-Set Description

The *Connectivity Map* (cMAP) data-set is a large public database of wide-genome expression data from five different lines of cultured human cells, treated with 164 compounds at different concentrations [2]. The data is organized in experiments (batches) composed by two or more micro-array hybridizations of the treated cell line and one or more hybridizations of the untreated cell line as negative control. The number of treatments and controls per batch can vary as the number of total treatments across batches per single drug. The change in expression of a cell line after a treatment is computed considering the differential expression values of a treated hybridization respect those of the untreated one (or the set of untreated ones). So for each treatment with a drug in a batch we have a wide-genome differential expression profile (GEP). A tool allowing users to find connections between a well defined set of genes (a signature) and the GEPs of the cMAP has been realized by the authors of this data-set [5]. The tool makes use of a non parametric pattern matching strategy based on the Kolmogorov-Smirnov statistic and, basically, is an extension of the Gene Set Enrichment Analysis (GSEA) tool [3]. Moreover, a mechanism to build *internal signatures*, deriving well characterized set of genes, from the GEPs of the cMAP itself is also provided [5].

3   A Map of Modes of Action

In the following subsections we show how, starting by the Enrichment Score (with cMAP internal signatures) we improved it and we derived a metric, suitable to build a metric space hence a map of GEPs as proxy of a map of MOA.

3.1   Notations

- $D = \{d_1, \ldots, d_n\}$: the set of Drugs contained in the Connectivity Map Dataset;
- $X = \{x_1, \ldots, x_m\}$: the set of Wide-Genome Differential Expression Profiles (GEPs) contained in the Connectivity Map Dataset;
- $f_T : D \rightarrow \mathcal{P}(X)$: a function from D to all the possible subsets of X assigning to a drug the set of GEPs obtained treating cells with that drug;
- k : the number of probe set in each GEP (note that $\forall i = 1, \ldots, m : |x_i| = k$);
- S_k : The set of all the possible permutation of $[1, k]$;
- $f_R : X \rightarrow S_k$: a ranking function assigning to each GEP x_i a list $\sigma_i \in S_k$ in which $\sigma_i(j)$ is the rank of the j gene of x_i, considering x_i sorted according the differential expression values of the genes (in decreasing order).

3.2   Total Enrichment Score

Given two set of probe identifiers $p = \{p_1, \ldots, p_h\}$ and $q = \{q_1, \ldots, q_w\}$ we define the Total Enrichment Score, TES, of the **signature** $\{p, q\}$ respect to the GEP x_i, as follows:

$$TES_i^{\{p,q\}} = \frac{(ES_i^p - ES_i^q)}{2}. \tag{1}$$

where ES_i^d is the enrichment score of the genes whose identifiers are in the set $d \in \{p, q\}$ respect to the GEP x_i. ES_i^d is a measure based on the Kolmogorov-Smirnov statistic and it quantifies how much a set of genes lies at the top of the list obtained sorting x_i according the differential expression values of its genes in decreasing order. The more this measure is large and positive the more the genes lie at the top of the list. The more this measure is large and negative the more the genes lie at the bottom of the list. $TES_i^{\{p,q\}}$ takes into account of two set of genes and it checks how much the genes in the first set are placed at the top of the list and how much the genes in the second set are placed at the bottom.

3.3   Optimal Signature Creation

For each drug $d \in D$, we constructed a prototype list aggregating the rankings $\{\sigma_i | x_i \in f_T(d)\}$. The aggregation was made with the Borda Method [6]. To create the **optimal signature** $\{U_d, D_d\}$ we selected the first 250 genes of the prototype list and the last 250 ones as U_d and D_d, respectively. We considered this *signature* of genes as a general cellular response to the drug that approximates its mode of action. In other words, we isolated sets of genes that seemed to vary in response to the drug across different experimental conditions (different cell lines, different dosages, etc.).

3.4   A Metric Space Based on the Enrichment Score

For each optimal signature $\{U_d, D_d\}$, $d \in D$, we computed a vector of Total Enrichment Scores

$$\mathbf{T}_d = [TES_1, \ldots, TES_m], \tag{2}$$

such that the element TES_i is the Total Enrichment Score of the signature $\{U_d, D_d\}$ respect to the GEP x_i, $i = 1,\ldots,m$. Basically, we fixed a mono-dimensional space in which the position of each x_i is given by the value of TES_i. Then for each $x_i \in f_T(d)$, we rescaled the vector $\mathbf{T}_d$ obtaining the distances of all the GEPs from x_i along our mono-dimensional space. This rescaling was realized subtracting to each element of $\mathbf{T}_d$ the value of TES_i:

$$\mathbf{T}_d^i = |\mathbf{T}_d - TES_i|. \tag{3}$$

Then, grouping each of these $\mathbf{T}_d^i$ in a column, we obtained a $|f_T(d)| \times m$ matrix $\mathbf{M}_d$ in which there is a row for each GEP x_i, obtained treating cells with the drug d. Each row of this matrix contains the distances from x_i to all the other GEPs in the mono-dimensional space relative to the drug d.

$$\mathbf{M}_d = \begin{bmatrix} \mathbf{T}_d^{i_1} \\ \vdots \\ \mathbf{T}_d^{i_{|f_T(d)|}} \end{bmatrix} \tag{4}$$

Finally, we grouped in a single matrix all the $\mathbf{M}_d$ builded for each drug $d \in D$:

$$\mathbf{M} = \begin{bmatrix} \mathbf{M}_1 \\ \vdots \\ \mathbf{M}_{|D|} \end{bmatrix} \tag{5}$$

Note that this is an $m \times m$ matrix in which we have exactly one entry equal to zero per row and per column, so with a proper permutation of the rows, we can place all these zeros entries on the diagonal. To obtain the final space we builded the Enrichment Score Distance matrix, by simply providing symmetry to $\mathbf{M}$ (its permuted version):

$$\mathbf{M}_{ES} = \frac{\mathbf{M} + \mathbf{M}^T}{2}. \tag{6}$$

The entry i, j of this matrix contains the distance between x_i and x_j. Note that, for the construction process explained above, this distance is given by:

$$\mathbf{M}_{ES}^{i,j} = \frac{\left|TES_j^{Di} - TES_i^{Di}\right| + \left|TES_i^{Dj} - TES_j^{Dj}\right|}{2}, \tag{7}$$

where $Di \in D : x_i \in f_T(Di)$ and $Dj \in D : x_j \in f_T(Dj)$.

Then this distance is the average value between the distance of the two GEP in the ES-induced mono-dimensional space relative to the drug whose response is represented by x_i and the distance of the two GEP in the ES-induced mono-dimensional space relative to the drug whose response is represented by x_j. It can be shown that this final total distance respects the metric-properties so, giving $\mathbf{M}_{ES}$ as input to a Multi-dimensional Scaling (MDS) procedure, an euclidean embedding can be obtained.

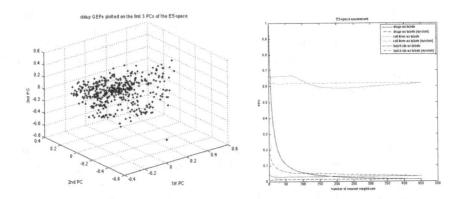

Fig. 1. All the cMap GEPs plotted on the first 3 principal components of the scaled space (left side) and assessment results of the embedding (right side). As we can see, data seem to have an underlying non-random structure and the embedding quality is obviously optimal.

3.5   Space Analysis, Assessment and Visualization

The output of the MDS showed that a space of 232 dimensions is sufficient to represent the GEPs according the ES-distance matrix computed. In Fig. 1 (on the left) the whole cMap dataset is plotted on the first 3 principal components of the obtained space. In order to assess the quality of this space let introduce the following additional notations.

Let be $f_{kn}^{\pi} : X \times K \rightarrow \mathcal{P}(X)$ a function assigning to each GEP x_i the set of its k-nearest neighbors in the mapping π (with $K = \{n \in N | n < |X|\}$). Then for a given map π, we define, $\forall k \in K$, $\forall x_i \in f_T(d)$ and $\forall d \in D$:

$P(\pi, x_i, k) = $ k (the number of predictions, the order of the neighborhood considered);

$TP(\pi, x_i, k) = \left| f_{kn}^{\pi}(x_i, k) \cap f_T(d) \right|$ (the number of GEPs in the k-neighborhood of x_i obtained with the same drug, number of true positive prediction);

$ppv(\pi, x_i, k) = TP(\pi, x_i, k)/P(\pi, x_i, k)$ (the number of true positive prediction on the number of total predictions).

Then we compute the average value of the *ppv* for a given value of k, in a given mapping π as follow:

$$PPV_k^{\pi} = \left(\sum_{d \in D} \sum_{x \in f_T(d)} ppv(\pi, x, k) \right) \Big/ |X|.$$

On the right side of Fig. 1 the assessment results on the ES-space are shown. The blue line is the average PPV in function of the cardinality of the considered neighbors, while the red line and the green line are the average PPV considering the treated cell line and the batch identifiers as supervised labeling, respectively.

This validation have been performed considering only the following subset of GEPs:

$$f_T(\{d \in D : |f_T(d)| \geq 2\}).$$

Obviously, due to the supervised nature of the space construction, GEPs obtained with the same drug lies in close positions. By other hand the proximity of the GEPs due to the similarities of the experimental conditions (same cell line or same batch) are substantially random. The interesting property of this space embedding lies in the fact that clusters of GEPs relative to drugs that share the same therapeutic application or the same set of molecular targets are very close and, in same cases, they are overlapped.

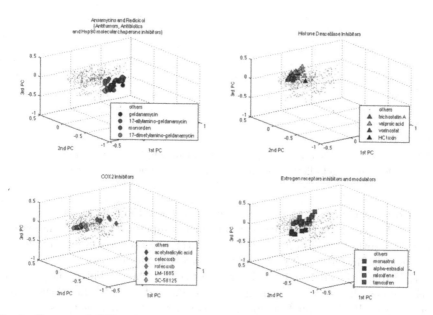

Fig. 2. Clusters of GEPs obtained with drugs sharing the therapeutic applications or a set of molecular targets lie in close and overlapping regions

In Fig. 2 and Fig. 3 some examples are provided. So we can conclude that, starting by a reference dataset of wide-genome differential expression data, obtained treating different cell lines with different drugs, our method provides an almost continuous and visualizable mapping of drug modes-of-action avoiding normalization procedures on the expression data across conditions. Note that no prior knowledge about the similarity of different drugs has been used.

Obviously, the more two drugs are represented in the dataset (with treatments on a sufficiently large variety of conditions) the more their similarity is reflected by our euclidean embedding. However (as in the case of the immunosuppressants in which we have an average number of treatments per cell line less than 0.35) this method seems to be sufficiently robust also when a class of drugs is not over-represented across conditions. In the rest of this section the

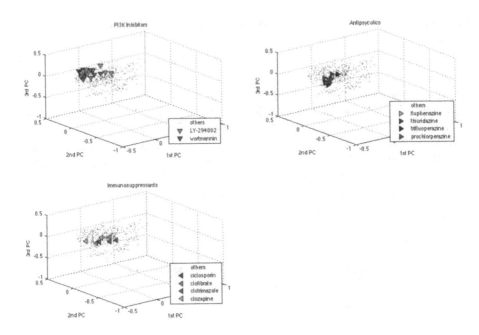

Fig. 3. Clusters of GEPs obtained with drugs sharing the therapeutic applications or a set of molecular targets lie in close and overlapping regions

explained property is assessed. We chose, as *test-set*, 156 GEPs composing the subset of cMap obtained with drugs upon which a supervised labeling, indicating their therapeutic application (or a subset of their molecular target), induces a partition of sufficiently *well-separated* and *uncorrelated* behaviors.

In the table 1 the GEPs composing the test set are shown. This table contains the considered drugs; **n** is the number of GEPs that are present in the cMap for a drug; **Cells H** is an Entropy value computed considering the set of the different cell lines treated with the drug as a random variable with an observation per treatment. The more this value is high the more the drug is represented in the cMap trough treatments across cell lines. Similarly, **Doses H** is computed considering the set of all the different dosages used to treat cells with the drug. The column **Labels** induces a partition on the considered set of GEPs. The table shows the heterogeneous nature of the test-set, composed by drugs used on all the cell lines, drugs used on some of them and drugs used only on a unique cell line. In order to evaluate the property explained above we made a hierarchical clusterization of the test set in the ES-space. The algorithm used was the classic Single-linkage with euclidean distance as metric. Then we cut the hierarchy of clusters at different level in order to obtain clusterings of k clusters, $\forall k = 1, \ldots, 156$.

In Fig. 4 a similarity measure between the different clustering obtained and the supervised labeling is shown. This measure is, basically, a normalized mutual information score defined as follow:

Table 1. Composition of the test set

n	Compound Type	Cells H	Doses H	Description	Label
18	'17-allylamino-geldanamycin '	0.54	0.31	ansamycin antibiotic that disrupts the chaperone function of heat shock protein (Hsp) 90	1
6	'geldanamycin '	0	0	"	1
10	'monorden'	0.79	0	Antifugal Metabolites	1
2	'17-dimethylamino -geldanamycin'	0	0	ansamycin antibiotic that disrupts the chaperone function of heat shock protein (Hsp) 90	1 1
12	'trichostatin A'	0.60	0.92	histone deacetylase inhibitor	2
18	'valproic acid'	0.67	0.85	"	2
2	'vorinostat'	0	0	"	2
1	'HC toxin'	0	0	toxin	2
3	'acetylsalicylic acid'	0	0	Aspirin, Ecotrin, and others	3
5	'celecoxib'	0.72	0	COX2 inhibitor	3
6	'rofecoxib'	0.79	0	"	3
3	'LM-1685'	0	0	"	3
4	'SC-58125'	0.95	0	"	3
8	'monastrol'	0.54	0.81	inhibitor of the mitotic kinesin, Eg5, arrests cells in mitosis with monoastral spindles	4
6	'alpha-estradiol'	0.78	0	inhibitor of estrogen receptors	4
3	'raloxifene'	1	0	estrogen receptors modulator	4
3	'tamoxifen'	0.92	0	"	4
17	'LY-294002'	0.62	0.67	PI3K inhibitor	5
8	'wortmannin'	0.77	0.81	"	5
4	'fluphenazine'	0.81	0	antipsycotic	6
4	'thioridazine'	0	0.81	"	6
3	'trifluoperazine'	0	0	"	6
3	'prochlorperazine'	0	0	"	6
2	'ciclosporin'	0	0	immunosuppressant	7
2	'clofibrate'	0	1	"	7
1	'clotrimazole'	0	0	"	7
2	'clozapine'	0	0	"	7

$$MI_s(L, C) = 2 \left(\frac{H(L) + H(C) - H(L,C)}{H(L) + H(C)} \right).$$

Where $H(L)$ is an entropy value computed considering the supervised labeling as a random variable with an observation per label. Similarly, $H(C)$ is an entropy value computed on a clustering (in this case the realization of the associated random variable are the cluster identifier). Finally $H(L,C)$ is the joint entropy between these two random variables. The probability functions of these random variables are obtained as frequencies starting by the *confusion matrix* between a clustering and the supervised labeling [4].

This similarity measure ranges in $[0,1]$ and it is equal to 0 when the clustering and the labeling are completely independents while is equal to 1 when they exactly induce the same partition on the data. As we can see in Fig. 4, the curve relative to the dendrogram obtained in our ES-space is not monotonic and it

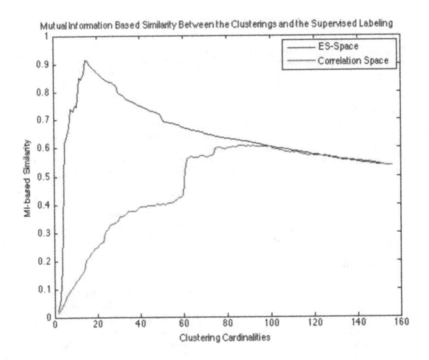

Fig. 4. Mutual Information score between clusterings of different number of clusters (different cardinality) and the supervised labeling

Table 2. Clustering composition

Cluster ID	n GEPs	Drug Composition	Classes
1	25	68% LY-294002, 32% wortmannin	100% label 5
2	14	21% trifluperazine, 28% thioridazine, 21%prochlorperazine, 28%fluphenazine	100% label 7
4	13	46% rofecoxib, 23% LM-1685, 30% SC-58125	100% label 3
7	32	37% trichostatin A, 56% valproic acid, 0.6% vorinostat	100% label 2
12	36	16% geldanamycin, 50% 17-allylamino-geldanamycin 27% monorden, 0.5% 17-dimethylamino-geldanamycin	100% label 1
3	6	100% alpha-estradiol	-
5	2	100% ciclosporin	-
6	2	100% clofibrate	-
8	1	100% HC toxin	-
9	5	100% celecoxib	-
10	6	100% tamoxifen	-
11	3	100% acetylsalicylic acid	-
13	8	100% monastrol	-
14	2	100% clozapine	-
15	1	100% clotrimazole	-

has a global maximum (0.91) for $k = 15$. This means that, properly cutting the obtained dendrogram in the ES-space, we can obtain a clusterisation that is very close to the supervised labeling. This result is not provided by applying this process in a space obtained using the classical correlation metric to the GEPs of the test-set in the domain of all their probes (the red curve of Fig. 4).

In the table 2 the composition of the most reliable clustering ($k = 15$) is shown. As we can see, even if there are singleton clusters, five classes of drug have been detected (with high *sensitivity* score: 100% for two of them, $\sim 80\%$ for the others) and drugs sharing the therapeutic application (or a subset of the molecular targets) have been grouped together.

4   Conclusions

We developed a data-mining and visualization approach that provides a Map of drug-modes-of-action from gene expression data. Our method starts from a compendium of gene expression data obtained treating five different human cell lines with more than 150 different compounds. To obtain the map we combine a supervised version of the Enrichment Score and a Rank aggregation method. Results shows that, in the obtained map, wide-genome expression profiles obtained with drugs that share the therapeutic application or a set of molecular targets lies in clusters placed in close position. Our method avoid complex normalization procedure across conditions and provide a map that is not achievable with classical similarity measures.

References

1. di Bernardo, D., Thompson, M.J., Gardner, T.S., Chobot, S.E., Eastwood, E.L., Wojtovich, A.P., Elliott, S.J., Schaus, S.E., Collins, J.J.: Chemogenomic profiling on a genome-wide scale using reverse-engineered gene networks. Nat. Biotechnol. 23(3), 377–383 (2005)
2. Lamb, J., Crawford, E.D., Peck, D., Modell, J.W., Blat, I.C., Wrobel, M.J., Lerner, J., Brunet, J.P., Subramanian, A., Ross, K.N., Reich, M., Hieronymus, H., Wei, G., Armstrong, S.A., Haggarty, S.J., Clemons, P.A., Wei, R., Carr, S.A., Lander, E.S., Golub, T.R.: The Connectivity Map dataset: Using gene-expression signatures to connect small molecules, genes and diseases. Science 313, 1929–1935 (2006)
3. Subramaniana, A., Tamayoa, P., Moothaa, V.K., Mukherjeed, S., Eberta, B.L., Gillettea, M.A., Paulovichg, A., Pomeroyh, S.L., Golub, T.R., Landera, E.S., Mesirova, J.P.: Gene set enrichment analysis: A knowledge-based approach for interpretating genome-wide expression profiles. P. Natl. Aacad. Sci. USA 102(43), 15545–15550 (2005)
4. Ciaramella, A.: Interactive data analysis and clustering of genomic data. Neural Networks 21(2-3), 368–378 (2008)
5. http://www.broad.mit.edu/cmap/
6. Parker, J.R.: Voting methods for multiple autonomous agents. In: Proceedings of the Third Conference on Intelligent Information Systems, ANZIIS 1995, Australian, New Zealand, 27 November 1995, pp. 128–133 (1995)

In Silico Evaluation of Osmoprotectants in *Eucalyptus* Transcriptome

Petra dos Santos Barros, Nina M. Soares-Cavalcanti, Gabriela S. Vieira-Mello,
Ana C. Wanderley-Nogueira, Tercílio Calsa-Junior,
and Ana M. Benko-Iseppon*

Universidade Federal de Pernambuco, Center for Biological Sciences (CCB),
Genetics Department, Laboratory for Plant Genetics and Biotechnology,
Av. Prof. Moraes Rêgo s/no. Recife, PE, Brazil
Tel.: +55-81-2126-8520
ana.benko.iseppon@pesquisador.cnpq.br

Abstract. The synthesis of osmoprotectants is one of the main mechanisms to dribble harmful effects caused by abiotic stresses. In the present work, osmoprotectant genes were selected and used in a search for orthologs in the FORESTs database using *in silico* procedures. The studied genes were here analyzed for the first time in *Eucalyptus*, including 51 identified orthologs from nine gene families (*P5CS*=3, *P5CR*=4, *TPS1*=10, *TPPB*=8, *SAT*=6, *OASTL*=13, *CMO*=1, *BADH*=3 and *INPS1*=3). In general, the most represented genes were those involved in the trehalose and cysteine synthesis, probably due to their vital importance in the maintenance of cellular homeostasis. Overall expression pattern revealed significant amounts of reads in most libraries. Dendrograms based on analysis of selected genes from other flowering plants revealed no clear separation between mono and dicots, since diversity regarding the evolution and diversity in gene signatures may be more related to fitness and adaptation, than evolutionary patterns.

Keywords: Salinity and Drought Stress, Osmolytes, Data Mining, Crop Evolution.

1 Introduction

Despite of significant information regarding response to abiotic stresses in herbaceous plants, as rice and arabidopsis, scarce information is available for understanding such mechanisms in woody plants, principal components of arid and semi-arid regions of the word [1]. Physiological characters widely associated with drought resistance in eucalyptus include greater osmotic adjustment, tissue elasticity, higher leaf water content and higher ability to resist desiccation [2]. Drought and salt stresses are often interconnected and cause cellular damages, such as osmotic and oxidative stresses. To cope with these environmental influences, plants have developed arrays of physiological and biochemical strategies

---

* Corresponding author.

F. Masulli, R. Tagliaferri, and G.M. Verkhivker (Eds.): CIBB 2008, LNBI 5488, pp. 66–77, 2009.

to adapt to the adverse conditions. One of the most common metabolic adaptations is the accumulation of certain organic solutes (known as osmoprotectants) that protect proteins and membranes against damage by high concentrations of inorganic ions [3].

Trehalose helps to stabilize enzymes, membrane proteins and lipids, acting also as protective of biological structures for several stress types. It is synthesized by a two-step process in which trehalose-6-phosphate synthase (TPS1) synthesizes T6P from UDP-glucose and glucose-6-phosphate, followed by dephosphorylation of trehalose by trehalose-6-phosphate phosphatase (TPPB) [4].

Myo-inositol is a cyclic alcohol formed from the glucose-6-phosphate by an internal reaction of oxy-reduction involving NAD^+ and INPS1 enzyme (myo-inositol-1-phosphate synthase), a reaction catalyzer. Methylated forms of myo-inositol accumulate in plants in response to salinity stress to act during osmotic adjustment [5].

Proline (Pro) is one of the most common compatible osmolytes in water-stressed plants. The accumulation of Pro in dehydrated plants is caused both by activation of the Pro biosynthesis and by inactivation of the Pro degradation. In plants, L-Pro is synthesized from L-glutamic acid (l-Glu) via Δ^1-pyrroline-S-carboxylate (P5C) by two enzymes, P5C synthetase (P5CS) and P5C reductase (P5CR) [6].

The final steps of cysteine biosynthesis are catalyzed by a bi-enzyme complex composed of serine acetyltransferase (SAT) and *O*-acetyl-serine(thiol)lyase (OASTL). Some previous studies revealed an increased synthesis of cysteine when the plants are exposed to high salinity levels. The authors suggested that the synthesis of this amino acid is essential for the protection and/or adaptation of the plants to the stress condition [7].

In higher plants the glycine-betaine (GB) is synthesized from two steps, with the first is catalyzed by the enzyme CMO (choline mono-oxygenase), which carries out the oxidation of choline in aldehyde betaine and the second by the enzyme BADH, which converts aldehyde betaine in GB. Under salt stress, GB helps to maintain the osmotic potential in cytosol, also protecting macromolecules, including proteins [8].

None of the above described enzymes have been described before in eucalyptus. The *Eucalyptus* Genome Sequencing Consortium (FORESTs project) generated 112,857 expressed sequence tags (ESTs) from 19 libraries regarding specific tissues and stages or conditions using five *Eucalyptus* species [9]. The present work performed a data-mining based identification of osmoprotectant genes in the FORESTs database, using well known arabidopsis sequences as seed sequences, also evaluating their expression in FORESTs libraries, and comparing them with sequences from public databases and literature data.

2   Methods

Descriptions of the FORESTs cDNA libraries, sequence nomenclature and clustering methods are described in Vicentini *et al.* (2005) [9]. Complete *A. thaliana* full length cDNA sequences of osmoprotectants were blasted against the FORESTs database using the tBLASTn tool [10]. Only sequences with e-value e^{-05} or less were annotated. *Eucalyptus* ESTs were translated with the aid of ORFfinder program and screened for similarity to entries in GenBank (NCBI) using BLASTx. Moreover, Blink function, was used to obtain similar sequences containing the searched conserved domains (CDs) in other organisms. The RPS-BLAST tool was used to inquire the presence and integrity of the CDs. A multiple alignment using CLUSTALx [11] program was generated including only sequences with complete CDs, aiming to compare the eucalyptus' and other osmoprotectants. The MEGA software (V 4.0) [12] allowed the construction of dendrograms using the Neighbor Joining method, gamma distance and bootstrap test (with 1,000 replications) to infer the relationship among sequences. To evaluate the ESTs expression pattern in eucalyptus sequences, a normalized data matrix including annotated reads for each library was constructed and used for a hierarchical clustering analysis using CLUSTER_v.2.20 program [13], with data displayed in the TreeView_v.1.60 program [14].

3   Results

3.1   Osmoprotectants in Eucalyptus

The search for candidates osmoprotectant gene in the FORESTs database revealed the presence of several clusters with high degree of similarity, with best e-value for each gene ranging from 0.0 to $2e^{-74}$ (Table 1). In general, 90.6% sequences of eucalyptus presented higher similarity to dicots, while only 5.66% were more similar to monocots and in few cases (3.74%) with other taxa. This was the case of *TPS1*, where one of the 13 clusters identified showed similarity with a bryophyte (*Physcomitrella patens*). In the case of *INPS1*, one of the three orthologs clusters presented high similarity with the fungus *Aspergillus terreus* and *Giberella zeae*.

Table 1. *Clusters* of *Eucalyptus* (best matches) in the FORESTs database as compared with osmoprotectants known from *A. thaliana*. Abbreviations: nt, nucleotide; aa, amino acid; Fr, Frame; Gi, GenBank identifier.

Gene	Results from FORESTs					Results from BLASTx in NCBI				
	FORESTs code	e-value	Size nt	Fr	# Hits	Gi number	Organism	e-value	Size aa	Fr
P5CS	P5CS_RT3103C01.g	0.0	2666	+2	3	6225816	*Actinidia chinensis*	0.0	717	+2
P5CR	P5CR_CL1286B03.g	$2e^{-84}$	851	+2	4	130972	*Glycine max*	$3e^{-84}$	274	+2
CMO	CMO_CL1347G03.g	e^{-91}	815	+1	1	33300598	*Oryza sativa*	$1e^{-90}$	410	+1
BADH	BADH_RT3200D09.g	0.0	1837	+1	3*	40850676	*Gossypium hirsutum*	0.0	503	+1
SAT	SAT_ST2201B08.g	e^{-93}	1309	+2	7	23343585	*Nicotiana tabacum*	$3e^{-116}$	300	+2
OASTL	OASTL_ST2203B10.g	e^{-143}	1333	+2	13	148562451	*Glycine max*	$2e^{-163}$	372	+2
TPPB	TPPB_RT5001B05.g	$2e^{-74}$	1059	+2	8	51458330	*Nicotiana tabacum*	$6e^{-101}$	384	+1
TPS1	TPS1_RT3140G03.g	0.0	3089	+2	13	15224213	*Arabidopsis thaliana*	0.0	862	+2
INPSI	INPSI_LV2201G02.g	0.0	2106	+3	3	14548095	*Sesamum indicum*	0.0	510	+3

*From 30 clusters with high e-values only three sequences were, in fact, orthologs to the BADH.

Moreover, the results in FORESTs revealed the presence of 13 and seven clusters similar to the *OASTL* and *SAT* genes respectively, which jointly take role in cysteine synthesis. From the 13 candidates to *OASTL*, five had the full conserved domain PALP, while in the other eight this domain was incomplete. Regarding the *SAT*, three clusters showed the full domains SATase_N and LbetaH while these domains were absent in the other four. For *TPS1* and *TPPB* 13 and eight candidates were identified, respectively. After reverse alignment most ortholog candidates of these genes showed similarity with *A. thaliana*. In conserved domain analyses for each gene, two sequences (e^{-74} and e^{-107}, respectively) were found to contain the complete procured domains (glico-transferase and TPPase) in the first case, and TPPase, in the second.

The search for similar sequences to *P5CS* gene revealed three clusters, with sequences ranging from 829 to 2,666 nt, the last sequence with e-value 0.0. The BLASTx results showed that eucalyptus sequences presented high similarity with the same gene from kiwi fruit (*Actinidia chinensis*; gi:6225816), *Saccharum arundinaceum*; (gi:157061835) and from ice-plant (*Mesembryanthemum crystallinum*; gi:6225818). Regarding the CD analysis, only one eucalyptus sequence had the desired domains AA_Kinase and ProA (the latter incomplete). In the search for orthologs of *P5CR*, two of the four clusters found presented a partial *P5CR* domain.

The search for ortholog candidates to *CMO* gene revealed one sequence with high similarity (e-value e^{-91}), but without the procured CD. In the search for *BADH*, 30 clusters were observed; however the analysis of these candidates in NCBI database, showed that three of them were in fact ortholog to the gene in question, one with the searched domain in its entirety, while the other clusters showed similarity to aldehyde dehydrogenase (ALDH) genes of the same family than BADH.

3.2   Library Specific Expression

625 osmoprotectant transcripts were identified at the FORESTs database and included in the evaluation. The exceptions were the candidates to *P5CR* and *CMO* genes, where the number of reads was not sufficient to build a representative analysis of the expression profile.

Considering all identified osmoprotectants, most reads came from stems (ST) and seedlings (SL) libraries, the two largest FORESTs libraries (with 26,318 and 27,437 reads, respectively). However, they were abundant also in smaller libraries, nominally 100 reads in callus libraries (CL, 12,533 reads), and 91 in root tissues (RT, 20,129 reads) of eucalyptus seedlings from nurseries. 625 identified transcripts were used for a hierarchical clustering analysis, to allow the identification of expression distribution and intensity, as well as co-expressed sequences (co-regulated), pertaining gene categories (pink dendrogram, in vertical) or libraries (gray dendrogram, in horizontal) (Figure 1).

Expression was evidently higher in stem of plants submitted to water deficit (ST) regarding all gene families studied. Increased but less strong expression of osmoprotectants could be observed in seedlings grown in the dark (SL) and

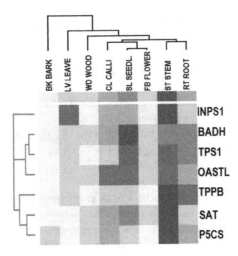

Fig. 1. Graphic representation of expression pattern analysis of seven osmoprotectants in the FORESTs database. Light yellow means no expression, and degrees from yellow to orange and red regard increased expression intensity. Abbreviations for libraries: BK (bark, duramen, pith and alburnum of trees of *E. grandis* with eight years); CL (calli grown in the dark), FB (buttons, flowers and fruits); LV (leaves of seedlings); RT (roots of seedlings from nursery), SL (seedlings grown in the dark), ST (stem seedlings subjected to water deficit) and WD (wood of *E. grandis*).

roots in the absence of stress (RT). Associated spatial co-expression was stronger among ST/RT, while SL had higher association with FB (buttons, flower and fruits) and CL (calli) libraries. Co-expression of genes regarded mainly BADH/-TPS1+OASTL and SAT/P5CS+TPPB (Figure 1).

3.3   Phenetic Analysis

Most dendrograms (Figure 2 and Figure 3), grouped proteins from monocots and dicots together in the same clades, with exception of BADH (Figure 2A), where

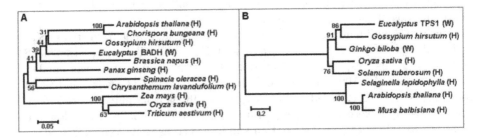

Fig. 2. Dendrograms generated after Neighbor Joining analysis illustrating relationship revealed by conserved domains similarity in (A) BADH-like and (B) TPS1-like sequences and putative *Eucalyptus* orthologs. The bar represents genetic distance, while numbers in the base of clades regard bootstrap values. The letters in parenthesis mean herbaceous (H) and woody (W) plants.

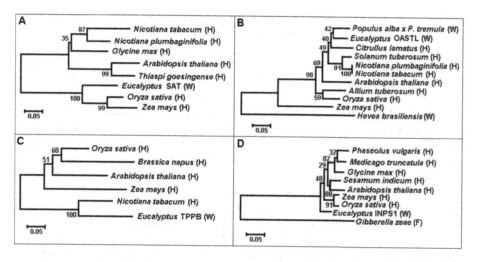

Fig. 3. Dendrograms generated after Neighbor Joining analysis, showing the relationships between the consensus sequences of *A. thaliana*, *Eucalyptus* and other organisms regarding protein sequences: (A) SAT; (B) OASTL; (C) TPPB and (D) INPS1. Bar represents genetic distance, while numbers in the base of clades indicate bootstrap values. The letters in parenthesis mean herbaceous (H) and woody (W) plants or fungi (F).

members of the two angiosperm classes grouped into different branches. However, within dicots classic taxonomic grouping was not observed for BADH sequences. With respect to TPS1 (Figure 2B), the dendrogram included primitive plant members (Lycopodiophyta and Ginkgophyta) together with Angiosperms.

Additional dendrograms for SAT, OASTL and TPPB comprised only flowering plants (Figures 3A, 3B and 3C), with exception of INPS1 (Figure 3D), where a fungus sequence (genus *Giberella*) was included.

4   Discussion

4.1   Osmoprotectants in Eucalyptus

All procured gene classes were present in the FORESTs database bearing structural conserved regions as compared to arabidopsis seed sequences, exhibiting also considerable divergence, sometimes even regarding the domain structure, enough to be indicative of long divergent evolution. Our findings confirmed observations [15] concerning to Resistance (R) genes analysis in eucalyptus, which revealed considerable discrepancies among genes of eucalyptus and those available in databases for angiosperms, especially considering the procured conserved domains. The authors suggested that the low number of sequences from woody plants available in public databases can be one of the factors that justify novel structural forms in eucalyptus.

The fact that OASTL gene family presents a larger number of representatives in eucalyptus may be due to the important role of this protein in the cysteine synthesis. The OASTL activity is higher than that of SAT [16], since SAT is generally found only in association with OASTL, forming the complex CSC, while OASTL may be also found in its free form in the cell.

The presence of restricted number of *TPS1* and *TPPB* representatives may be explained by previous evidences in arabidopsis [17] where some plants did not accumulate trehalose effectively when subjected to stresses, probably due to the high activity of the trehalase enzyme (found in all tissues of *A. thaliana* and also in *Glycine max*) that hydrolyses trehalose in two glucose molecules. With reference to *P5CR*, none of the candidates presented complete domains. Regardless that, the high similarity of the identified sequences (e-value 0.0 to *A. thaliana*) indicates the existence of this route in *Eucalyptus* transcriptome. Previous identification of this protein in many organisms, from prokaryotes to eukaryotes (bacteria, algae, higher plants, mammals, etc.) provide strong evidences that the processes involved in the proline formation are conserved [18]. Physiological experiments performed with *Eucalyptus tereticornis* showed that when subjected to salt stress, the leaves presented increased proline content [19]. Since the FORESTs project presents only a library subjected to conditions of water deficit (regarding the stem tissue of seedlings), it is possible that additional libraries from different tissues and treatments (as drought and salinity) would allow the observation of a greater number of representatives of these genes.

Considering glycine-betaine, it was expected to find at least one representative of the *CMO* with its complete CD. Despite of that, BLASTx search revealed that the identified CMO eucalyptus cluster showed high similarity with the same gene in rice, indicating that even in the absence of the CD the eucalyptus transcriptome included orthologs of this gene. Evidences in favor to this supposition are given by the effective accumulation of glycine-betaine in *Eucalyptus* and in many woody plants as *Pinus*, *Larix* and *Fraxinus* [20].

Besides the already mentioned compatible osmolytes, the *INPS1* also participates in cellular protection against dehydration and salt accumulation. Three sequences with significant similarity were found in eucalyptus for this gene category. The first two aligned significantly (e-values 0.0 and e^{-95}) with *Sesamum indicum* after BLASTx. Together these results point to the presence of all important gene classes for the synthesis of compatible osmolytes in *Eucalyptus* transcriptome, considering the knowledge available to *A. thaliana* model plant.

4.2   In Silico Expression

The observed abundance of osmoprotectants in stems of seedlings subjected to water deficit library was expected, since stem tissues are responsible for the support of young plants, preventing its falling down when under drought or salinity [21]. The existence of a higher expression in two libraries subjected to other abiotic stresses (SL=seedlings grown in the dark and CL=calli grown in the dark) confirms that these genes are recruited under additional adverse conditions, other than those related to drought and salinity, in the present case, the absence of light. It is interesting to note that co-expression of osmoprotectants was evident in the above mentioned libraries, with closer co-expression of *BADH*,

TPS1 and *OASTL*, in contrast to *SAT*, *P5CS* and *TPPB*. It is worth noticing that besides higher expression in some libraries and tissues, almost all tissues presented some expression of compatible osmolytes, even considering that only one library was built under abiotic stress. This indicates a relative importance of some basal osmoprotection in plants, even when the environmental conditions are not stressful.

4.3 Dendrograms

It is noteworthy that none of the dendrograms followed a strict taxonomic grouping forming often mixed non taxonomic grouping. This should be surprising, considering that alignments included mainly conserved domains. As emphasized by Weston *et al.* [22], despite the advances made possible by "omics"-based technologies, we still struggle to accurately associate the genes, transcriptional cascades, and signaling networks with physiological performance and ecological fitness. This is especially true for osmoprotectants, with gene evolution and functional structure probably more related to fitness and adaptation, than to stochastic evolutionary patterns, revealing groupings that seem to associate different gene signatures to physiological performance. Results regarding genes related to abiotic stress responses have shown that dendrograms very often group sequences in functional arrangements, with common signatures probably reflecting environmental driven adaptation or selection [23]. This may be interpreted as the result of common ancestry, partially lost (or silenced/rarely expressed) genes in some taxa, but also to convergent mutation (or reverse mutation) associated to ecological fitness.

The separation of the BADH dendrogram (Figure 2A) in two major groups (dicots and monocots) can be explained by the difference of RNA processing pattern of BADH homologs among both groups. Moreover, poaceous species as *Zea mays*, *Triticum aestivum* and *Oryza sativa* accumulate less glycinebetaine (GB) in comparison with other families. In dicots clade, spinach is a chenopod member that accumulates far more GB than do cereal crops in response to osmotic stresses [24]. In our dendrogram rice and wheat grouped together, maybe because the rice sequence used came from salt stressed plants fed with betaine aldehyde for one week, leading to expression of some new candidates and producing GB levels close to that observed in wheat.

The presence of trehalose-6-phosphate synthase (Figure 2B) in Ginkgophyta, Magnoliophyta and Algae indicates that TPS1 represents an arqueomorphic character. The monocots, as well as Rosidae, also formed a polyphyletic group while the branch composed by eucalyptus, cottons and *G. biloba* formed a paraphyletic group. *TPS1* appear to be an essential gene in plants, present in primitive plants including *Selaginella*, a known desert plant from the spikemoss Lycopodiophyta group (known for its ability to survive almost complete desiccation) and also Ginkgo that is remarkably resistant to cold and drought, being regarded as one of the first terrestrial plants.

Considering the SAT dendrogram (Figure 3A), monocot plants (rice and maize) grouped together, both sensible to abiotic factors. In the second clade all dicots grouped (again with exception of eucalyptus) with both Brassicaceae joining a distinct sub-clade with high bootstrap value (99). The here aligned gene from *A. thaliana* was an ortholog to a *Salmonella typhimurium* gene [25], also revealing its ancestral character. Furthermore, previously observations revealed that elevated glutathione, driven by constitutive activation of *SAT* gene, plays a role in the Ni, Co and Zn tolerance of *T. goesingense* (Brassicaceae) [26]. The soybean SAT was positioned in an intermediary position between the two Brassicaceae and two *Nicotiana* species. Furthermore, it was shown [27] that a serine acetyltransferase from soybean (GmSerat2;1) is a substrate of CDPK (calcium dependent protein kinase), so it seems that SAT, at least in *Glycine max*, is strongly involved with tolerance to different kinds of stresses.

After analysis using OASTL sequence, the distribution of species in this dendrogram may be explained by the existence of different isoforms of this gene (*OASTLA*, cytosolic; *OASTLB*, plastid; and *OASTLC*, mitochondrial), a fact already observed in arabidopsis and spinach [28]. In fact previous analysis [29] showed that the here used sequences of *Z. mays*, *O. sativa*, *N. plumbaginifolia*, *Citrullus lanatus*, *Populus alba* X *P. tremula* and *Allium tuberosum*, are the cytosolic OASTL isoforms and therefore are grouped together in the dendrogram. It seems clear that the here analyzed eucalyptus sequence represents the cytosolic isoform, *Hevea brasiliensis* may represent another isoform, considering its behavior as an out-group in the dendrogram.

Regarding the few complete TPPB proteins from data banks, a junction of several organisms from different taxonomic groups are mixed in the formed clades, with eucalyptus composing an outgroup together with tobacco (Figure 3C) with high bootstrap value (100). The availability of a larger number of orthologs may bring useful evidences concerning diversity and also regarding complex regions in conserved/non conserved segments of this gene category.

With exception of the eucalyptus positioning, the INPS1 dendrogram (Figure 3D) grouped most species according to their taxonomic affinities. Three legumes joined a clade, with *Phaseolus* and *Medicago* in the same subclade together with soybean in a separate subclade, what may also reflect adaptive features. Beans and alfalfa are more tolerant to drought than to saline soils [30] and also share the ability to develop in soils with high levels of aluminum [31], while soybean appear to be more tolerant to salinity than to drought [32]. *Arabidopsis* and sesame also joined a clade being both characterized by their ability to grow in arid and semi-arid deserts [33]. Rice and maize are annual grasses that need high temperatures, both sensible to abiotic factors as day length for maize and availability of flooded environments in the case of rice [34]. As for other osmoprotectants, the eucalyptus position near the fungus *Giberella zeae* was unexpected, again confirming their slow evolving manner as woody most primitive plant group, as compared with all other herbaceous species. It is interesting to note that this fungus presents tolerance to heavy metals, a feature also present in *Neurospora crassa* [35].

5   Concluding Remarks

The present study allowed the identification of all procured osmoprotectants in eucalyptus (FORESTs database). Nevertheless different aspects of structure, abundance, function and pattern of expression in all tissues could be observed and reported by the first time.

Osmoprotectant sequences were identified in low levels in most tissues, but were more abundant (and co-expressed) in libraries produced under abiotic stresses, providing important evidences for physiological studies, expression assays and germplasm evaluation using quantitative real-time PCR assays.

Dendrograms generated using eucalyptus osmoprotectants, as compared to other organisms bearing procured conserved domains revealed similarities to plants adapted to different kinds of abiotic stresses and also to model plants, abundant in public databases. Divergent patterns in eucalyptus sequences were also identified, probably due to the prevalence of herbaceous plants in databases, as compared with the woody (primitive) habit and taxonomic position of eucalyptus, indicating urgent need of more information regarding basal Angiosperm and also other primitive plants (especially that adapted to dry environments) to clarify structure, functionality and evolutionary patterns regarding osmoprotectants.

Acknowledgments

The authors thank CNPq (Conselho Nacional de Desenvolvimento Científico e Tecnológico) and FACEPE (Fundação de Amparo à Pesquisa do Estado de Pernambuco) for the concession of fellowships. To FAPESP (Fundação de Amparo à Pesquisa do Estado de São Paulo) and FORESTs coordination we thank for the access to the *Eucalyptus* EST data bank.

References

1. Lemcoff, J.H., Guarnaschelli, A.B., Garau, A.M., Bascialli, M.E., Ghersa, C.M.: Osmotic Adjustment and its Use as a Selection Criterion in *Eucalyptus* Seedlings. Can. J. For. Res. 24(12), 2404–2408 (1994)
2. Ladiges, P.Y.: Variation in Drought Tolerance in *Eucalyptus viminalis* Labill. Aust. J. Bot. 22(3), 489–500 (1974)
3. Paston, G.M., Foyer, C.H.: Common components, Networks, and Pathways of Cross-Tolerance to Stress. The Central Role of Redox and Abscisic Acid-mediated Controls. Plant Physiol. 129, 460–468 (1999)
4. Elbein, A.D., Pan, Y.T., Pastuszak, I., Carroll, D.: New Insights on Trehalose: a Multifunctional Molecule. Glycobiology 13, 17–27 (2003)
5. Hegeman, C.E., Good, L.L., Grabau, E.A.: Expression of D-Myo-Inositol-3-Phosphate Synthase in Soybean. Implications for Phytic Acid Biosynthesis. Plant Physiol. 125, 1941–1948 (2001)
6. Hu, C.A., Delauney, A.J., Verma, D.P.: A Bifunctional Enzyme (Delta 1-Pyrroline-5-Carboxylate Synthetase) Catalyzes the First Two Steps in Proline Biosynthesis in Plants. Proc. Natl. Acad. Sci. USA 89(19), 9354–9358 (1992)

7. Sirko, A., Blaszczyk, A., Liszewska, F.: Overproduction of SAT and/or OASTL in Transgenic Plants: a Survey of Effects. J. Exp. Bot. 55(404), 1881–1888 (2004)
8. Sakamoto, A., Murata, N.: The role of Glycine-Betaine in the Protection of Plants from Stress: Clues from Transgenic Plants. Plant Cell Environ. 25(2), 163–171 (2002)
9. Vicentini, R., Sassaki, F.T., Gimenes, M.A., Maia, I.G., Menossi, M.: In Silico Evaluation of the *Eucalyptus* Transcriptome. Gen. Mol. Bio. 28(3), 487–495 (2005)
10. Altschul, S.F., Gish, W., Miller, W., Myers, E.W., Lipman, D.J.: Basic Local Alignment Search Tool. J. Mol. Biol. 251, 403–410 (1990)
11. Thompson, J.D., Higgins, D.G., Gibson, T.J.: CLUSTALX, Multiple Sequence Alignment Program Version 1.63b. EMBO Rep. (1997)
12. Tamura, K., Dudley, J., Nei, M., Kumar, S.: MEGA4: Molecular Evolutionary Genetics Analysis (MEGA) Software Version 4.0. Mol. Biol. Evol. 24, 1596–1599 (2007)
13. Eisen, M.B., Spellman, P.T., Brown, P.O., Botstein, D.: Cluster Analysis and Display of Genome-Wide Expression Patterns. Proc. Natl. Acad. Sci. USA 95, 14863–14868 (1998)
14. Page, R.D.: Treeview Program Version 161. Comput. Appl. Biosci. 12, 357–358 (1996)
15. Barbosa-da-Silva, A., Wanderley-Nogueira, A.C., Silva, R.R.M., Berlarmino, L.C., Soares-Cavalcanti, N.M., Benko-Iseppon, A.M.: In Silico Survey of Resistance (R) Genes in *Eucalyptus* Transcriptome. Gen. Mol. Biol. 28(3), 562–574 (2005)
16. Hell, R., Wirtz, F., Berkowitz, O., Droux, M.: The Cysteine Synthase Complex from Plants. Mitochondrial Serine Acetyltransferase from *Arabidopsis thaliana* Carries a Bifunctional Domain for Catalysis and Protein-protein Interaction. Eur. J. Biochem. 268, 683–686 (2001)
17. Müller, J., Aeschbacher, R.A., Wingler, A., Boller, T., Wiemken, A.: Trehalose and Trehalase in Arabidopsis. Plant Physiol. 125, 1086–1093 (2001)
18. Verbruggen, N., Villarroel, R., Montagu, M.V.: Osmoregulation of a Pyrroline-5-Carboxylate Reductase Gene in *Arabidopsis thaliana*. Plant Physiol. 103, 771–781 (1993)
19. Marsh, N.R., Adams, M.A.: Decline of *Eucalyptus tereticornis* near Bairnsdale, Victoria: Insect Herbivory and Nitrogen Fractions in Sap and Foliage. Aust. J. Bot. 43, 39–50 (1995)
20. Adams, M.A., Richter, A., Hill, A.K., Colmer, T.D.: Salt Tolerance in *Eucalyptus* spp.: Identity and Response of Putative Osmolytes. Plant Environ. 28, 772–787 (2005)
21. Hazen, S.P., Wu, Y., Kreps, J.A.: Gene Expression Profiling of Plant Responses to Abiotic Stress. Funct. Integr. Genomics 3(3), 105–111 (2003)
22. Weston, D.J., Gunter, L.E., Rogers, A., Wullschleger, S.D.: Connecting Genes, Co-expression Modules and Molecular Signatures to Environmental Stress Phenotypes in Plants. BMC Syst. Biol. 2, 1–17 (2008)
23. Benko-Iseppon, A.M., Soares-Cavalcanti, N.M., Wanderley-Nogueira, A.C., Berlarmino, L.C., Silva, R.R.M., Almeida, P.M.L., Brunelli, K.R., Kido, L.M.H., Kido, E.A.: Genes Associated to Biotic and Abiotic Stresses in Cowpea [*Vigna unguiculata* (L.) Walp.] and Other Angiosperms. In: Nogueira, R.J.M.C., Araújo, E.L., Willadino, L.C., Cavalcante, U.M.T. (eds.) Environmental Stresses: Damages and Benefits to Plants, pp. 350–359. UFRPE University Press Recife, Brazil (2005)
24. Ishitani, M., Arakawa, K., Mizuno, K., Kishitani, S., Takabe, T.: Betaine Aldehyde in Leaves of both Betain-Accumulating and Non-Accumulating Cereal Plants. Plant Cell Physiol. 34, 493–495 (1993)

25. Kawashima, C.G., Berkowitz, O., Hell, R., Noji, M., Saito, K.: Characterization and Expression Analysis of a Serine Acetyltransferase Gene Family Involved in a Key Step of the Sulfur Assimilation Pathway in Arabidopsis. Plant Physiol. 137, 220–230 (2005)
26. Freeman, J.L., Salt, D.E.: The Metal Tolerance Profile of *Thlaspi goesingense* is Mimicked in *Arabidopsis thaliana* Heterologously Expressing Serine Acetyl-Transferase. BMC Plant Biol. 7, 63 (2007)
27. Liu, F., Byung-Chun, Y., Jung-Youn, L., Pan, W., Harmon, A.C.: Calcium-Regulated Phosphorylation of Soybean Serine Acetyltransferase in Response to Oxidative Stress. J. Biol. Chem. 281(37), 27405–27415 (2006)
28. Saito, K.: Isolation and Characterization of a cDNA that Encodes a Putative Mitochondrion-Localized Isoform of Cysteine Synthase (O-Acetylserine(Thiol)-lyase) from *Spinacia oleracea*. J. Biol. Chem. 269(28), 187–192 (1994)
29. Liszewska, F., Gaganidze, D., Sirko, A.: Isolation of *Nicotiana plumbaginifolia* cD-NAs Encoding Isoforms of Serine Acetyltransferase and O-Acetylserine (Thiol)lyase in a Yeast Two-Hybrid System with *Escherichia coli* CysE and CysK Genes as Baits. Acta Biochim. Pol. 52(1), 117–128 (2005)
30. González, E.M., Ladrera, R., Larrainzar, E., Arrese-Igor, C.: Response of *Medicago truncatula* to Drought Stress. In: Mathesius, U., Journet, E.P., Sumner, L.W. (eds.) The Samuel Roberts Noble Foundation, Ardmore, OK, pp. 1–6 (2007)
31. Chandran, D., Sharopova, N., Ivashuta, S., Gantt, J., VandenBosch, K.A., Samac, D.A.: Transcriptome Profiling Identified Novel Genes Associated with Aluminum Toxicity, Resistance and Tolerance in *Medicago truncatula*. Planta 228, 151–166 (2008)
32. Waldren, R.P., Teare, I.D.: Free Proline Accumulation in Drought-Stressed Plants under Laboratory Conditions. Plant Soil 40(3), 689–692 (1974)
33. Weiss, E.A. (ed.): Oilseed Crops. Longman, London (1983)
34. Yoshida, K.T., Wada, T., Koyama, H., Mizobuchi-Fukuoka, R., Naito, S.: Temporal and Spatial Patterns of Accumulation of the Transcript of Myo-Inositol-1-Phosphate Synthase and Phytin-Containing Particles During Seed Development in Rice. Plant Physiol. 119(1), 65–72 (1999)
35. Kiranmayi, P., Mohan, P.M.: Metal Transportome of *Neurospora crassa*. Silico Biol. 6(3), 169–180 (2006)

Mining Association Rule Bases from Integrated Genomic Data and Annotations

Ricardo Martinez[1], Nicolas Pasquier[1], and Claude Pasquier[2]

[1] Laboratoire I3S, Université de Nice Sophia-Antipolis/CNRS UMR-6070,
06903 Sophia Antipolis, France
[2] IDBC, Université de Nice Sophia-Antipolis/CNRS UMR-6543,
Parc Valrose, 06108 Nice, France

Abstract. During the last decade, several clustering and association rule mining techniques have been applied to highlight groups of co-regulated genes in gene expression data. Nowadays, integrating these data and biological knowledge into a single framework has become a major challenge to improve the relevance of mined patterns and simplify their interpretation by biologists. GenMiner was developed for mining association rules from such integrated datasets. It combines a new nomalized discretization method, called NorDi, and the JClose algorithm to extract condensed representations for association rules. Experimental results show that GenMiner requires less memory than Apriori based approaches and that it improves the relevance of extracted rules. Moreover, association rules obtained revealed significant co-annotated and co-expressed gene patterns showing important biological relationships supported by recent biological literature.

1 Introduction

Gene expression technologies are powerful methods for studying biological processes through a transcriptional viewpoint. Since many years these technologies have produced vast amounts of data by measuring simultaneously expression levels of thousands of genes under hundreds of biological conditions. The analysis of these numerical datasets consists in giving meaning to changes in gene expression to increase our knowledge about cell behavior. In other words, we want to interpret gene expression data via integration of gene expression profiles with corresponding biological knowledge (gene annotations, literature, etc.) extracted from biological databases. Consequently, the key task in the interpretation step is to detect the present co-expressed (sharing similar expression profiles) and co-annotated (sharing the same properties such as function, regulatory mechanism, etc.) gene groups.

Several approaches dealing with the interpretation problem have recently been reported. These approaches can be classified in three axes [18]: *expression-based approaches, knowledge-based approaches* and *co-clustering approaches*. The most currently used interpretation axis is the *expression-based* axis that gives more weight to gene expression profiles. However, it presents many well-known drawbacks. First, this approach cluster genes by similarity in expression profiles across

F. Masulli, R. Tagliaferri, and G.M. Verkhivker (Eds.): CIBB 2008, LNBI 5488, pp. 78–90, 2009.
© Springer-Verlag Berlin Heidelberg 2009

all biological conditions. However, gene groups involved in a biological process might be only co-expressed in a small subset of conditions [2]. Second, many genes have different biological roles in the cell, they may be conditionally co-expressed with different groups of genes. Since almost all clustering methods used place each gene in a single cluster, that is a single group of genes, his relationships with different groups of conditionally regulated genes may remain undiscovered. Third, discovering biological relationships among co-expressed genes is not a trivial task and requires a lot of additional work, even when similar gene expression profiles are related to similar biological roles [23].

The use of association rule mining (ARM), that is another unsupervised data mining technique, was proposed to overcome these drawbacks. ARM aims at discovering relationships between sets of variable values, such as gene expression levels or annotations, from very large datasets. Association rules identify groups of variable values that frequently co-occur in data lines, establishing relationships with the form: $A \Rightarrow B$ between them. This rule means that when a data line contains variable values in A it is also likely to contain variable values in B. It has been shown in several research reports that ARM has several advantages. First, ARs can contain genes that are co-expressed in a subset of the biological conditions only. From this viewpoint, it and can be considered as a *bi-clustering* technique. Second, a gene can appear in several AR, if its expression profile fulfills the assignation criteria. That means, if a gene is involved in several co-expressed gene groups, it will appear in each and every one of these groups. Third, association rules are orientated knowledge patterns with the form *if condition then consequent* that describe directed relationships. This enables the discovery of any type of relationships between gene expression measures and annotations as they can be premisses or consequents of association rules. Fourth, since all types of data are considered in the same manner with ARM, several heterogeneous biological sources of information can be easily integrated in the dataset. These features make ARM a technique that is complementary to clustering for gene expression data analysis.

The GenMiner principle was introduced, with preliminary experimental results, in [19]. In this paper, we present a new Java implementation of GenMiner and new experimental results on the biological significance of extracted rules, the applicability and scalability of the algorithm and performance comparisons with other ARM approaches. This paper is organized as follows. Section 2 and 3 present ARM basics and related works respectively. The GenMiner approach is described in section 4 and the integrated dataset constituted for the experiments is presented in section 5. Experimental results are presented in section 6 and the paper ends with a discussion and conclusion in section 7.

2   Association Rule Mining

Association rules (ARs) express correlations between occurrences of variable values in the dataset as directed relationships between sets of variable values. In the data mining literature, variable values are called *items* and sets of items are called *itemsets*. For each AR, statistical measures assess the scope, or frequency, and the precision of the rule in the dataset. The classical statistics for this are respectively the *support* and the *confidence* measures. For instance, an AR *Event(A)*, *Event(B)* ⇒ *Event(C)*, *support=20%*, *confidence=70%* states that when events A and B occur, event C also occurs in 70% of cases, and that all three events occur together in 20% of all situations. This AR is extracted from a dataset containing *Event(A)*, *Event(B)* and *Event(C)* as items and data lines of the dataset describe co-occurred events, that is known situations. Since all ARs are not useful or relevant, depending on their frequency and precision, only ARs with support and confidence exceeding some user defined minimum support (*minsupp*) and minimum confidence (*minconf*) thresholds are extracted.

Extracting ARs is a challenging problem since the search space, i.e. the number of potential ARs, is exponential in the size of the set of items and several dataset scans, that are time expensive, are required. Several studies have shown that ARM is a NP-Complete problem and that a trivial approach, considering all potential ARs, is unfeasible for large datasets. The first efficient approach proposed to extract ARs is the Apriori algorithm [1]. Several optimisations of this approach have been proposed since, but all these algorithms give response times of the same order of magnitude and have similar scalability properties. Indeed, this approach was conceived for the analysis of sales data and is thus efficient when data is weakly correlated and sparse but performances drastically decrease when data are correlated or dense [5]. Moreover, with such data, a huge number of ARs are extracted, even for high *minsupp* and *minconf* values, and a majority of these rules are redundant, that is they cover the same information. For instance, consider the following five rules that all have the same support and confidence and the item *annotation* in the antecedent:

1. *annotation* ⇒ *gene1*↑
2. *annotation* ⇒ *gene2*↑
3. *annotation* ⇒ *gene1*↑, *gene2*↑
4. *annotation, gene1*↑ ⇒ *gene2*↑
5. *annotation, gene2*↑ ⇒ *gene1*↑

The most relevant rule from the user's viewpoint is rule 3 since all other rules can be deduced by inference from this one, including support and confidence (but the reverse does not hold). Information brougth by all other rules are summed up in rule 3, that is a *non-redundant association rule with minimal antecedent and maximal consequent*, or *minimal non-redundant ARs* for short. This situation is usual when mining correlated or dense data, such as genomic data, and to address this problem the GenMiner ARM approach integrates the JClose algorithm to extract minimal non-redundant ARs only.

3   Related Works

Several applications of ARM to the analysis of gene expression data have been recently reported [7,24,13]. These applications aimed at discovering frequent gene patterns among a subset of biological conditions. These patterns were represented as ARs such as: *gene1*↓ ⇒ *gene2*↑, *gene3*↓. This rule states that, in a significant number of biological conditions, when *gene1* is under-expresssed, we also observe an over-expression of *gene2* and an under-expression of *gene3*. These applications successfully highlighted correlations between gene expression profiles, avoiding some drawbacks of classical clustering techniques [13]. However, in these applications, biological knowledge was not taken into account and the task of discovering and interpreting biological similarities hidden within gene groups was left to the expert.

Recently, an approach to integrate gene expression profiles and gene annotations to extract rule with the form *annotations ⇒ expression profiles* was proposed in [6]. However, this approach presents several weaknesses. First, it uses the Apriori ARM algorithm [1] that is time and memory expensive in the case of correlated data. Moreover, it generates a huge number of rules among which many are redundant thus complexifying the interpretation of results. This is a well-known major limitation of the Apriori algorithm for correlated data [6,24]. Second, extracted rules are restricted to a single form: Annotations in the left-hand-side and expression profiles in the right-hand-side. However, all rules containing annotations and/or expression profiles, regardless of the side, bring important information for the biologist. Third, it uses the two-fold change cut-off method for discretizing expression measures in three intervals, a dangerous simplification that presents several drawbacks [21].

Discretization, which is needed for most of ARM implementations, is a delicate issue. According to the criteria used, there may be drastic changes on the rules generated. A recent paper proposed a way around this problem by running a biclustering algorithm over the gene expression matrix and then, by associating genes with the groups to which they belong [17]. The authors claim that the main advantage of this approach is that it reduces drastically the number of columns in the matrix and thus, that it simplify both the processing of the data and the interpretation of the rules. However, this depends mainly on the number of biclusters generated. In order to obtain very specific rules with low support, one needs to generate a huge number of small biclusters. Thus, the use of an efficient ARM algorithm is still needed and the interpretation of the resultant rules will still be very difficult.

GenMiner was developed to address these weaknesses and fully exploit ARM capabilities. It enables the integration of gene annotations and gene expression profile data to discover intrinsic associations between them. We chose to keep every colum from gene expression data but we use the novel NorDi method for discretizing gene expression measures. GenMiner takes advantage of the JClose algorithm [22] that can efficiently generate low support and high confidence non-redundant association rules, thus reducing the number of ARs and facilitating their interpretation by the biologist. With these features, GenMiner is an ARM approach that is adequate to biologists requirements for genomic data analysis.

4   GenMiner Approach

GenMiner follows the classical three steps of ARM approaches: (1) data selection and preparation, (2) ARs extraction and (3) ARs interpretation. It uses the NorDi algorithm for discretizing gene expression data during phase (1) and the

JClose algorithm for extracting minimal non-redundant ARs during phase (2). It is a co-clustering approach that discovers co-expressed and co-annotated gene groups at the same time according to co-ocurrences of gene expression profiles and annotations. It is a bi-clustering approach that finds co-annotated and co-expressed gene groups even in a small subset of biological conditions.

The whole process of GenMiner is deterministic and extracted ARs are not constrained in their form and their size in order to ensure that all kinds of relationships between gene expression profiles and annotations are discovered. The actual implementation of GenMiner does not integrate graphical visualization tools and complementary programs must be used to manipulate the results.

4.1 NorDi Algorithm

The *Normal Discretization* (NorDi) algorithm was developed to improve gene expression measures discretization into items. This algorithm is based on statistical detection of outliers and the continuous application of normality tests for transforming the initial sample distribution "almost normal" to a "more normal" one. The term "almost" means that the sample distribution can be normally distributed without the outlier's presence.

Let us assume that the expression data measures are presented as an nXm matrix: E with n genes (rows) and m samples or biological conditions (columns). Each matrix entry, $e_{i,j}$ represents the gene expression measure of gene i in sample j where $e_{i,j}$ is continuous in all real numbers. Let's suppose that the gene expression matrix E accomplishes the following assumptions:

1. All data is well cleaned (minimal noise).
2. Number of genes is largely enough.
3. The samples of the matrix S_j for every $j = 1, 2, ..., m$ are independent from each other and they are "almost" normally distributed $S_j \sim N(\mu_j, \sigma_j)$.
4. Missing values are no significant regarding the number of genes.

The NorDi algorithm is based on the observation that every sample of the expression matrix S_j can be "more" normally distributed $S_j^k \sim N(\mu_j, \sigma_j)$ if all outliers of each sample are momentarily removed (that is keeping a list of the k removed outliers for each sample, i.e. L_j^k) by Grubbs outliers method [14]. Each time an outlier k is removed, a Jaque-Bera normality test [3] has to be accomplished for the remaining sample S_j^k, where k is the number of removed outliers at each step in sample S_j and $k = 0, 1, 2, ..., clean$ ($k = clean$ means that there are no more outliers in the sample according to the Grubbs criterium). So, for every sample, we obtain the remaining sample S_j^{clean} that is "more normally" distributed than the original sample S_j. To verify this assertion we compare S_j^{clean} against S_j using the QQ-plot [20] and Lilliefors [16] normality tests. Then, we calculate the over-expressed, Ot, and under-expressed, Ut, cutoff thresholds using the $z - score$ methodology [25] over the cleaned sample S_j^{clean}.

Supposing the four precedent assumptions with $S_j^{clean} \sim N(\mu_j, \sigma_j)$ normal distributed and a $1 - \alpha$ predetermined confidence degree, the $z - score$ threshold cutoffs for three intervals are defined as:

- $Z_j = \frac{e_{i,j} - \mu_j}{\sigma_j} \geq z_{\alpha/2} = Ot \Rightarrow e_{i,j}$: over-expressed ($\uparrow$),
- $Z_j = \frac{e_{i,j} - \mu_j}{\sigma_j} \leq z_{\alpha/2} = Ut \Rightarrow e_{i,j}$: under-expressed ($\downarrow$),
- $Ut < e_{i,j} > Ot \Rightarrow e_{i,j}$: unexpressed,

where $z_{\alpha/2} = \Phi^{-1}(1 - \alpha/2)$, if the cumulative distribution function is $\Phi(z_{\alpha/2}) = P(S_j^{clean} \leq z_{\alpha/2}) = 1 - \alpha/2$.

It is important to notice that this procedure for computing the threshold cutoffs is done over all the m cleaned samples S_j^{clean} contained in the expression matrix $\mathbf{E}$. Once the computation of threshold cutoffs is done, the k elements in each sample's outliers list L_j^k are integrated to the original sample S_j and the discretization procedure is calculated for all values in S_j. The main reason is that outliers values cannot be removed from the analysis because they may contain relevant information of the biological experiment.

4.2 JClose Algorithm

JClose is a *frequent closed itemsets* based approach [22] for extracting minimal non-redundant AR defined as follows. An AR is *redundant* if it brings the same or less general information than is brought by another rule with identical support and confidence [8]. Then, an AR R is a minimal non-redundant AR if there is no AR R' with same support and confidence, which antecedent is a subset of the antecedent of R and which consequent is a superset of the consequent of R. JClose first extracts equivalence classes of itemsets, defined by *generators* and *frequent closed itemsets*, and generates from them the *Informative Basis* containing only minimal non-redundant ARs. This basis (minimal set) is a generating set for all ARs that captures all information brought by the set of all rules in a minimal number of rules, without information loss [8]. Experiments conducted on benchmark datasets show that the rule number reduction factor varies from 5 to 400 according to data density and correlation [22]. Moreover, when data is dense or correlated, JClose reduces extraction time and memory usage since the search space of frequent closed itemsets based approaches is a subset of the search space of Apriori based approaches. Several algorithms for extracting frequent closed itemsets, using complex data structures to improve efficiency, have been proposed since JClose. However, they do not extract generators, precluding the Informative Basis generation, and their response times, that depends mainly on data density and correlation, are of the same order of magnitude.

5   Annotations Enriched Eisen *et al.* Dataset

To validate the GenMiner approach we applied it to the well-known genomic dataset used by Eisen *et al.* [10]. This dataset contains expression measures of 2 465 *Saccharomyces cerevisiae* genes under 79 biological conditions extracted from a collection of four independent microarray studies during several biological processes: Cell cycle, Sporulation, Temperature shock and Diauxic shift experiments. Gene expression measures were discretized using NorDi algorithm at a 95% confidence level.

Each yeast gene was annotated with its associated terms in *Yeast GO Slim* (a yeast-specific cut-down version of Gene Ontology), its associations with research papers, the KEGG pathways in which it is involved, its phenotypes and the transcriptional regulators that bind its promoter regions.

The resulting dataset is a matrix of 2 465 lines representing yeast genes and 737 columns representing expression levels (discretized gene expression measures) over the 79 biological conditions and at most 658 gene annotations (24 GO annotations, 14 KEGG annotations, 25 transcriptional regulators, 14 phenotypes and 581 pubmed keywords). On the whole, the dataset contains 9 839 items (variable values). This dataset and the GenMiner application are available on the GenMiner web site [12] and on the KEIA web site [15].

6   Experimental Results

We conducted several experiments to evaluate the biological significance of extracted ARs, to compare the applicability of GenMiner and Apriori based approaches and to evaluate the scalability of GenMiner when mining very large dense biological datasets. For these experiments, the Java implementation of GenMiner was applied to the annotations enriched Eisen *et al.* dataset. All types of rules, containing gene annotations or gene expression levels either or both in the antecedent and the consequent, were extracted for *minsupp*=0.003 (at least 7 lines) and *minconf*=30%.

6.1   Biological Interpretation of Extracted Association Rules

Table 1 to 3 show some examples of the different form of rules extracted by GenMiner. In these tables, supports are given in number of transactions and confidences are given in percentages; the prefixes *go:*, *path:*, *pmid:*, *pr:*, *phenot:* are used to identify GO terms, KEGG pathways, Pubmed identifiers, promoters and phenotypes respectively; the labels *heat*, *diau* and *spo* refer to the different time points of the Heat shock, Diauxic shift and Sporulation experiments respectively; ↑ denotes an over-expression while ↓ denotes an under-expression.

ARs with the form *annotations* ⇒ *expression levels* (Table 1) show groups of genes associated with the same annotations that are over-expressed or under-expressed in a set of biological conditions. Rules 1 and 2 highlight a general reduction of transcription and protein synthesis following a heat shock, leading to cellular damages. This is confirmed by rule 3 which shows that genes regulated by RAP1 and FHL1, which are two key regulators of ribosomal protein genes, are under-expressed in this experiment. This last rule reflects the known fact that RAP1 recruits FHL1 to activate transcription [26]. A reduction of protein synthesis in the last time point of the Diauxic shif experiment is highlighted by rule 4. Additionally, rules 5 to 7 show that the genes involved in *oxidative phosphorylation, citrate cycle* and *glyoxylate and dicarboxylate metabolism* were also mainly over-expressed at the last time points. These rules reflect the main metabolic changes associated to the diauxic shift in yeast, manually identified in [9].

Table 1. Associations *annotations* ⇒ *expression levels*

Rule Antecedent	Consequent	supp.	conf.
1 go:0006412 (translation) go:0005840 (ribosome)	heat3↓	103	51
2 go:0005840 (ribosome) go:0003723 (RNA binding)	heat3↓	12	57
3 pr:RAP1 pr:FHL1	heat3↓	71	62
4 path:sce03010 (ribosome)	diau7↓	121	92
5 path:sce00190 (oxidative phosphorylation)	diau7↑	18	33
6 path:sce00020 (citrate cycle)	diau6↑ diau7↑	18	60
7 path:sce00630 (glyoxylate/dicarboxylate metabolism)	diau7↑	8	53

ARs with the form *expression levels* ⇒ *annotations* (Table 2) show groups of genes that are over-expressed or under-expressed in a set of biological conditions and have the corresponding gene annotations. Selected rules show information related to the Sporulation experiment (rules 1 and 2), the Heat shock process (rules 3 and 4) and the Diauxic shift process (rule 5) reported in the corresponding biological literature.

Table 2. Associations *expression levels* ⇒ *annotations*

Rule Antecedent	Consequent	supp.	conf.
1 spo4↓ spo5↓ spo6↓	go:0005975 (carbohydrate metabolism)	12	52
2 spo3↓ spo4↓ spo5↓	path:sce00010 (Glycolysis)	13	52
3 heat3↓ heat4↓ heat5↓	go:0006412 (translation)	35	88
4 heat2↓	go:0042254 (ribosome biogenesis)	39	66
5 diauxic6↓ diauxic7↓	go:0006412 (translation)	21	66

ARs with the form *annotations* ⇒ *annotations* (Table 3) contain gene annotations both in the antecedent and consequent. They highlight existent relationships among gene annotations, independently from gene expression levelsRules 1 and 2 identify associations between annotations from different sources like the relationship between the KEGG term *cell cycle* and the Gene Ontology term *cell cycle*, or the less obvious one between the KEGG term *purine metabolism* and the GO term *cytoplasm*. Rules 3 and 4 confirm the strong relationship between promoters $FHL1$ and $RAP1$. Rule 5 highlight a relationship between genes cited in a scientific article (which presents a review of the essential yeast genes) with the phenotype *inviable*. Rules 6 and 7 are two examples of a special group of rules that simply reflect the hierarchical structure of the bio-ontologies used. They represent an important proportion of rules that either depict the hierarchical links or represent identical relationships at different levels of abstraction corresponding to the hierarchically linked annotations. Such kind of rules can be filtered during a post-processing phase without information loss.

Table 3. Associations *annotations* ⇒ *annotations*

Rule Antecedent	Consequent	supp.	conf.
1 path:sce04111 (cell cycle)	go:0007049 (cell cycle)	67	78
2 path:sce00190 (purine metabolism)	go:0005737 (cytoplasm)	49	91
3 pr:FHL1	pr:RAP1	114	86
4 pr:RAP1	pr:FHL1	114	61
5 pmid:16155567	phenot:inviable	168	93
6 go:0016192 (vesicle transport)	go:0006810 (transport)	171	100
7 go:0005739 (mitochondrion)	go:0005737 (cytoplasm)	532	100

6.2 Execution Times and Memory Usage

These experiments were conducted to assess the applicability of GenMiner to very large dense biological datasets and to compare its results with Apriori based approaches. They were performed on a PC with one Pentium IV processor running at 2 GHz and 1 GB of RAM was allocated for the execution of GenMiner and implementations of Apriori based approaches. We tested several implementations of Apriori based approaches (Apriori, FP-Growth, Eclat, LCM, DCI, etc.). Execution times presented in Table 4 are these of Borgelt's implementation [4], available on the FIMI web site [11], that is globally the most efficient for mining ARs (and not only frequent itemsets). We can see in this table that execution times of GenMiner and the Apriori implementation are similar for *minsupp* between 0.02 (2%) and 0.007 (0.7%). However, executions of Apriori based approaches for lower *minsupp* values were interrupted as they required more than 1 GB of RAM. GenMiner could be run for *minsupp* = 0.003, i.e. rules supported by at least 7 data lines (genes), but the execution for *minsupp* = 0.002 was interrupted as more than 1 GB of RAM was required.

Table 4. Execution times and number of rules (*minconf*=0.3)

	GenMiner		Apriori	
minsupp (#)	Time (s)	Number of rules	Time (s)	Number of rules
0.020 (50)	10	10 028	5	65 312
0.015 (37)	21	28 492	16	325 482
0.010 (25)	72	110 989	76	3 605 486
0.009 (22)	101	147 966	110	6 115 366
0.008 (19)	187	230 255	182	12 138 561
0.007 (17)	289	315 090	264	21 507 415
0.006 (14)	673	542 746	Out of Memory	-
0.005 (12)	1 415	824 518	Out of Memory	-
0.004 (9)	5 353	1 675 811	Out of Memory	-
0.003 (7)	18 424	2 883 710	Out of Memory	-
0.002 (4)	Out of Memory	-	Out of Memory	-

Table 5. Execution times of GenMiner (in seconds)

minsupp (#)	minconf = 0.9	minconf = 0.5	minconf = 0.3
0.020 (50)	9.18	10.40	10.88
0.015 (37)	16.47	19.58	21.21
0.010 (25)	47.50	63.47	72.63
0.009 (22)	65.10	87.68	101.49
0.008 (19)	118.78	162.17	187.33
0.007 (17)	182.27	249.60	289.41
0.006 (14)	435.41	595.23	673.27
0.005 (12)	974.14	1 274.57	1 415.38
0.004 (9)	4 065.05	4 937.74	5 353.63
0.003 (7)	14 163.02	17 412.65	18 424.72
0.002 (4)	Out of Memory	Out of Memory	Out of Memory

We can see that for *minsupp* between 0.02 (2%) and 0.007 (0.7%), the Informative Basis is from 6 to 68 times smaller than the set of all ARs, that contains up to more than 21 millions of rules. However, the number of ARs in the Informative Basis is important for low *minsupp* values and it cannot be manually explored without tools to select subsets of ARs.

6.3   GenMiner Scalability

Experimental results presented in Table 5 were conducted to evaluate execution times and memory usage of GenMiner when the *minsupp* and *minconf* thresholds vary. Three series of executions were run for *minconf* equals to 0.9 (90%), 0.5 (50%) and 0.3 (30%). For each serie, *minsupp* was varied between 0.02 (2%) and 0.002 (0.2%). As in the previous experiment, GenMiner could not be run for *minsupp* lower than 0.003, independently from the *minconf* value. We can also see that the longest executions, for *minsupp* equals to 0.003, took from 4 to 5 hours depending on the *minconf* value.

7   Discussion and Conclusion

GenMiner was developed for mining association rules from very large dense datasets containing both gene expression data and annotations. Contrarily to most approaches for gene expression interpretation, as well *expression-based* as *knowledge-based*, in which biological information and gene expression profiles are incorporated in an independent manner, with GenMiner both data sources are integrated in a single framework.

GenMiner implements a new discretization algorithm, called NorDi, that was designed for processing data generated by gene expression technologies in the case of independent biological conditions. Experiments conducted on the Eisen *et al.* dataset show that its results are relevant. However, the discretization issue is delicate when using data mining methods such as ARM. We thus propose to use several discretization scenarios, analyzing the pertinence of obtained results against expected results, to validate the discretization method. As pointed out in [21]: "The robustness of biological conclusions made by using microarray analysis should be routinely assessed by examining the validity of the conclusions by using a range of threshold parameters issued from different discretization algorithms". Unfortunately, to our knowledge no discretization algorithm, specially designed for time process data, can integrate the time variable without an important loss of temporal information.

GenMiner also integrates the JClose algorithm developed to extract condensed representations of ARs from dense and correlated data. JClose is based on the frequent closed itemsets framework that allows to reduces both the search space and the number of dataset accesses, and thus the memory usage, for dense and correlated data. It extracts a minimal set of non-redundant ARs called Informative Basis [22] in order to reduce the number of extracted ARs and improve the result's relevance. In this basis, all information is summarized in a minimal number of ARs, each rule bringing as much information as possible, without information loss. The GenMiner, JClose and NorDi applications are available on the KEIA web site [15].

To evaluate the efficiency and scalability of GenMiner, it was run on a dataset combining the Eisen *et al.* gene expression data [10] and annotations of these genes. Experimental results show that GenMiner can deal with such large datasets and that its memory usage, as well as the number of ARs generated, are significantly smaller than these of Apriori based approaches. Moreover, ARs extracted by GenMiner are not constrained in their form and can contain both gene annotations and gene expression profiles in the antecedent and the consequent. The analyze of these ARs has shown important relationships supported by recent biological literature. These results show that GenMiner is a promising tool for finding meaningful relationships between gene expression patterns and gene annotations. Furthermore, it enables the integration of thousands of gene annotations from heterogenous sources of information with related gene expression data. This is an essential feature as the integration of different types of biological information is indispensable to fully understand the underlying biological processes. In addition, qualitative variables such as gender, tissue and age could easily be integrated in order to extract ARs among these features and gene expression patterns. In the future, we plan to integrate in GenMiner tools to filter, select, compare and visualize ARs during the interpretation phase to simplify these manipulations.

References

1. Agrawal, R., Srikant, R.: Fast Algorithms for Mining Association Rules. In: Proceedings of the VLDB international conference, pp. 478–499 (1994)
2. Altman, R., Raychaudhuri, S.: Whole-Genome Expression Analysis: Challenges Beyond Clustering. Current Opinion Structural Biology 11, 340–347 (2001)

3. Bera, A., Jarque, C.: Efficient Tests for Normality, Homoscedasticity and Serial Independence of Regression Residuals: Monte Carlo Evidence. Economics Letters 7, 313–318 (1981)

4. Borgelt, C.: Recursion Pruning for the Apriori Algorithm. In: Proceedings of the FIMI international workshop (2004)

5. Brin, S., Motwani, R., Ullman, J.D., Tsur, S.: Dynamic Itemset Counting and Amplication Rules for Market Basket Data. In: Proceedings of the ACM SIGMOD international conference, pp. 255–264 (1997)

6. Carmona-Saez, P., Chagoyen, M., Rodriguez, A., Trelles, O., Carazo, J., Pascual-Montano, A.: Integrated Analyis of Gene Expression by Association Rules Discovery. BMC Bioinformatics 7, 54 (2006)

7. Creighton, C., Hanansh, S.: Mining Gene Expression Databases for Association Rules. Bioinformatics 19, 79–86 (2003)

8. Cristofor, L., Simovici, D.A.: Generating an Informative Cover for Association Rules. In: Proceedings of the ICDM international conference, pp. 597–600 (2002)

9. DeRisi, J., Iyer, L., Brown, V.: Exploring the Metabolic and Genetic Control of Gene Expression on a Genomic Scale. Science 278, 680–686 (1997)

10. Eisen, M., Spellman, P., Brown, P., Botsein, D.: Cluster Analysis and Display of Genome Wide Expression Patterns. Proc. Nat. Aca. Sci. 95, 14863–14868 (1998)

11. FIMI: Frequent Itemset Mining Implementations Repository, http://fimi.cs.helsinki.fi

12. GenMiner: Genomic Data Miner, http://bioinfo.unice.fr/publications/genminer_article

13. Georgi, E., Richter, L., Ruckert, U., Kramer, S.: Analyzing Microarray Data using Quantitative Association Rules. Bioinformatics 21, 123–129 (2005)

14. Grubbs, F.: Procedures for Detecting Outlying Observations in Samples. Technometrics 11, 1–21 (1969)

15. KEIA: Knowledge Extraction, Integration and Applications, http://keia.i3s.unice.fr

16. Lilliefors, H.: On the Kolmogorov-Smirnov Test for Normality with Mean and Variance Unknown. Journal of the American Statistical Association 62 (1967)

17. Lopez, F.J., Blanco, A., Garcia, F., Cano, C., Marin, A.: Fuzzy Association Rules for Biological Data Analysis: A Case Study on Yeast. BMC Bioinformatics 9, 107 (2008)

18. Martinez, R., Collard, M.: Extracted knowledge: Interpretation in Mining Biological Data, a Survey. Int. J. of Computer Science and Applications 1, 1–21 (2007)

19. Martinez, R., Pasquier, N., Pasquier, C.: GenMiner: Mining Informative Association Rules from Genomic Data. In: Proceedings of the IEEE BIBM international conference, pp. 15–22 (2007)

20. NIST: e-Handbook of Statistical Methods. SEMATECH (2007), http://www.itl.nist.gov/div898/handbook/

21. Pan, K., Lih, C., Cohen, N.: Effects of Threshold Choice on Biological Conclusions Reached During Analysis of Gene Expression by DNA Microarrays. Proc. Nat. Aca. Sci. 102, 8961–8965 (2005)

22. Pasquier, N., Taouil, R., Bastide, Y., Stumme, G., Lakhal, L.: Generating a Condensed Representation for Association Rules. Journal of Intelligent Information Systems 24(1), 29–60 (2005)

23. Shatkay, H., Edwards, S., Wilbur, W., Boguski, M.: Genes, Themes, Microarrays: Using Information Retrieval for Large-Scale Gene Analysis. In: Proceedings of the ISMB international conference, pp. 340–347 (2000)
24. Tuzhilin, A., Adomavicius, G.: Handling Very Large Numbers of Association Rules in the Analysis of Microarray Data. In: Proceedings of the SIGKDD international conference, pp. 396–404 (2002)
25. Yang, I., Chen, E., Hasseman, J., Liang, W., Frank, B., Sharov, V., Quackenbush, J.: Within the Fold: Assesing Differential Expression Measures and Reproducibility in Microarray Assays. Genome Biology 3, 11 (2002)
26. Zhao, Y., McIntosh, K., Rudra, D., Schawalder, S., Shore, D., Warner, J.: Fine-Structure Analysis of Ribosomal Protein Gene Transcription. Molecular Cellular Biology 26(13), 4853–4862 (2006)

Stability and Performances in Biclustering Algorithms

Maurizio Filippone[1], Francesco Masulli[2,3,4], and Stefano Rovetta[2,3]

[1] Department of Computer Science, University of Sheffield, United Kingdom
[2] Department of Computer and Information Sciences, University of Genova, Genova, Italy
[3] CNISM Genova Research Unit, Genova, Italy
[4] Center for Biotechnology, Temple University, Philadelphia, USA
{rovetta,masulli}@disi.unige.it

Abstract. Stability is an important property of machine learning algorithms. Stability in clustering may be related to clustering quality or ensemble diversity, and therefore used in several ways to achieve a deeper understanding or better confidence in bioinformatic data analysis. In the specific field of fuzzy biclustering, stability can be analyzed by porting the definition of existing stability indexes to a fuzzy setting, and then adapting them to the biclustering problem. This paper presents work done in this direction, by selecting some representative stability indexes and experimentally verifying and comparing their properties. Experimental results are presented that indicate both a general agreement and some differences among the selected methods.

1 Introduction

Many bioinformatic data sets come from DNA microarray experiments and are normally given as a rectangular m by n matrix $X = (x_{ij})_{mn}$, where each column represents a feature (e.g., a gene) and each row represents a data point or condition (e.g., a patient), and value x_{ij} is the expression of i-th gene in j-th condition. The analysis of microarray data sets can provide valuable information on the biological relevance of genes and correlations among them.

Biclustering (also known under other names like *co-clustering* and *two-way clustering*) [17] is a methodology allowing for feature set and data points clustering simultaneously, i.e., to find clusters of samples possessing similar characteristics together with features creating these similarities. In other words, biclustering answers the question: *What characteristics make "similar" objects similar to each other?*

The output of biclustering is not a partition or hierarchy of partitions of either rows or columns, but a partition of the whole matrix into sub-matrices or patches. We can obtain different biclustering structures: single bicluster, different non-overlapping structures (as exemplified in Fig. 1), and overlapping with or without structure.

The goal of biclustering is to find as many patches as possible, and to have them as large as possible, while maintaining strong homogeneity within patches. This task is reported to be an NP-complete task [17,21].

In gene expression microarray data analysis biclustering methods allow us to identify genes with similar behavior with respect to different conditions. A single patch represents a given subset of genes in a given subset of conditions.

F. Masulli, R. Tagliaferri, and G.M. Verkhivker (Eds.): CIBB 2008, LNBI 5488, pp. 91–101, 2009.

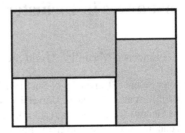

Fig. 1. Example of biclusters, shown as "patches"

Biclustering algorithms able to find largest biclusters from DNA microarray data that do not exceed an assigned homogeneity constraint [3] are necessary as they convey relevant biological information able to support important tasks, such as:

- *identification of coregulated genes* and/or *specific regulation processes* by identifying sets of genes that, under specific conditions, exhibit coherent activations;
- *gene functional annotation* by extending the class label shared by the the majority of genes in the bicluster to the remaining non-annotated genes of the same bicluster;
- *sample* and/or *tissue classification*, since considering the diagnosis of a specific pathology biclusters identify the different responses to treatment, and then the group of genes to be used as the most effective probe.

In the next section we introduce three non-pairwise indexes suited to studying the stability of biclustering algorithms. In Sect. 3 after stating the fuzzy framework for biclustering, we give a short overview of two biclustering algorithms based on this approach. Sect. 4 presents the experimental study of their stabilities. In Sect. 5 we draw the conclusions.

2   Stability Indexes

2.1   Stability of Learning Machines

In machine learning *stability* among solutions has been related to some important properties of learners, e.g., generalization [15,5].

When learning is formulated as an optimization process, the reliability of a solution can be inferred from its robustness with respect to perturbations in the data, parameters, or training process. A learning algorithm is *stable* if it produces robust solutions. If the analysis of a cost function landscape is possible, this relationship can be proved, but often we deal with difficult cost functions and then we must estimate the stability of a learning machine from empirical observations.

In case of biclustering algorithms we can evaluate their stability by means of indexes measuring the degree of similarity or overlap among solution sets, as for clustering, but preferring computationally parsimonious indexes, as the dimension of the solution space is very large (product of the row- and column-dimensions of the data matrix). We shall now describe the stability indexes we have considered.

2.2   Normalized Mutual Information between Partitions

The Normalized Mutual Information between partitions is a pairwise index defined as [9]:

$$
NMI = \frac{-2 \sum_{i=1}^{|A|} \sum_{j=1}^{|B|} N_{ij} \log \left(\frac{N_{ij}N}{N_i N_j} \right)}{\sum_{i=1}^{|A|} N_i \log \left(\frac{N_i}{N} \right) + \sum_{j=1}^{|B|} N_j \log \left(\frac{N_j}{N} \right)}
\tag{1}
$$

where A and B are two partitions; $|A|$ is the number of biclusters in partition A; $|B|$ is the number of biclusters in partition B ($|A|$ and $|B|$ can be different); i a bicluster in A; j a bicluster in B; N_i and N_j are respectively the cardinalities of bicluster i and j; N_{ij} the cardinality of the intersection of bicluster i and bicluster j (i.e., number of data points which are assigned to bicluster i in A and to bicluster j in B); and N is the cardinality of the dataset.

Note that computation of NMI doesn't involve relabeling of biclusters, moreover we have NMI $= 1$ for perfect overlap and NMI $= 0$ (asymptotically) for completely independent partitions.

2.3   Jaccard Coefficient

Jaccard coefficient is another pairwise index and is defined as the ratio of cardinalities of the intersection of two sets to their union [12]:

$$
J(A,B) = \frac{|A \cap B|}{|A \cup B|}.
\tag{2}
$$

After Jaccard's paper, this coefficient and some related ones have been proposed in several occasions, see, e.g., [18,2,4].

2.4   Entropy of Consensus Matrix

The *co-association matrix* [20] is a $N \times N$ matrix $\mathbf{M}^{(s)}$. Their elements $m_{ij}^{(s)}$ indicates whether two matrix elements i and j are in the same bicluster in experiment number s.

The *consensus matrix* $\mathbf{M}$ averages co-association matrixes over all experiments [20]. Then: (a) if all the clusterers agree on joining objects i and j in the same bicluster, $m_{ij} = 1$; (b) if all clusterers agree that objects i and j are in different clusters, $m_{ij} = 0$; (c) otherwise, if there is disagreement on joint membership of the two objects, m_{ij} between 0 and 1. Note that in the case of the largest disagreement, where i and j are in the same biclusters in exactly $L/2$ of the partitions $P_1, \ldots, P_L$, $m_{ij} = 0.5$.

We can define a global index called the *Entropy of Consensus Matrix* as the *averaged entropy* of the cells of $\mathbf{M}$ as [15]:

$$
H(\mathbf{M}) = -\frac{1}{N^2} \sum_{i=1}^{N} \sum_{j=1}^{N} \left(m_{ij} \log (m_{ij}) + (1 - m_{ij}) \log (1 - m_{ij}) \right)
\tag{3}
$$

3 Fuzzy Biclustering

3.1 Fuzzy Framework for Biclustering

Let x_{ij} be the expression level of the i-th gene in the j-th condition. A *bicluster* is defined as a subset of the $m \times n$ data matrix X, i.e., a bicluster is a pair $(\mathbf{g}, \mathbf{c})$, where $\mathbf{g} \subset \{1, \dots, m\}$ is a subset of genes and $\mathbf{c} \subset \{1, \dots, n\}$ is a subset of conditions [3,11,16,24].

The *size* (or volume) n of a bicluster is usually defined as the number of elements in the gene expression matrix X belonging to it, that is the product of the cardinalities $n_g = |\mathbf{g}|$ and $n_c = |\mathbf{c}|$:

$$n = n_g \cdot n_c \tag{4}$$

The *bicluster mean*, *bicluster row mean*, and *bicluster column mean* are defined as:

$$x_{IJ} = \frac{1}{n} \sum_{i \in \mathbf{g}} \sum_{j \in \mathbf{c}} x_{ij} \qquad x_{iJ} = \frac{1}{n_c} \sum_{j \in \mathbf{c}} x_{ij} \qquad x_{Ij} = \frac{1}{n_g} \sum_{i \in \mathbf{g}} x_{ij} \tag{5}$$

Hartigan [11] proposed the following definition for a biclustering *residue*:

$$d_{ij} = x_{ij} - x_{IJ}. \tag{6}$$

A residue suited for general "trends" of DNA microarray data analysis has been proposed by Cheng and Church [3,7]:

$$d_{ij} = x_{ij} - (x_{IJ} + \alpha_i + \beta_j) = x_{ij} + x_{IJ} - x_{iJ} - x_{Ij}. \tag{7}$$

This definition takes into account that for ideal constant row biclusters, each element x_{ij} is equal to the bicluster mean x_{IJ} plus an offset $\alpha_i = x_{iJ} - x_{IJ}$, and similarly, for ideal constant column biclusters, each element x_{ij} is equal to the bicluster mean x_{IJ} plus an offset $\beta_j = x_{Ij} - x_{IJ}$.

Following [3], we are interested in the largest biclusters from DNA microarray data that do not exceed an assigned homogeneity constraint. To this aim we can utilize one of those definitions of biclustering homogeneity, or better, of *biclustering (crisp) heterogeneity*:

$$H = \sum_{i \in \mathbf{g}} \sum_{j \in \mathbf{c}} d_{ij}^2 \qquad \text{Sum-squared residue [7];} \tag{8}$$

$$G = \frac{1}{n} \sum_{i \in \mathbf{g}} \sum_{j \in \mathbf{c}} d_{ij}^2 = \frac{H}{n} \qquad \text{Mean Squared residue [3].} \tag{9}$$

G measures the *bicluster heterogeneity*, i.e., the *difference between the actual value of an element x_{ij} and its expected value* as predicted from the corresponding row mean, column mean, and bicluster mean.

Hartigan's residue (Eq. 6) can capture constant biclusters only, while Cheng and Church's residue (Eq. 7) can capture biclusters with constant rows, constant columns, and even coherent values (addictive models) and is then best suited for gene expression data analysis [17].

In order to obtain a fuzzy formulation of the biclustering problem, we should first set biclustering in a (crisp) set theory framework, and then we can extend this setting to a fuzzy set formulation. To this aim, for each bicluster we assign two membership vectors, one for the rows and one other for the columns, denoting them respectively **a** and **b**. In a *crisp sets framework* row i and column j can either belong to the bicluster or not. An element x_{ij} of X belongs to the bicluster if both $a_i = 1$ and $b_j = 1$, i.e., its membership to the bicluster is $u_{ij} = \text{and}(a_i, b_j)$. Therefore, we can define the cardinality of a bicluster as follows:

$$n = \sum_i \sum_j u_{ij} \tag{10}$$

To proceed toward a fuzzy set theory framework, we allow membership u_{ij}, a_i and b_j to belong in the interval $[0,1]$. The membership u_{ij} of an element x_{ij} of X to the bicluster can be obtained by the *aggregation* of row and column memberships, using, e.g..:

$$u_{ij} = a_i b_j \quad \text{(product)} \tag{11}$$

or

$$u_{ij} = \frac{a_i + b_j}{2} \quad \text{(average)}. \tag{12}$$

The fuzzy cardinality of the bicluster is defined as the sum of the memberships u_{ij} for all i and j and, formally, is still given in eq. 10. The same happens for *fuzzy residue* that is formally identical to the definitions in eq.s 6 and 7 (we will use the second one, as we will work with DNA microarray data), but variables included in them must be interpreted as *fuzzy bicluster mean, fuzzy bicluster row mean, fuzzy bicluster column mean*:

$$x_{IJ} = \frac{\sum_i \sum_j u_{ij} x_{ij}}{\sum_i \sum_j u_{ij}}, \quad x_{iJ} = \frac{\sum_j u_{ij} x_{ij}}{\sum_j u_{ij}}, \quad x_{Ij} = \frac{\sum_i u_{ij} x_{ij}}{\sum_i u_{ij}} \tag{13}$$

We can introduce now the definitions of *Fuzzy Sum-squared residue g* and *Fuzzy Mean Squared residue G* that generalize the *bicluster heterogeneity* concept:

$$g = \sum_i \sum_j u_{ij} d_{ij}^2 \qquad G = \frac{1}{n} \sum_i \sum_j u_{ij} d_{ij}^2 \tag{14}$$

3.2 Minimum Sum-Squared Residue for Fuzzy Co-clustering Algorithm

We shall now present two fuzzy biclustering algorithms, namely the *Minimum Sum-squared Residue for Fuzzy Co-clustering* (MSR-FCC) algorithm [23] and the Possibilistic Biclustering (PCB) [8], that are inspired from the fuzzy central clustering algorithms. In the versions applied in this paper, both methods employ the product aggregator (11) in the computation of the membership of an element of the data matrix X to a bicluster. The former imposes the *probabilistic constraint* on memberships, used, e.g., in the Fuzzy C-Means [1], according to which the sum of the membership values of a matrix element to all the biclusters must be equal to one. PCB, instead, applies more relaxed constraints to the memberships, following the so-called possibilistic clustering framework [13].

The *Minimum Sum-squared Residue for Fuzzy Co-clustering* (MSR-FCC) algorithm [23] is based on the constrained minimization of a generalization of the fuzzy central clustering objective function:

$$J = \sum_{cdij} a_{ci} b_{dj} d_{cdij}^2 + T_a \sum_{ci} a_{ci}^2 + T_b \sum_{dj} b_{dj}^2 +$$

$$\sum_i \lambda_i \left(\sum_c a_{ci} - 1 \right) + \sum_j \gamma_j \left(\sum_d b_{dj} - 1 \right), \qquad (15)$$

$$c,q \in \{1,...,C\}, \qquad d,p \in \{1,...,D\}, \qquad T_a, T_b, \lambda_i, \gamma_j \in \mathbb{R}$$

where C is the number of data clusters; D is the number of feature clusters; the 1st term is the Fuzzy Sum-squared residue g; the 2nd, 3rd terms are fuzzy Gini indexes; while the last two terms are Lagrange constraints due to the probabilistic constraints imposed for row- and column-memberships normalization.

By setting the derivatives of J with respect to the memberships a_i and b_j to zero we obtain these necessary conditions:

$$a_{ci} = \frac{1}{C} + \frac{1}{2T_a} \left(\frac{1}{C} \sum_{pdj} b_{dj} d_{pdij}^2 - \sum_{dj} b_{dj} d_{cdij}^2 \right) \qquad (16)$$

$$b_{dj} = \frac{1}{D} + \frac{1}{2T_b} \left(\frac{1}{D} \sum_{cqi} a_{ci} d_{cqij}^2 - \sum_{ci} a_{ci} d_{cdij}^2 \right) \qquad (17)$$

Fig. 2 shows a generic algorithm for fuzzy biclustering iterating these necessary conditions for minimizing the the objective function J. One thing that may be worth noticing is that, in the MSR-FCC optimization process, iterations are not necessarily contraction mappings, and hence convergence of the Picard iterations is not guaranteed as required by related fixed-point theorems [6,10].

---

1. Initialize ε and the memberships $a_{ci}\, b_{dj}\, \forall c,d,i,j$
2. Compute $d_{cdij}^2\, \forall c,d,i,j$
3. Update $a_{ci}\, \forall c,i$
4. Update $b_{dj}\, \forall d,j$
5. Compute $\Delta_{max} = \max \left\{ \{|a_{ci} - a_{ci}'|\forall c,i\} \cup \{|b_{dj} - b_{dj}'|\forall d,j\} \right\}$
6. **if** $\Delta_{max} < \varepsilon$ then **stop**
7. **else jump** to step 2

Fig. 2. Generic fuzzy biclustering algorithm

3.3 Possibilistic Biclustering Algorithm

The Possibilistic Biclustering (PCB), algorithm proposed by our group [8], is based on the possibilistic clustering framework proposed by Krishnapuram and Keller in 1993 [13], that relaxes the clustering probabilistic constraint to:

$$u_{pq} \in [0,1] \quad \forall p,q; \qquad (18)$$

$$0 < \sum_{q=1}^{r} u_{pq} < r \quad \forall p; \qquad \bigvee_p u_{pq} > 0 \quad \forall q. \tag{19}$$

These minimal constraints say that clusters cannot be empty and each pattern must be assigned to at least one cluster.

In PBC framework we can go to minimize this objective function [8,14]:

$$J = \sum_{ij} a_i b_j d_{ij}^2 + \lambda \sum_i (a_i \ln(a_i) - a_i) + \mu \sum_j (b_j \ln(b_j) - b_j) \tag{20}$$

where the first term is the *fuzzy squared residue H*, while the other two are penalization terms. The parameters λ and μ control the size of the bicluster. Setting the derivatives of J with respect to the memberships a_i and b_j to zero we obtain:

$$a_i = \exp\left(-\frac{\sum_j b_j d_{ij}^2}{\lambda}\right) \qquad b_j = \exp\left(-\frac{\sum_i a_i d_{ij}^2}{\mu}\right) \tag{21}$$

These necessary conditions for the minimization of J_B together with the definition of fuzzy residue d_{ij} can be used by an algorithm able to find a numerical solution for the optimization problem (Picard iteration), as show in Fig. 2.

4   Experimental Analysis

We studied the stability and the performances of the Minimum Sum-squared Residue for Fuzzy Co-clustering (MSR-FCC) algorithm and of the Possibilistic Biclustering (PBC) algorithm using the Yeast data set [22] that is DNA microarray data set measuring the gene expression of 2879 genes in 17 conditions.

We used this data set and some its modifications obtained by adding to it uniform random noise of different levels. We initialized the two algorithms with random memberships, setting the other parameters as follows: the threshold for the stopping criterion is $\varepsilon = 10^{-6}$, and the cutting threshold for defuzzification of data matrix elements' membership (α-cut) is 0.5. The number of requested biclusters was 2×2 for MSR-FCC, and one for PBC. Stability indexes were all evaluated in pairwise way, to assess the overlap between each individual experiment and a reference solution (the one obtained without noise). The heterogeneity index is defined as $\Omega \equiv G$ (Eq. 9).

Table 1 reports the experimental results of stability analysis. All presented results are averaged on 10 runs. The Normalized Mutual Information (H), the Jaccard Coefficient (NMI), and the Entropy of Consensus Matrix (J) show a good concordance, confirming in such a way their usefulness, but at the same time suggesting that the information they provide is redundant. Concordance is to be expected at the extreme values (partitions matching completely, partitions completely independent); however, experimental results show that this holds even for intermediate values.

In general, the MSR-FCC method appears to be very stable. The possibilistic version has a certain dependence from user-defined parameters, but Table 1 shows that heterogeneity is always better than for the competitive MSR-FCC method, even if stability is

Table 1. Results on yeast data. Average indexes with uniform noise added

	Noise	H	NMI	J	Largest n	avg.Ω
	0%	0.000	1.00	1.00	23518	493.44
	1%	0.002	0.96	0.99	23485	491.75
PBC ($\mu = 0.6, \lambda = 80$)	2%	0.022	0.83	0.91	23309	479.78
	4%	0.057	0.59	0.78	18576	431.26
	8%	0.136	0.20	0.28	6820	271.55
	0%	0.000	1.00	1.00	15496	330.5
	1%	0.004	0.93	0.98	15470	329.32
PBC ($\mu = 0.34, \lambda = 120$)	2%	0.010	0.87	0.95	15236	326.81
	4%	0.022	0.75	0.88	14196	313.27
	8%	0.100	0.28	0.34	6320	220.66
	0%	0.000	1.00	1.00	15228	935.43
	1%	0.000	0.99	1.00	15228	935.94
MSR-FCC ($T_u = 1000, T_v = 1000$)	2%	0.001	0.99	1.00	15264	935.74
	4%	0.001	0.98	0.99	15273	936.35
	8%	0.002	0.97	0.98	15291	936.63

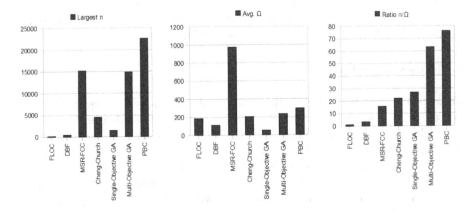

Fig. 3. Comparative performances of some biclustering techniques: heterogeneity Ω (left), size n (center), and synthetic indicator n/Ω (right)

slightly lower. We can see from the table that the largest biclusters are obtained by PBC with the first set of parameters ($\lambda = 0.6, \mu = 80$); the heterogeneity of solutions obtained with the other set of parameters is smaller, but this is obvious for smaller biclusters. On the other hand, MSR-FCC shows superior stability, but this is not associated to equally high performance.

Figure 3, based on data from [8], shows the relationship between the two quality criteria as obtained also on the other biclustering techniques outlined previously in this paper (namely, DBF [26], FLOC [25], Cheng-Church [3], Single-objective GA [19], and Multi-objective GA [19]). The first column graph shows the heterogeneity level (Ω) and the second graph shows bicluster size (n), computed, in case of fuzzy biclustering algorithms, after the final defuzzication. The third graph concerns the value of the

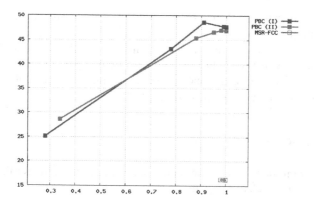

Fig. 4. Synthetic performance (n/Ω) as a function of Jaccard index J, computed from Table 1. PBC (I) and (II) indicate PBC with the first and second sets of parameters respectively. MSR-FCC is consistently worse.

performance index, defined as the ratio n/Ω: the largest the ratio, the better the result. Finally, as it is reasonable to expect, the value of the performance index shows low correlation with respect to stability.

The experiments reported in Figure 3 are different from those of the present paper, but the results on MSR-FCC and PBC are very similar. The competitive method is clearly the most stable. However, by taking into account performance along with stability, we see that PBC with both sets of parameters performs consistently better than MSR-FCC. This can be observed by computing the performance index n/Ω for the results in Table 1, as a function of stability. This is illustrated by the graphs in Figure 4, where n/Ω is plotted for the three methods as a function of J.

5   Conclusions

The stability of a learning algorithm is reported in the literature as related to is performance, e.g. to the generalization capabilities in supervised algorithms [5]. In ensemble methods diversity of base learners is often exploited to increase stability of the ensemble.

In standard clustering complex relationships among stability indexes and performances have been experimental shown in [15] using k-means algorithm. In this study we have studied the relationships between similarity and performances in biclustering that is a generalization of clustering as it is a methodology allowing for feature set and data points clustering simultaneously.

While performances in clustering can be related to the *representation error* (that is the count of data points in each cluster disagreeing with the majority label in that cluster, summed over all clusters and expressed as a percentage), as done, e.g., in [15], performances for biclustering can be related to bicluster cardinality and heterogeneity.

In this study we have employed the Normalized Mutual Information, the Jaccard Coefficient, and the Entropy of Consensus Matrix as stability indexes and we have considered two fuzzy biclustering algorithms, namely the *Minimum Sum-squared Residue for Fuzzy Co-clustering* (MSR-FCC) algorithm [23] and the *Possibilistic Biclustering* (PBC) algorithm [8].

We noticed that for PBC there is a relationship between all stability indexes and the performance index, while MSR-FCC is more stable, but shows a worse performance index that does not depend on any stability index. Moreover, in general, the three considered stability indexes are strongly correlated.

Acknowledgments

We thank Luca Zini for programming support and Giorgio Valentini for the useful suggestions provided while discussing this work.

References

1. Bezdek, J.C.: Pattern recognition with fuzzy objective function algorithms. Plenum, New York (1981)
2. Brivio, P.A., Binaghi, E., Rampini, A., Ghezzi, P.: A Fuzzy Set-Based Accuracy Assessment of Soft Classifications. Pattern Recognition Letters 20, 935–948 (1999)
3. Cheng, Y., Church, G.M.: Biclustering of expression data. In: Proc. Int. Conf. Intell. Syst. Mol. Bio., vol. 8, pp. 93–103 (2000)
4. Congalton, R.G.: Assessing the Accuracy of Remotely Sensed Data: Principles and Practices. CRC Press, Boca Raton (1999)
5. Bousquet, O., Elisseeff, A.: Stability and generalization. J. Mach. Learn. Res. 2, 499–526 (2002)
6. Browder, F.: Nonexpansive nonlinear operators in a banach space. Proc. Nat. Acad. Sci. U.S.A. 54, 1041–1044 (1965)
7. Cho, H., Dhillon, I.S., Guan, Y., Sra, S.: Minimum Sum-Squared Residue Co-clustering of Gene Expression Data. In: Proc. Fourth SIAM Int. Conf. on Data Mining, pp. 114–125 (2004)
8. Filippone, M., Masulli, F., Rovetta, S., Mitra, S., Banka, H.: Possibilistic approach to biclustering: An application to oligonucleotide microarray data analysis. In: Priami, C. (ed.) CMSB 2006. LNCS (LNBI), vol. 4210, pp. 312–322. Springer, Heidelberg (2006)
9. Fred, A.L.N., Jain, A.K.: Robust Data Clustering. In: IEEE Computer Society Conference on Computer Vision and Pattern Recognition (CVPR 2003), vol. 2, pp. 128–136 (2003)
10. Göhde, D.: Zum prinzip der kontraktiven abbildung. Math. Nachr. 30, 251–258 (1965)
11. Hartigan, J.A.: Direct clustering of a data matrix. Journal of the American Statistical Association 67(337), 123–129 (1972)
12. Jaccard, P.: Etude comparative de la distribution florale dans une portion des alpes et des jura. Bulletin de la Société Vaudoise des Sciences Naturelles 37, 547–579 (1901)
13. Krishnapuram, R., Keller, J.M.: A possibilistic approach to clustering. IEEE Transactions on Fuzzy Systems 1(2), 98–110 (1993)
14. Krishnapuram, R., Keller, J.M.: The possibilistic C-Means algorithm: insights and recommendations. IEEE Transactions on Fuzzy Systems 4(3), 385–393 (1996)
15. Kuncheva, L.I., Vetrov, D.P.: Evaluation of stability of k-means cluster ensembles with respect to random initialization. IEEE Trans. Pattern Anal. Mach. Intell. 28, 1798–1808 (2006)
16. Kung, S.Y., Mak, M.W., Tagkopoulos, I.: Multi-metric and multi-substructure biclustering analysis for gene expression data. In: Proceedings of the 2005 IEEE Computational Systems Bioinformatics Conference (CSB 2005) (2005)

17. Madeira, S.C., Oliveira, A.L.: Biclustering algorithms for biological data analysis: A survey. IEEE Transactions on Computational Biology and Bioinformatics 1, 24–45 (2004)
18. Masulli, F., Schenone, A.: A fuzzy clustering based segmentation system as support to diagnosis in medical imaging. Artificial Intelligence in Medicine 16, 129–147 (1999)
19. Mitra, S., Banka, H.: Multi-objective evolutionary biclustering of gene expression data. Pattern Recogn. 39(12), 2464–2477 (2006)
20. Monti, S., Tamayo, P., Mesirov, J., Golub, T.: Consensus clustering: A resampling-based method for class discovery and visualization of gene expression microarray data. Machine Learning 52(1-2), 91–118 (2003)
21. Peeters, R.: The maximum edge biclique problem is NP-Complete. Discrete Applied Mathematics 131, 651–654 (2003)
22. Tavazoie, S., Hughes, J.D., Campbell, M.J., Cho, R.J., Church, G.M.: Systematic determination of genetic network architecturem. Nature Genetics 22(3) (1999)
23. Tjhi, W.-C., Chen, L.: Minimum sum-squared residue for fuzzy co-clustering. Intelligent Data Analysis 10(3), 237–249 (2006)
24. Turner, H., Bailey, T., Krzanowski, W.: Improved biclustering of microarray data demonstrated through systematic performance tests. Computational Statistics and Data Analysis 48(2), 235–254 (2005)
25. Yang, J., Wang, H., Wang, W., Yu, P.: Enhanced biclustering on expression data. In: BIBE 2003: Proceedings of the 3rd IEEE Symposium on BioInformatics and BioEngineering, p. 321. IEEE Computer Society, Washington (2003)
26. Zhang, Z., Teo, A., Ooi, B.C.A.: Mining deterministic biclusters in gene expression data. In: Proceedings of the Fourth IEEE Symposium on Bioinformatics and Bioengineering (BIBE 2004), pp. 283–292 (2004)

Splice Site Prediction Using Artificial Neural Networks

Øystein Johansen, Tom Ryen, Trygve Eftesøl, Thomas Kjosmoen,
and Peter Ruoff

University of Stavanger, Norway

Abstract. A system for utilizing an artificial neural network to predict splice sites in genes has been studied. The neural network uses a sliding window of nucleotides over a gene and predicts possible splice sites. Based on the neural network output, the exact location of the splice site is found using a curve fitting of a parabolic function. The splice site location is predicted without prior knowledge of any sensor signals, like 'GT' or 'GC' for the donor splice sites, or 'AG' for the acceptor splice sites. The neural network has been trained using backpropagation on a set of 16965 genes of the model plant Arabidopsis thaliana. The performance is then measured using a completely distinct gene set of 5000 genes, and verified at a set of 20 genes. The best measured performance on the verification data set of 20 genes, gives a sensitivity of 0.891, a specificity of 0.816 and a correlation coefficient of 0.552.

1 Introduction

Gene prediction has become more and more important as the DNA of more organisms are sequenced. DNA sequences submitted to databases are often already characterized and mapped when they are submitted. This means that a molecular biologist has already used genetics and biochemical methods to find genes, promoters, exons and other meaningful subsequences in the submitted material. However, the number of sequencing projects are increasing, and a lot of DNA sequences have not yet been mapped or characterized. Having a computational tool to predict genes and other meaningful subsequences is therefore of great value, and can save a lot of expensive and time consuming experiments for biologists.

This study tries to utilize an *artificial neural network* to predict where the splice sites of a gene can be located. The splice sites are the transitions from exon to intron or from intron to exon. A transition from exon to intron is called a *donor splice site* and a transition from intron to exon is called *acceptor splice site*.

2 Neural Network

The main premise in this study is to use a window of nucleotides that moves stepwise over the sequence to be analysed. The inputs to the neural network

F. Masulli, R. Tagliaferri, and G.M. Verkhivker (Eds.): CIBB 2008, LNBI 5488, pp. 102–113, 2009.

are calculated from the *input calculator*. For each step of the sliding window the neural network will give an output score if it recognizes there is a splice site in the window. A diagram of the entire prediction system is shown in Fig. 1. The window size is chosen to be 60 nucleotides. This is hopefully wide enough to find significant patterns on both sides of the splice site. A bigger window will make the neural network bigger and thereby harder to train. Smaller window would maybe exclude important information around the splice site.

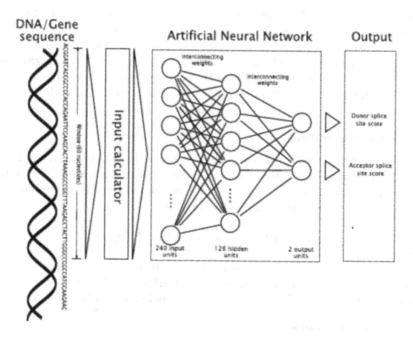

Fig. 1. The connection between the DNA sequence and the neural network system. A sliding window covers 60 nucleotides, which is calculated to 240 input units to the neural network. The neural network feedforward calculates a score for donor and acceptor splice site.

2.1   Network Topology

The neural network structure is a standard three layer feedforward neural network.[1] There are two output units corresponding to the donor and acceptor splice sites, 128 hidden layer units and 240 input units. The 240 input units were used since the orthogonal input scheme uses four inputs each nucleotide in the window. The neural network program code was reused from a previous study, and in this code the number of hidden units was hard coded and optimized for 128 hidden units. There is also a bias signal added to the hidden layer and the output layer. The total number of adjustable weights is therefore $(240 \times 128) + (128 \times 2) + 128 + 2 = 31106$.

---

[1] This kind of neural network has several names, such as multilayer perceptrons (MLP), feed forward neural network, and backpropagation neural network.

2.2 Activation Function

The activation function is a standard sigmoid function, shown in Eq. 1. The β values for the sigmoid functions are 0.1 for both the hidden layer activation and the output layer activation. Preliminary experiments were performed to test the effect of these values. These tests indicated that 0.1 was a suitable value.

$$f(x) = \frac{1}{1 + e^{-\beta x}} \tag{1}$$

When doing forward calculations and backpropagation, the sigmoid function is called repetitively. It is therefore important that this function has a high computational performance. A fast and effective evaluation of the sigmoid function can improve the overall performance considerably. To improve the performance of the sigmoid function, a precalculated table for the exponential function is used. This table lookup method is known to be very fast, and with an acceptable accuracy.

2.3 Backpropagation

The neural network was trained using a normal backpropagation method as described in Duda, Hart, Stork [3], Haykin [4] or Kartalopoulos [6]. There is no momentum used in the training. We have not implemented any second order methods to help the convergence of the weights.

3 Training Data and Benchmarking Data

Based on data from the The Arabidopsis Information Resource (TAIR) release 8 website [8], we compiled a certain set of genes. TAIR is an on-line database resource of genetic and molecular biology data of the model plant *Arabidopsis thaliana*.

3.1 Excluded Genes

All genes that contain unknown nucleotides were excluded from the data set. In addition, all single exon genes were excluded. Further, all genes with very short exons or introns were excluded. By "short" we mean 30 nucleotides or less. These genes were excluded to avoid very short exons or very short introns. Excluding these genes also simplifies the calculation of desired outputs, since it then can not be more than two splice sites in a window. For genes with alternative splicing, only one splicing variation was kept.

3.2 Training Data Set and Benchmark Data Set

The remaining data set, after exclusion of some genes, consists of 21985 genes. This set is divided into a training data set and a benchmarking data set. The training set and the benchmark set have 16965 and 5000 genes, respectively. The remaining 20 genes, four from each chromosome, was kept for a final verification. The number of genes in each set is chosen such that the benchmark set is large enough to achieve a reliable performance measure of the neural network. This splitting was done at random. Both data sets contains genes from all five chromosomes.

4   Training Method

The neural network training is done using standard backpropagation. This section describes how the neural network inputs were calculated and how the desired output was obtained.

4.1   Sliding Window

For each gene in the training set, we let a window slide over the nucleotides. The window moves one nucleotide each step, covering a total of $L_G - L_W + 1$ steps, where L_G is the length of the gene and L_W is the length of the window. As mentioned earlier, the length of the window is 60 nucleotides in this study.

4.2   Input to the Neural Network

For each nucleotide in the sliding window, we have four inputs to the neural net. The four inputs are represented as an orthogonal binary vector. (A=1000, T=0100, G=0010, C=0001). This input description has been used in several other studies [5], and is described in Baldi and Brunak [1]. This input system is called *orthogonal* input, due to the orthogonal binary vectors. According to Baldi and Brunak [1] this is the most used input representation for neural networks in the application of gene prediction. This input scheme also has the advantage that each nucleotide input is separated from each other, such that no arithmetic correlation between the monomers need to be constructed.

4.3   Desired Output and Scoring Function

The task is to predict splice sites, thus the desired output is 1.0 when a splice site is in the middle of the sliding window. There are two outputs from the neural network: One for indicating acceptor splice site and one for indicating donor splice site.

However, if it is only a 1.0 output when a splice site is in the middle of the window, and 0.0 when a splice site is not in the middle of the window, there will probably be too many 0.0 training samples that the neural network would learn to predict everything as 'no splice site'. This is why we introduce a score function which calculates a target output not only when the splice site is in the middle of the window, but whenever there is a splice site somewhere in the window. We use a weighting function where the weight of a splice site depends on the distance from the respective nucleotide to the nucleotide at the window mid-point. The further from the mid point of the window this splice site is, the lower value we get in the target values. The target values decrease linearly from the mid point of the window. This gives the score function as shown in Eq. 2

$$f(n) = 1 - |1 - \frac{2n}{L_W}| \tag{2}$$

If a splice site is exactly at the mid point, the target output is 1.0. An example window is shown in Fig. 2.

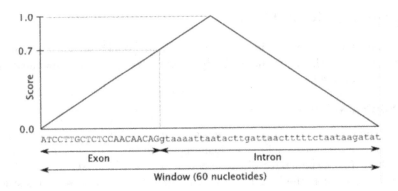

Fig. 2. Score function to calculate the desired output of a sliding window. The example window in the figure has a splice site after 21 nucleotides. This makes the desired output for the acceptor splice site 0.7.

In some cases there may be two splice sites in a window. It is then one acceptor splice site and one donor splice site.

4.4   Algorithm for Training on One Single Gene

The algorithm for training on a single gene is very simple. The program loops through all the possible window positions. For each position, the program computes the desired output for the current window, computes the neural net input and then calls BACKPROPAGATION() with these computed data. See Algorithm 1.

Algorithm 1. Training the neural net on one single gene

1: **procedure** TRAINGENE(NN, G, η)   ▷ Train the network on gene G
2:     $n \leftarrow \text{length}[G] - L_W$
3:     **for** $i \leftarrow 0$ **to** n **do**
4:         $W \leftarrow G[i..(i + L_W)]$ ▷ Slice the gene sequence
5:         $desired \leftarrow$ CALCULATEDESIRED(W)
6:         $input \leftarrow$ CALCULATEINPUT(W)
7:         BACKPROPAGATION($NN, input, desired, \eta$)
8:     **end for**
9: **end procedure**

In this algorithm, NN is a composite data structure that holds the neural network data to be trained. G is one specific gene that the neural network should be trained on. The first integer variable n is simply the total number of positions of the sliding window. L_W is the number of nucleotides in the sliding window, which in this study, is set to 60 nucleotides.

The desired output is calculated as described in Section 4.3, and is listed in Algorithm 2.

Algorithm 2. Calculating the desired output based on a given window

---

1: **function** CALCULATEDESIRED(W)   ▷ Calculate desired output
2:      $D \leftarrow [0.0, 0.0]$ ▷ Initialize the return array
3:      $prev \leftarrow$ ISEXON($W[0]$)   ▷ Boolean value
4:      **if** $prev$ **then**
5:          $j \leftarrow 0$
6:      **else**
7:          $j \leftarrow 1$
8:      **end if**
9:      **for** $i \leftarrow 1$ **to** L_W **do**
10:         $this \leftarrow$ ISEXON($W[i]$)
11:         **if** $prev \neq this$ **then**
12:             $D[j] \leftarrow D[j] + 1 - |1 - \frac{i}{30}|$ ▷ Score function
13:             $j \leftarrow 1 - j$ ▷ Flip the index 0 to 1, 1 to 0
14:         **end if**
15:         $prev \leftarrow this$
16:     **end for**
17:     **return** D
18: **end function**

---

5   Evaluation Method

The evaluation of a gene is simply the forward calculation performed for all the window positions in that gene. The neural network outputs are accumulated as an indicator value for each nucleotide in the gene.

5.1   Sliding Window

Because the neural network is trained to recognize splice sites in a 60 nucleotides wide window, the forward calculation process is also performed on the same sized window. The window slides over the gene in the same way as in the training procedure.

5.2   Cumulative Output and Normalization

As the sliding window moves over the gene and forward calculates whenever there is a splice site in the window. A nucleotide gets a score contribution from from 60 outputs corresponding to the sliding window passing over it. All these outputs are accumulated.

The accumulated output is then normalized. Most of the nucleotides will get a contribution from 60 different window positions, and these nucleotides are normalized by dividing the cumulative output by the area under the score function (30.0). These normalized cumulative scores are called *acceptor splice site indicator* and *donor splice site indicator*.

5.3   Algorithm for Evaluating One Single Gene

The pseudo code of the evaluation of a gene is given in Algorithm 3. The algorithm contains two loops. The first loop a slides a window over all positions in the gene and adds up all the predictions from the neural network. The second loop normalizes the splice site indicators.

Algorithm 3. Evaluation of gene

---

function EVALUATEGENE(NN, G)   ▷ Calculate splice site indicators
in: Neural network (NN), Gene (G)
out: Two arrays, D and A, which contains the donor and acceptor splice site indicator.
$\quad n \leftarrow \text{length}[G] - L_W$
$\quad$**for** $i \leftarrow 0$ **to** n **do**
$\quad\quad W \leftarrow G[i..(i + L_W)]$ ▷ Slice the gene sequence
$\quad\quad input \leftarrow$ CALCULATEINPUT(W)
$\quad\quad pred \leftarrow$ EVALUATE($NN, input$)   ▷ Gets predicted output
$\quad\quad$**for** $j \leftarrow i$ **to** $L_W + i$ **do**
$\quad\quad\quad D[j] \leftarrow D[j] + pred[0]$
$\quad\quad\quad A[j] \leftarrow A[j] + pred[1]$
$\quad\quad$**end for**
$\quad$**end for**
$\quad$**for** $i \leftarrow 0$ **to** $\text{length}[G] - 1$ **do**   ▷ Normalizing loop
$\quad\quad D[i] \leftarrow 2D[i]/L_W$
$\quad\quad A[i] \leftarrow 2A[i]/L_W$
$\quad$**end for**
$\quad$**return** D, A
end function

The normalizing loop in the implemented code also takes into account that the nucleotides close to the ends of the gene gradually gets evaluated by less window positions.

6   Measurement of Performance (Benchmark)

For monitoring the learning process and knowing when it is reasonable to stop the training, it is important to have a measurement of how well the neural network performs. This measurement process is also called *benchmarking*.

6.1   Predicting Splice Sites

As mentioned earlier the transition from exon to intron is called a *donor splice site*. The algorithm for predicting exons and introns in the gene is more or less done as a finite state machine with two states – *exon* state and *intron* state. The gene sequence starts in *exon* state. The algorithm then searches for the first high value in the donor splice site indicator. When the algorithm finds a significant top in the donor splice site indicator, the state switches to *intron*. The algorithm continues to look for a significant top in the acceptor splice site indicator, and the state is switched back to *exon*. This process continues until the end of the gene. The gene must end in the *exon* state.

In the above paragraph, it is unclear what is meant by a *significant top*. To indicate a top in a splice site indicator, the algorithm first finds a indicator value above some threshold value. It then finds all successive indicator data points that are higher than this threshold value. Through all these values, a second order polynomial regression line is fitted, and the maximum of this parabola is used to indicate the splice site. This method is explained with some example data in Fig. 3. In this example the indicator value at 0 and 1 is below the threshold. The value at 2 is just above the threshold and the successive values at 3,4,5 and 6 is also above the threshold and these five values are used in the curve fitting. The rest of the data points are below the threshold and not used in the curve fitting.

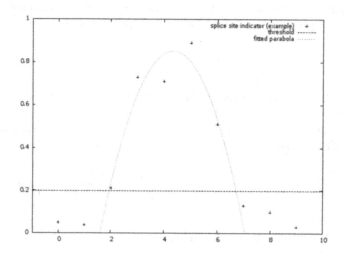

Fig. 3. Predicting a splice site based on the splice site indicator. When the indicator reaches above the threshold value, 0.2 in the figure, all successive data points above this threshold are used in a curve fitting of a parabola. The nucleotide closest to the parabola maxima is used the splice site.

Finding a good threshold value is difficult. Several values have been tried. We have performed some simple experiments with dynamically computed threshold values based on average and standard deviation. However, the most practical threshold value was found to be a constant at 0.2.

6.2 Performance Measurements

The above method is used to predict the locations of the exons and introns. These locations can be compared with the actual exon and intron locations.

There are four different outcomes of this comparison, true positive (TP), false negative (FN), false positive (FP) and true negative (TN). The comparison of actual and predicted location is done at nucleotide level.

The count of each comparison outcome are used to compute standard measurement indicators to benchmark the performance of the predictor. The *sensitivity*, *specificity* and *correlation coefficient* has been the de facto standard way of measuring the performance of prediction tools. These prediction measurement values are defined by Burset and Guigó [2] and by Snyder and Stormo [7].

The sensitivity (Sn) is defined as the ratio of correctly predicted exon nucleotides to all actual exon nucleotides as given in Eq. 3.

$$Sn = \frac{TP}{TP + FN} \tag{3}$$

The higher the ratio, the better prediction. As we can see, this ratio is between 0.0 and 1.0, where 1.0 is the best possible.

The specificity (Sp) is defined as the ratio of correctly predicted exon nucleotides to all predicted exon nucleotides as given in Eq. 4.

$$Sp = \frac{TP}{TP + FP} \tag{4}$$

The higher the ratio, the better prediction. As we can see, this ratio is between 0.0 and 1.0, where 1.0 is the best possible.

The correlation coefficient (CC) combines all the four possible outcomes into one value. The correlation coefficient is defined as given in Eq. 5.

$$CC = \frac{(TP \times TN) - (FN \times FP)}{\sqrt{(TP + FN)(TN + FP)(TP + FP)(TN + FN)}} \tag{5}$$

6.3 The Overall Training Algorithm

The main loop of the training is very simple and is an infinite loop with two significant activities. First, the infinite loop trains the neural network on all genes in the training data set. Second, it benchmarks the same neural network on the genes in the benchmark data set. There are also some other minor activities in the main loop like reshuffling the order of the training data set, saving the neural network, and logging the results. The main training loop is shown in Algorithm 4.

In this algorithm, NN is the neural net to be trained, T is the data set of genes to be used for training and B is the data set of genes for benchmarking. In this algorithm the learning rate, η, is kept constant. The bm variable is a composite data structure to hold the benchmarking data.

The subroutine SAVE() saves the neural network weights, and the SHUFFLE() subroutine reorders the genes in the data set. LOGRESULT() logs the result to the terminal window and to a log file.

Algorithm 4. Main training loop

```
 1: procedure TRAIN(NN, T, B, η)              ▷ Train the neural network
 2:     repeat
 3:         for all g ∈ T do                  ▷ Train neural net on each gene in dataset
 4:             TRAINGENE(NN, g, η)
 5:         end for
 6:         SAVE(NN)
 7:         SHUFFLE(T)                         ▷ A new random order of the training set
 8:         for all g ∈ B do
 9:             BENCHMARK(NN, g, bm)
10:         end for
11:         LOGRESULT(bm)
12:     until break                           ▷ Manually break when no improvement observed
13: end procedure
```

---

7 Experiments and Results

The training data set of 16965 genes where then used to train a neural network. The training was done in three sessions, and for each session we chose separate, but constant, learning rates. The learning rate, η, was chosen to be 0.2, 0.1, and 0.02, respectively. For each epoch[2] through the training data set, the neural networks performance was measured with the benchmark data set.

7.1 Finding Splice Sites in a Particular Gene

The splice site indicators can be plotted for a single gene. To illustrate our results, we present an arbitrarily chosen gene, AT4G18370.1. The curves in Fig. 4 represent the donor and acceptor splice site indicators for an entire gene. The donor splice sites are marked using a red line, the acceptor splice sites using a green line, and the predicted and actual exons are marked with the upper and lower dashed lines, respectively. The shown indicators are computed using a neural network which has been trained for about 80 epochs, with a learning rate of 0.2. As noted in the header of Fig. 4, the prediction on this gene achieves a better than average CC of 0.835. The results are promising. Most splice sites match the actual data, and some of the errors are most likely due to the low-pass filtering effect of using a sliding window, causing ambiguous splice sites.

7.2 Measurements of the Best Neural Networks

The best performing neural network, achieved a correlation coefficient of 0.552. The correlation coefficients, as well as the sensitivity, specificity, and standard simple matching coefficient (SMC), are shown in Tab. 1. When calculating these performance measurements, the benchmark algorithm averages the sensitivities and specificities for all genes in the data set. In addition the specificity, sensitivity, and correlation coefficient for the entire dataset is reported.

---

[2] An *epoch* is one run through the data set of training data.

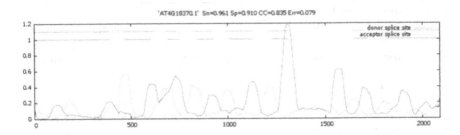

Fig. 4. The splice site indicators plotted along an arbitrary gene (AT4G18370.1) form the verification set. Above the splice site indicators, there are two line indicators where the upper line indicates predicted exons, and the other line indicates actual exons. The sensitivity, specificity and correlation coefficient of this gene is given in the figure heading. (*Err* is an error rate defined as the ratio of false predicted nucleotides to all nucleotides. $Err = 1 - SMC$.)

Table 1. Measurements of the neural network performances for each of the three training sessions. Numbers are based on a set of 20 genes which are not found in the training set nor the benchmarking set.

Session	*Average*		*All nucleotides in set*			
	Sn	*Sp*	*Sn*	*Sp*	*CC*	*SMC*
$\eta = 0.20$	0.864	0.801	0.844	0.802	0.5205	0.7761
$\eta = 0.10$	0.891	0.816	0.872	0.806	0.5517	0.7916
$\eta = 0.02$	0.888	0.778	0.873	0.777	0.4978	0.7680

8   Conclusion

This study shows an artificial neural networks used in splice site prediction. The best neural network trained in this study, achieve a correlation coefficient at 0.552. This result is achieved without any prior knowledge of any sensor signals, like 'GT' or 'GC' for the donor splice sites, or 'AG' for the acceptor splice sites. Also note that some of the genes in the data sets did not store the base case for splicing, but an alternative splicing, which may have disturbed some of the training. It is fair to conclude that artificial neural networks are usable in gene prediction, and the method used, with a sliding window over the gene, is worth further study. This method combined with other statistical methods, like General Hidden Markov Models, would probably improve the results further.

References

1. Baldi, P., Brunak, S.: Bioinformatics, The Machine Learning Approach, 2nd edn. MIT Press, Cambridge (2001)
2. Burset, M., Guigó, R.: Evaluation of gene structure prediction programs. Genomics 34(3), 353–367 (1996)
3. Duda, R.O., Hart, P.E., Stork, D.G.: Pattern Classification, 2nd edn. Wiley, New York (2001)

4. Haykin, S.: Neural Networks, A Comprehensive Foundation, 2nd edn. Prentice-Hall, Englewood Cliffs (1998)
5. Hebsgaard, S., Korning, P.G., Tolstrup, N., Engelbrecht, J., Rouzé, P., Brunak, S.: Splice site prediction in Arabidopsis thaliana pre-mRNA by combining local and global sequence information. Nucleic Acids Research 24(17) (1996)
6. Kartalopoulos, S.V.: Understanding Neural Networks and Fuzzy Logic. IEEE Press, Los Alamitos (1996)
7. Snyder, E.E., Stormo, G.D.: Identifying genes in genomic DNA sequences. In: DNA and Protein Sequence Analysis. Oxford University Press, Oxford (1997)
8. The Arabidopsis Information Resource, http://www.arabidopsis.org

Interval Length Analysis in Multi Layer Model

Vito Di Gesù, Giosuè Lo Bosco, and Luca Pinello

DMA - Università di Palermo, Italy
{vito.digesu,lobosco,pinello}@unipa.it

Abstract. In this paper we present an hypothesis test of randomness based on the probability density function of the symmetrized Kulback-Leibler distance estimated, via a Monte Carlo simulation, by the distributions of the interval lengths detected using the Multi-Layer Model (MLM). The MLM is based on the generation of several sub-samples of an input signal; in particular a set of optimal cut-set thresholds are applied to the data to detect signal properties. In this sense MLM is a general pattern detection method and it can be considered a preprocessing tool for pattern discovery. At the present the test has been evaluated on simulated signals which respect a particular tiled microarray approach used to reveal nucleosome positioning on Saccharomyces cerevisiae; this in order to control the accuracy of the proposed test of randomness. It has been also applied to real biological data. Results indicate that such statistical test may indicate the presence of structures in the signal with low signal to noise ratio.

1 Introduction

The Multi-Layers Model (MLM)[1,2,3] is a general pattern detection method and it can be considered a preprocessing tool for pattern discovery, its main advantages are the computational cost and a better structural view of the input data. Informally, it consists in the generation of several sub-samples from the input signal and, in previous works [1,2,3] it has been applied to both synthetic and real microarray data for the identification of particular patterns in the Yeast DNA called *nucleosome* and *linker* regions. MLM is strongly related to the class of methods successfully used in the analysis of very noisy data which, by using several views of the input data-set are especially able to recover statistical properties of a signal. In this paper a test of randomness, based on the distance of the interval lengths p.d.f's detected by the Multi-Layer Model (MLM) is presented. Such p.d.f's are estimated for each cut-set and the hypothesis test is performed against random signals generated via a Monte Carlo simulation. At this end the symmetrized Kullback-Leibler measure has been used to estimate the distribution distances.

Here, we present the evaluation of the test using simulated and real data respecting a particular tiled microarray approach able to reveal nucleosome positioning information on the DNA [4]. Results indicate that such statistical test may indicate the presence of structures in real and simulated biological signals,

F. Masulli, R. Tagliaferri, and G.M. Verkhivker (Eds.): CIBB 2008, LNBI 5488, pp. 114–122, 2009.
© Springer-Verlag Berlin Heidelberg 2009

showing also its robustness to data noise and its superiority to the Wilcoxon rank sum test. In Fig.s 1(a),1(b),1(c) three examples of input signals with signal to noise ratio $SNR = 1, 1.5, 10$, are given. This allows us to control the accuracy of the proposed test of randomness and perform the calibration of the methodology. The same test has been applied to the data used in the simulation phase.

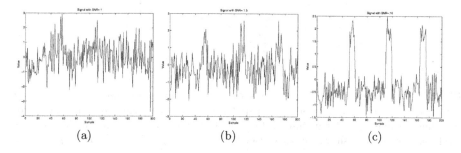

Fig. 1. Examples of input signals: (a) input signal $SNR = 1$; (b) input signal $SNR = 1.5$; (b) input signal $SNR = 10$

The paper is organized as it follows in Section 2 the Multi-Layer Model is shortly described; Section 3 introduces the hypothesis test based on Kullback-Leibler distance; the estimation of the probability density function, starting from the MLM interval information is provided in Section 4; Section 5 provides the assessment of the statistical test of randomness on simulated and real data; final remarks are given in Section 6.

2   Multi-Layer Model Outline

The MLM analysis is performed on a generic mono-dimensional signal **S**, and, in particular, it consists in the following distinct processing steps:

– *Preprocessing and Pattern Model construction*. The input signal is subject to a preprocessing phase (i. e. convolution) in order to smooth the signal and reduce the noise. Then, starting from information about the specific problem to solve, a *model* of the pattern to study is built. For instance, in the case of nucleosome spacing, starting from the assumption that the interesting information of the signal is located around the local maxima, the *well positioned nucleosome* model is built by looking locally around these maxima.

– *Interval Identification and Pattern construction*. Given the convolved signal **S**, its intersections with K equally spaced horizontal lines is found, each one representing a threshold t_k. For each t_k a set of intervals $R_k = \left\{I_k^1, I_k^2, \cdots, I_k^{n_k}\right\}$ is obtained; where, $I_k^i = [b_k^i, e_k^i]$ and $\mathbf{S}(b_k^i) = \mathbf{S}(e_k^i) = t_k$ (see Fig. 2). Then, patterns are formed by grouping together those intervals; the way of grouping them depends on the problem to solve.

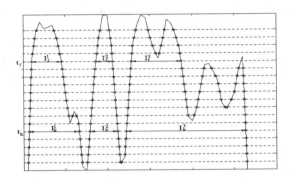

Fig. 2. An example of input signal representing Saccharomyces cerevisiae microarray data portion. Each x value represents a spot (probe) on the microarray and the corresponding y value is the logarithmic ratio of its Green and Red values. The intervals at level k and r are also shown, in this case, $R_k = \{I_k^1, I_k^2, I_k^3\}$, $R_r = \{I_r^1, I_r^2, I_r^3\}$.

- *Pattern identification and feature extraction.* In this step the *interesting patterns*, which represent the one satisfying a particular criterion, are identified, assigning them also a feature vector. For instance, in the case of nucleosome spacing the interesting pattern contains m nested intervals and its feature vector is a matrix depending on such intervals.
- *Dissimilarity function.* A measure of dissimilarity between patterns is defined in order to make decisions about them.

Papers [1,2,3] show more details about the MLM on the specific problem of *nucleosome* and *linker* identification on the Yeast DNA.

3   Hypothesis Test

In order to detect the presence of structures in the signal an hypothesis test based on the expected probability distribution function (p.d.f.) of the segments length is proposed. The hypothesis 0 ($H0$) represents a *random signal* and it is accepted if the p.d.f. of the segment lengths, p_1, is compatible with a random signal distribution, p_0; the hypothesis $H1$ represents a *structured signal* and it is accepted if the p.d.f. of the segment lengths is not compatible with a random signal, p_0. It follows that we need to measure the similarity (dissimilarity) of two p.d.f.'s and set a confidence level α to perform the decision.

The symmetric Kullback-Leibler measure, SKL, has been considered to evaluate the dissimilarity of the two distributions p_0, and p_1 [5]:

$$SKL(p_0, p_1) = \frac{KL(p_0, p_1) + KL(p_1, p_0)}{2}$$

where, KL is the no-symmetric Kullback-Leibler measure. In the continuous case, p.d.f.'s are defined in a dominion $I \subseteq \mathbb{R}$ and the KL measure is defined as:

$$KL(p, q) = \int_I p(x) log \frac{p(x)}{q(x)} dx$$

In the discrete case $I \subseteq \mathbb{N}$ and the KL become:

$$KL(p, q) = \sum_{i \in I} p_i log \frac{p_i}{q_i}$$

In order to perform the hypothesis test we should know the p.d.f. of the SKL in the case of a random signal. The derivation of analytical form of this p.d.f. is usually an hard problem that has been solved by a Monte Carlo simulation. For example in [6] a goodness-of-fit test for normality is introduced; it is based on Kullback-Leibler information and a Monte Carlo simulation is performed to derive and estimate the p.d.f.'s. In [7] an extension of the previous test is described for s-normal, exponential, and uniform distributions and also in this work a Monte Carlo simulation has been used to estimate the p.d.f. of the measure KL.

4   Probability Density Functions Estimation

In this section we provide a description of the simulation performed to estimate the p.d.f.'s of both the intervals length, IL_k (PIL_k), and the SKL_k ($PSKL_k$), at a given threshold, t_k; here, SKL_k is the distance between the p.d.f.'s of two interval length.

To estimate the p.d.f. of IL_k we have generated $n = 1, ..., N$ signals, RS_n, of length l according to a normal distribution with $\hat{\mu}$ and $\hat{\sigma}$ estimated from an input signal S of length L. Each signal, RS_n, is then used to evaluate experimentally $PIL_k^{(n)}$ ($n = 1, 2, ..., N$).

In our simulation, for each threshold, t_k, we derive the experimental distributions of the IL_k in R_k. Therefore we obtain $k = 1, 2, ..., K$ normalized p.d.f. $PIL_k^{(n)}$ with nb bins. Fig.3(a) shows examples of PIL_k for a simulation using $l = 20000$, $L = 200000$, $N = 1000$, $K = 9$, $nb = 100$.

The estimation of the p.d.f. of SKL_k, $PSKL_k$, is carried out by computing the SKL_k between the pairs $\left(PIL_k^{(m)}, PIL_k^{(n)} \right)$, with $m \neq n$. In our case we draw the evaluation of $PSKL_k$ from a sample of $\frac{N \times (N-1)}{2}$ elements by using a density estimation with Gaussian kernel. Fig.3(b) shows examples of $PSKL_k$ for a simulation using $l = 20000$, $L = 200000$, $N = 1000$, $K = 9$, $nb = 100$.

5   Experimental Results

In the following the assessment of the proposed test of randomness are shown on both synthetic and real data.

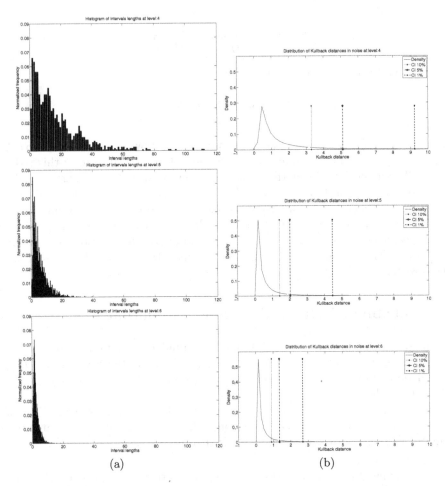

Fig. 3. Examples of PIL_k (a), and $PSKL_k$ (b) for $k = 4$, 5, *and* 6

5.1   Assessment on Synthetic Data

The input signals used to evaluate the test, are synthetically generated following the procedure described in [3] and represent signals which emulate the nucleosome positioning data of Saccharomyces cerevisiae. In particular, such signals represents microarray data where each spot is a *probe i* of 50 base pairs that overlaps every 20 base pairs with probe $i+1$. Such microarray data are collected by spanning the chromosome moving a window (probe) i of width 50 base pairs from left to right, measuring both the percentage of mononucleosomal DNA G_i (*green channel*) and whole genomic DNA R_i (*red channel*) within such window, respecting also that two consecutive windows (probes) have an overlap of 30 base pairs. The resulting signal $S(i)$ for each probe i is the logarithmic ratio of the *green channel* G_i to *red channel* R_i. Intuitively, nucleosomes presence is related to peaks of S which correspond to higher logarithmic ratio values, while

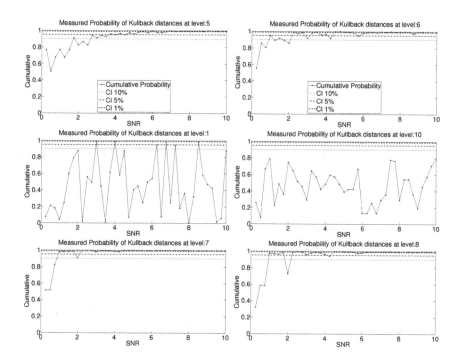

Fig. 4. Examples of hypothesis test at different SNR and thresholds

lower ratio values show nucleosome free regions called *linker regions*. Detailed information about the microarray labeling and hybridization protocols of the Saccharomyces cerevisiae can be found in [4].

In the following we perform an assessment of the proposed hypothesis test to guess the presence of structures in an unknown input signal. In this sense these can be considered part of an exploratory data analysis procedure. This experiment has been carried out generating 40 synthetic test signals of length $L = 200000$ (base pairs), with signal to noise ratio ranging from $SNR = 0$ to $SNR = 10$ by steps of 0.25 and N random samples of length $l = 20000$ (base pairs). The simulation used to estimate the p.d.f. of SKL, has been done using the synthetic signals of length L (base pairs), $N = 1000$ random samples of length l (base pairs), $K = 9$ thresholds and $nb = 100$ bins. The number of bins has been set as a good compromise among different sample size at different thresholds. For each test signal S its SKL_k from a random sample drawn from the RS_n samples is computed and used to verify the test of hypothesis on the $PSKL_k$. In Fig. 4 some results of the test are provided for increasing SNR, for confidence level $\alpha = 99\%$, 95%, 90% and at different thresholds. In the abscissa is represented the SNR while in the ordinate the probability that the symmetrized Kullback-Leibler distance is in the interval $[0, SKL_k]$. If the ordinate value is greater than the confidence α the random test is rejected. From previous results it can be seen that the test is not reliable for lower and the higher thresholds while it is quite sensitive for intermediated thresholds. For example, for $t_k = 5, 6, 7, 8$ and $\alpha \geq$

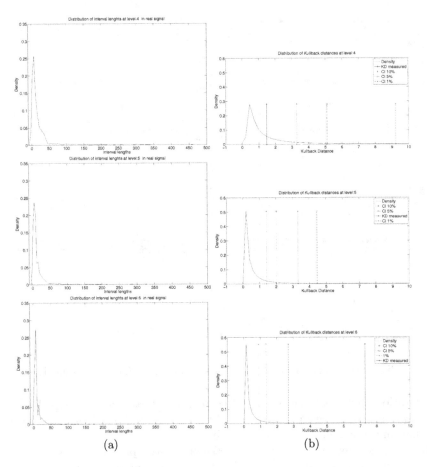

(a)   (b)

Fig. 5. PIL_k (a) and $PSKL_k$ and hypothesis test results (b) of the real signal for $k = 4$, 5, and 6

90% the random hypothesis is rejected for $SNR \geq 3.0$, 1.5, 1.25, 1.5 respectively. Intuitively, this can be explained because the number of intersections decrease for higher and lower thresholds.

5.2   Assessment on Real Data

The test of randomness has been applied to real biological data derived from a tiled microarray approach able to reveal nucleosome positioning information on the Saccharomyces cerevisiae DNA [4]. The input microarray data, **S**, are organized in T contiguous fragments $S_1, \cdots, S_T$ which represents DNA sub-sequences.

In the experiment we set $K = 10$, $nb = 100$, for each signal fragment S_i the corresponding intervals INT_{ik} are extracted for each threshold t_k.

Finally, the set of intervals $INT_k = \bigcup_{i=1}^{T} INT_{ik}$ are used to compute the interval distribution length PIL_k. Then, the SKL_k from a random sample R drawn

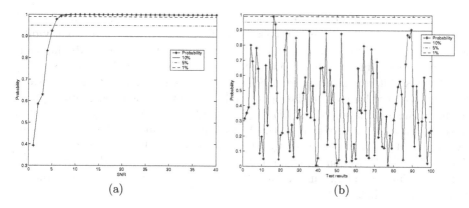

Fig. 6. Mann-Whitney rank sum test results for different signal to noise ratio (a) and for the real signal (b)

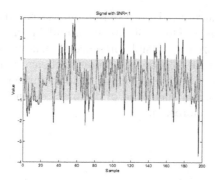

Fig. 7. The gray strep indicates the useful part of the input signal in order to perform the test of randomness

from the RS_n samples is computed and used to verify the test of hypothesis on the $PSKL_k$. Note that, in this experiment, the length of the real signal and of the random sample is 20000 base pairs. Figures 5(a) show PIL_k for $k = 4, 5, 6$.

The experiment indicates that the hypothesis test is rejected at confidence level 95% for $k = 5$, while for $k = 6, 7, 8, 9$ is rejected at a confidence level $\geq 99\%$. In Figure 5(b) we show the result of the test of randomness for $k = 4, 5, 6$. Moreover, the test of randomness is quite unstable for $k \leq 4$ and $k = 10$; this property highlights that the central part of the signal contains the majority of the useful information for the test of randomness (see Figure 7).

5.3   Comparison with Wilcoxon Rank Sum Test

We have also compared the results of our test with the Wilcoxon rank sum test, both on synthetic and real data. Such test, is an hypothesis test that can be applied when no assumption about sample distribution can be made, condition which falls in our case.

We have first verified if each synthetically generated S (a total amount of 40 signal) and $N = 100$ random samples drawn from the RS_n are significantly different by using a Wilcoxon rank sum test. Figure 6(a) shows the results that can be summarized affirming that S and a generic random signal R are at least 90% significantly different starting from $SNR = 1.25$. This reveal that the Wilcoxon test and our test have quite the same predictive power when considering intermediate threshold levels of the MLM (k=6,7,8,9).

In the case of a real signal S, the Wilcoxon rank sum test has rejected the hypothesis of randomness on S only 3 times over $N = 100$ tests (see figure 6(b)). This makes the Wilcoxon rank sum test not reliable for such kind of data, while our test, as already shown in section 5.2, confirms his predictive power on intermediate thresholds.

6   Conclusions

In this paper we have introduced a test of randomness based on the symmetrized Kullback-Leibler distance. This test can be useful in the case of exploratory analysis in order to verify the possible presence of structures in an input signal. Finally, the method is able to guess structures in the case of real and simulated data with low SNR (1.5) while a simple Mann-Whitney rank sum test has not shown full reliability on such kind of data. An extension of the method to multidimensional data is foreseen.

References

1. Corona, D.F.V., Di Gesù, V., Lo Bosco, G., Pinello, L., Collesano, M., Yuan, G.-C.: A Multi-layer method to study Genome-Scale Positions of Nucleosomes. In: Di Gesù, V., Maccarone, M.C., Lo Bosco, G. (eds.) Proc. of Data Analysis in Astronomy, Modelling and Simulation in Science, Erice, Italy, pp. 169–177 (2007)
2. Corona, D.F.V., Di Gesù, V., Lo Bosco, G., Pinello, L., Yuan, G.C.: A new Multi-Layers Method to Analyze Gene Expression. In: Apolloni, B., Howlett, R.J., Jain, L. (eds.) KES 2007, Part III. LNCS (LNAI), vol. 4694, pp. 862–869. Springer, Heidelberg (2007)
3. Di Gesù, V., Lo Bosco, G., Pinello, L.: A one class classifier for Signal identification: a biological case study. In: Lovrek, I., Howlett, R.J., Jain, L.C. (eds.) KES 2008, Part III. LNCS (LNAI), vol. 5179, pp. 747–754. Springer, Heidelberg (2008)
4. Yuan, G.-C., Liu, Y.J., Dion, M.F., Slack, M.D., Wu, L.F., Altschuler, S.J., Rando, O.: Genome-Scale Identification of Nucleosome Positions in S. cerevisiae. Science 309, 626–630 (2005)
5. Johnson, D.H., Sinanović, S.: Symmetrizing the Kullback-Leibler Distance. Technical report, Rice University (March 18, 2001)
6. Arizono, I., Ohta, H.: A Test for Normality Based on Kullback-Leibler Information. The American Statistician 43(1), 20–22 (1989)
7. Senoglu, B., Surucu, B.: Goodness-of-Fit Tests Based on Kullback-Leibler Information. IEEE TRANS. on Reliability 53(3), 357–361 (2004)

A Multivariate Algorithm for Gene Selection Based on the Nearest Neighbor Probability

Enrico Ferrari and Marco Muselli

Institute of Electronics, Computer and Telecommunication Engineering
Italian National Research Council
via De Marini, 6 - 16149 Genoa, Italy
{ferrari,muselli}@ieiit.cnr.it

Abstract. Experiments performed with DNA microarrays have very often the aim of retrieving a subset of genes involved in the discrimination between two physiological or pathological states (e.g. ill/healthy). Many methods have been proposed to solve this problem, among which the *Signal to Noise ratio* (*S2N*) [5] and SVM-RFE [6]. Recently, the complementary approach to RFE, called *Recursive Feature Addition* (*RFA*), has been successfully adopted. According to this approach, at each iteration the gene which maximizes a proper ranking function ϕ is selected, thus producing an ordering among the considered genes. In this paper an RFA method based on the nearest neighbor probability, named *NN-RFA*, is described and tested on some real world problems regarding the classification of human tissues. The results of such simulations show the ability of NN-RFA in retrieving a correct subset of genes for the problems at hand.

1 Introduction

In the last few years, the development of larger and larger datasets in different scientific fields has given rise to an increasing interest in *feature selection* methods. In particular, DNA-microarray datasets involve up to one hundred of thousands genes, which can be regarded as input variables of a classification problem. Moreover, the cost required for a single DNA-microarray experiment makes very difficult to obtain a large number of examples for performing the desired classification. Due to these motivations, researchers are very interested in reducing the input space (i.e. the number of considered genes) while keeping the maximum information about the classification.

Feature selection can be performed as a pre-processing step before the learning procedure, in order to obtain lighter datasets to be used to save computational time, or as a learning method itself in order to retrieve information about the input variables.

Many methods have been proposed in literature, a review of which can be found in [1]. Some of the proposed algorithms are *univariate*, i.e. they consider each variable separately; others are *multivariate*, i.e. they take into account correlations among features. Usually univariate methods permit a simpler description

F. Masulli, R. Tagliaferri, and G.M. Verkhivker (Eds.): CIBB 2008, LNBI 5488, pp. 123–131, 2009.

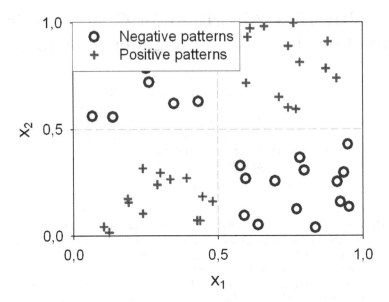

Fig. 1. Graphical representation of the XOR problem

and a faster execution; however, in many practical situations, the correlation among features is essential to correctly classify patterns.

For example, consider the XOR (or *chessboard like*) problem, represented in Fig. 1. It is clear that neither the input variable x_1 nor x_2 is able to independently perform the classification, whereas the combination of x_1 and x_2 points out the regularities in the data.

Moreover, suppose that the input variables include many other irrelevant or noisy features besides x_1 and x_2 so that the above shown regularity cannot be found through statistical analysis. In such case, it should be important to adopt a smart algorithm capable of retrieving the two relevant features thus simplifying the classification task.

2   The Proposed Algorithm

The problem can be formalized as follows: d-dimensional continuous valued vectors $\boldsymbol{x} \in X$ have to be assigned to one of two classes labelled by a binary value $y \in \{0, 1\}$. This classification task must be inferred by examining a training set S including N examples $\{\boldsymbol{x}_k, y_k\}$, $k = 1, \ldots, N$.

Let $I_d = \{1, \ldots, d\}$ be the set of indexes for the input variables; the feature selection task consists in finding a subset $G \subset I_d$ of M indexes associated with the most relevant features. The set G can be constructed by considering a ranking F of I_d, obtained according to a suitable criterion, and by selecting the first M elements of F.

A possible way to define a multivariate model for feature selection consists in the definition of a proper *ranking function* ϕ, which depends on a subset $H \subset I_d$

of input variables. The function $\phi(H)$, which measures the relevance of the subset of features defined by H, is used to incrementally generate the ranking F.

A general algorithm, called *Recursive Feature Addition (RFA)* [2], is shown in Fig. 2: it describes how the ranking F can be generated according to the analysis of the function ϕ. At each iteration the variable which scores the highest value of ϕ (together with the features already selected) is added to the ordered set F.

1. Set $F = \emptyset$ and $J = I_d$.
2. While J is not empty do
   (a) For $i \in J$ compute $\psi(i) = \phi(F \cup \{i\})$.
   (b) Select $\tilde{\imath} = \arg\max_i \psi(i)$.
   (c) Set $F = F \cup \{\tilde{\imath}\}$.
   (d) Set $J = J \setminus \{\tilde{\imath}\}$.

Fig. 2. A recursive algorithm to generate the ranking F

In many situations the discriminant function can be associated with the accuracy of a classifier. For instance, a cross-validation test can be performed on the available data in order to find the best classifier obtained through the analysis of the features included in the set $F \cup \{i\}$. According to the different choices of the function space where such classifier is searched for (e.g. multilayer perceptrons, support vector machines...), different kinds of feature selection algorithms can be obtained [2].

A possible choice for the function ϕ is based on the *nearest neighbor probability* P_{nn} [3]. Consider a vector $x \in X$ and a distance $D : X \times X \to \mathbb{R}^+$; then the nearest neighbor of x in the training set S is defined by

$$\mathcal{N}(x) = \arg\min\{D(x, x_k) \mid \{x_k, y_k\} \in S, \, x_k \neq x\} \ . \tag{1}$$

P_{nn} is the probability that a pattern has the same output value as its nearest neighbor. If x_k is an input pattern in the training set, suppose that $x_r = \mathcal{N}(x_k)$ is the nearest neighbor of x_k and define $\delta_k = 1$ if $y_r = y_k$, $\delta_k = 0$ otherwise, being y_k (resp. y_r) the output value associated with x_k (resp. x_r). Thus, the nearest neighbor probability can be approximated as the average of the δ_k on the whole training set S:

$$P_{nn} \simeq \frac{1}{N} \sum_{k=1}^{N} \delta_k \ .$$

Now, consider the nearest neighbor classifier [3], which assigns to a point x the label of its nearest neighbor; it is easy to notice that P_{nn} is just the accuracy of such classifier obtained through a *leave-one out* test.

In order to employ P_{nn} in the evaluation of the classification ability of a reduced group of features, a *reduced* distance can be introduced using only a subset of variables.

Let $D_H : X \times X \to \mathbb{R}^+$ be the distance obtained by employing only the subset of variables defined by H. For instance, if D is given by the Euclidean distance, the reduced distance D_H is defined by

$$D_H(x, x') = \sqrt{\sum_{i=1}^{|H|} (x_{h_i} - x'_{h_i})^2}$$

where h_i is the i-th element of the set H and $|H|$ is the number of elements in H.

Starting from D_H a *reduced* nearest neighbor can be found for each pattern in the training set, as in (1), and a *reduced* nearest neighbor probability P_{nn}^H can be computed in an analogous way.

A simple analysis shows that, when H is empty, P_{nn}^H corresponds to the probability of randomly extracting two patterns having the same label (disregarding any metric information about the input space). In the case of a two class problem, such probability is given by $P_{nn}^{(rand)} = P_0^2 + P_1^2$, where P_0 (P_1) is the fraction of patterns having label 0 (1). If the two classes are uniformly represented, i.e. if $P_0 = P_1 = 0.5$, $P_{nn}^{(rand)} = 0.5$.

It is easy to notice that it is possible to adopt P_{nn}^H as the discriminant function $\phi(H)$. The resulting algorithm will be called *Nearest Neighbor Recursive Feature Addition (NN-RFA)*.

According to this algorithm, at each iteration the input variable which is able to score the maximum improvement of the nearest neighbor classifier is selected.

Notice that NN-RFA is a very convenient method from the computational point of view since only N^2 elementary operations have to be performed at each iteration. Since N^2 is usually small (about a hundred examples are available in most real world problems), NN-RFA is able to retrieve a good set of genes in a very short time (at most few minutes).

Smart criteria can be adopted to decide when the algorithm must be stopped, if the number M of features is not fixed by the user. In general, the probability P_{nn} should increase at each iteration of the RFA algorithm since a new information is carried by the added feature. Nevertheless, if an irrelevant variable is selected, P_{nn} decreases since only noise is added to the actual information.

Therefore, as the size of H increases, the behavior of P_{nn}^H, is usually characterized by the following phases:

1. At the beginning, when H is empty, $P_{nn}^H = P_{nn}^{(rand)} = P_0^2 + P_1^2$.
2. Then, when relevant features are added, P_{nn}^H increases up to its maximum value $P_{nn}^{(max)}$ (which depends on the regularity of the available data). Such maximum value, obtained with a set of d_{min} genes is kept until the added noise becomes prevalent, in correspondence with a set of d_{max} genes.
3. At last, when many redundant features are added, P_{nn}^H decreases to the global nearest neighbor probability $P_{nn}^{I_d}$.

Consider the example in Fig. 1, with the addition of other 8 noisy features; a direct computation allows to obtain the values

$$P_{nn}^{(\text{rand})} \simeq 0.5 \,, \quad P_{nn}^{(\text{max})} = P_{nn}^{\{1,2\}} \simeq 0.95 \,, \quad P_{nn}^{I_d} \simeq 0.59 \,.$$

whereas the complete behavior of P_{nn}^H is shown in Fig. 3. In general, the following inequalities hold: $P_{nn}^{(\text{rand})} < P_{nn}^{I_d} < P_{nn}^{(\text{max})}$.

Due to these considerations, the feature extraction procedure can be stopped when the accuracy of the classifier starts decreasing. If a more compact set of features is needed, only the genes which cause an increase of the accuracy can be selected. However, no theoretical reason exists in order to prefer a subset of features with respect to another scoring the same level of accuracy. Only biological validation can ensure the soundness of the extracted subset of features; otherwise a computational convenience criterion can be adopted to choose among different set of genes.

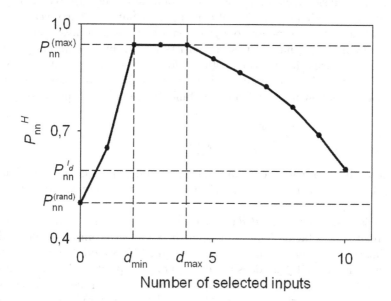

Fig. 3. The plot shows the behavior of the nearest neighbor probability P_{nn}^H when elements are added to the set H of considered features for the XOR problem (see Fig. 1)

3   Simulation Results

In order to evaluate the performances of the NN-RFA algorithm, two different kinds of experiment have been carried out. The first series of tests concerns the ability of NN-RFA in retrieving a subset of the input space which is able to classify the data within a good level of accuracy. In particular three datasets have been considered, regarding the colon cancer [7], leukemia [5] and lymphoma [8] diseases. The characteristics of these datasets are summarized in Table 1.

Table 1. The real world datasets used for the tests on the NN-RFA algorithm

Dataset	Number of examples	Number of genes
COLON	62 (40 ill, 22 healthy)	2000
LEUKEMIA	72 (47 affected by ALL, 25 by AML)	7129
LYMPHOMA	81 (43 affected by DLBL, 38 by BCL or FL)	4682

Since the genes really involved in the disease are not known *a priori*, it is impossible to evaluate whether the method has retrieved all the relevant features or just a minimal subset suitable to correctly classify most patterns.

For this reason, the NN-RFA algorithm has been tested on some datasets simulating the behavior of human tissues, obtained through the *TAGGED* (*Technique for the Artificial Generation of Gene Expression Data*) procedure [4].

The TAGGED approach is based on the definition of *expression signature*, introduced in [8]. An expression signature is a group of genes which coordinately determines a physiological or pathological condition. The *expression profile* of a patient, i.e. the whole collection of expression signatures that reveal different biological features, corresponds to a *functional state* (e.g. ill, healthy).

A model simulating human tissues consists in defining a function $f : \mathbb{R}^d \rightarrow \{0,1\}$, which returns the functional state y for a given vector $\boldsymbol{x}$ of gene expressions. Due to biological variability and errors in measurements, noise is included in the model by allowing a switch of the functional state with probability ε:

$$y = \begin{cases} f(\boldsymbol{x}) & \text{with probability } 1 - \varepsilon \\ 1 - f(\boldsymbol{x}) & \text{with probability } \varepsilon \ . \end{cases}$$

Suppose that for each gene, labeled by an integer value i, a modulation threshold t_i exists such that i is *overexpressed* (resp. *underexpressed*) if $x_i > t_i$ (resp. $x_i < -t_i$).

A gene is said to be *modulated* if it is either overexpressed or underexpressed. Therefore, it is possible to define a mapping $\boldsymbol{b}$, whose i-th component depends on the modulation thresholds t_i and returns the value 1 if i is modulated, 0 otherwise. Our assumption is that $f(\boldsymbol{x})$ does not depend on the specific value of $\boldsymbol{x}$, but only on the state of each gene (modulated or not). For this reason the function specifying the model can be written as $f(\boldsymbol{x}) = \varphi(\boldsymbol{z})$, where $\boldsymbol{z} = \boldsymbol{b}(\boldsymbol{x})$ and φ is a Boolean function defined on $\{0,1\}^d$. Thus, after the modulation mapping $\boldsymbol{b}$ has been specified, the model is completely determined by the definition of φ. According to the TAGGED model, the choice of φ is performed in the set of positive Boolean functions, which maintain ordering between binary strings. In fact, it can be shown [4] that positive Boolean functions satisfy all the biological requirements for f thus constituting a good model for simulating the behavior of human tissues.

In order to write the function φ in a more convenient form, consider a collection of sets $G_i(q_i) = \{z_{j_{i1}}, \ldots, z_{j_{il_i}}, j_{ir} \neq j_{is} \text{ if } r \neq s\}$, $i = 1, \ldots, h$, each including l_i distinct components of $\boldsymbol{z}$ and characterized by an integer value $q_i \leqslant l_i$. $G_i(q_i)$ is said to be *active* if at least q_i of its components have value 1. It can be shown [4] that a positive Boolean function can always be written in its *m-of-n* form:

$$\varphi(z_1, \ldots, z_d) = \begin{cases} 1 & \text{if at least } p \text{ of the } h \text{ sets } G_1(q_1), \ldots, G_h(q_h) \text{ are active} \\ 0 & \text{otherwise .} \end{cases}$$

where p is an integer value such that $p \leqslant h$.

In practice, a group of genes is said to be active if at least a fixed number of them is modulated, whereas a functional state is present if at least p groups of genes are active. The parameters of the model (ε, p, h, $G_i(q_i)$ for $i = 1, \ldots, h$) are tuned so that the artificial tissue has a statistical behavior as close as possible to the corresponding real one.

Since the datasets are artificially generated it is known which are the genes involved in the classification. In such a way it is possible to evaluate a given feature selection method by calculating the percentage of relevant genes selected by the algorithm.

For each dataset the accuracy of the nearest neighbor classifier has been evaluated through a leave-one out test for different dimensions of the input space.

Table 2. The values of $P_{nn}^{(rand)}$, $P_{nn}^{(max)}$, d_{min}, d_{max}, $P_{nn}^{I_d}$ for some real and artificial datasets

Dataset	$P_{nn}^{(rand)}$	$P_{nn}^{(max)}$	d_{min}	d_{max}	$P_{nn}^{I_d}$
Real COLON	0.545	1	18	587	0.742
Artificial COLON	0.545	1	2	1843	0.935
Real LEUKEMIA	0.545	1	3	6157	0.875
Artificial LEUKEMIA	0.545	1	303	6863	0.847
Real LYMPHOMA	0.502	1	8	3238	0.937
Artificial LYMPHOMA	0.502	1	15	3588	0.917

The behavior of P_{nn}^{H} is similar to that of Fig. 3 for all the considered datasets. The maximum value $P_{nn}^{(max)}$ is reached in correspondence with a set of d_{min} genes, whereas the performances starts decreasing when more than d_{max} genes are considered. The classification accuracy is constant in the range between d_{min} and d_{max}. The values of $P_{nn}^{(rand)}$, $P_{nn}^{(max)}$, d_{min}, d_{max} and $P_{nn}^{I_d}$ are reported in Table 2.

An analysis of Table 2 shows that an optimal classification ability is achieved for small subspaces of the input domain (almost always less than 20 genes) and maintained also for a high number of genes. Only in one case (real COLON) the performances get worse considering less than the 75% of the genes. In the other cases, some thousands variables may be considered without perturbing the classification ability. Notice that the global nearest neighbor probability is significantly higher than the random one $P_{nn}^{(rand)}$, thus pointing out that some regularity exists in the data.

The behavior of P_{nn} relative to a set of examples not employed in the feature selection task has been studied, too. To this aim, each dataset has been divided into two subsets, each including the 50% of the total number of examples. Then, one of these subsets was used to select a set of genes which will be tested on the remaining data (*test set*). Usually, the subsets of genes selected by NN-RFA are able to obtain a better level of classification accuracy with respect to the whole set of genes. In particular, Table 3 shows the number of genes $\bar{d}$ which must be considered in order to obtain in a robust way a nearest neighbor probability (on the test set) higher than $P_{nn}^{I_d}$. Notice that the maximum nearest neighbor $P_{nn}^{(max)}$, obtained in the test set, is significantly higher than $P_{nn}^{I_d}$, thus confirming the good level of classification accuracy of the selected set of genes.

Table 3. The number of genes which must be considered in order to obtain a nearest neighbor probability higher than $P_{nn}^{I_d}$ and the maximum P_{nn} (obtained on the test set) for different real and artificial datasets

Dataset	$\bar{d}$	$P_{nn}^{(max)}$	$P_{nn}^{I_d}$
Real COLON	187	0.839	0.710
Artificial COLON	263	1	0.903
Real LEUKEMIA	611	0.861	0.778
Artificial LEUKEMIA	613	0.889	0.778
Real LYMPHOMA	198	0.875	0.812
Artificial LYMPHOMA	330	1	0.958

The percentage of overlap between the selected set of genes and the actual one has been measured for each artificial dataset. The results obtained by NN-RFA on the artificial datasets are compared with those achieved by two other feature selection algorithms: *S2N* (the Signal to Noise ratio, a variation of the t-test) [5] and *SVM-RFE* (a recursive feature elimination method based on the training of a support vector machine) [6]. Table 4 shows the percentage of the genes correctly selected by the three methods in three different artificial benchmarks associated with the COLON, LEUKEMIA and LYMPHOMA datasets.

Table 4. Percentage of correctly selected genes for three different feature selection algorithms in three artificial datasets

Method	COLON	LEUKEMIA	LYMPHOMA
S2N	91.5%	40.5%	100%
SVM-RFE	30%	18.5%	31%
NN-RFA	96%	63.6%	100%

The results reported in Table 4 show that the NN-RFA algorithm scores a higher percentage of correct genes with respect to the other methods, selecting almost all the relevant genes in two datasets (COLON and LYMPHOMA).

At last, the quality of NN-RFA has been evaluated as a function of the number of examples in the training set. To this aim NN-RFA has been tested on training sets including the 20%, 40%, 60% and 80% of the total available data. The percentage of correctly selected features for different sizes of the training set is

Table 5. The percentage of correctly selected genes in three artificial datasets as a function of the fraction of examples employed by the NN-RFA algorithm

% of patterns	COLON	LEUKEMIA	LYMPHOMA
20	53.0%	54.5%	100%
40	75.6%	60.3%	100%
60	91.3%	64.5%	100%
80	96.5%	61.2%	100%
100	97.3%	63.6%	100%

shown in Table 5. The results for LYMPHOMA does not depend on the percentage of patterns, whereas the performances on the other datasets decrease when the number of examples is reduced. However, the percentage of correctly selected genes is satisfactory also for a small number of examples (20%, 40%).

4   Conclusions

In this paper, the problem of gene selection has been considered. A general approach to feature selection, named Recursive Feature Addition (RFA), has been introduced in order to generate a ranking of the input variables of a classification problem. The NN-RFA algorithm, based on the nearest neighbor probability, has been described and used to select the relevant genes in three real world problems regarding the discrimination of different kinds of diseases. In order to better evaluate the ability of NN-RFA in retrieving a good subset of genes, the TAGGED method has been employed for generating artificial data similar to the real ones. In such a way it has been possible to prove that NN-RFA is able to retrieve a very high fraction of the genes really involved in the discrimination of the disease.

References

1. Guyon, I., Elisseeff, A.: An introduction to variable and feature selection. Journal of Machine Learning Research 3, 1157–1182 (2003)
2. Liu, Q., Sung, A.H.: Recursive feature addition for gene selection. In: Proceedings of IJCNN 2006, Vancouver, BC, Canada, pp. 1360–1367 (2003)
3. Duda, R.O., Hart, P.E., Stork, D.G.: Pattern Classification. Wiley-Interscience, Hoboken (2000)
4. Ruffino, F., Muselli, M., Valentini, G.: Gene expression modeling through positive Boolean functions. International Journal of Approximate Reasoning 47, 97–108 (2008)
5. Golub, T.R., et al.: Molecular classification of cancer: class discovery and class prediction by gene expression monitoring. Science 286(5439), 531–537 (1999)
6. Guyon, I., et al.: Gene selection for cancer classification using support vector machines. Machine learning 46(1–3), 389–422 (2002)
7. Alon, U., et al.: Broad patterns of gene expressions revealed by clustering analysis of tumor and normal colon tissues probed by oligonucleotide arrays. Proceedings of the National Academy of Science USA 96, 6745–6750 (1999)
8. Alizadeh, A.A., et al.: Different types of diffuse large B-cell lymphoma identified by gene expression profiling. Nature 403, 503–511 (2000)

Control of Cellular Glycolysis by Perturbations in the Glucose Influx

Jaime Arturo de la Torre, Maria Carmen Lemos, and Antonio Córdoba*

Departamento de Física de la Materia Condensada,
Universidad de Sevilla,
P. O. Box 1065,
41080 Seville, Spain
cordoba@us.es

Abstract. For certain metabolic routes of living organism oscillations in the concentration of the reactant substances appear. During the glycolysis, when the glucose is broken to obtain energy, oscillations can be seen in many steps of the process. These oscillations are damped down, in general, due to the feedback that keeps stable the cellular mechanism. However, certain values of the flux of glucose input keep the oscillations of the metabolic system. In this paper, the influence of a periodic modulation added to the influx of glucose inside the cell in a kinetic model which simulates the glycolysis is analyzed. Simulations of the perturbed glycolytic model exhibit periodic and quasiperiodic behaviors depending on the amplitude and the frequency of the perturbation. The next-maximum maps obtained distinguish between periodic behavior and quasiperiodic one.

Keywords: Glycolysis, metabolic oscillations, glygolytic model, periodic perturbation, quasiperiodic behavior.

1 Introduction

Researchers in the field of Biology have felt attracted to the implications of nonlinear dynamical systems since its appearance. In particular, studies of the oscillatory behavior of metabolic systems have produced significant developments as a result of the application of the principles of nonlinear dynamics. Metabolic systems go from the strictly monoenzymatics (the case of the catalyzed reaction by the peroxidase), to those composed by many enzymes like it is the case of the metabolic route known as glycolysis. Glycolysis, a series of chemical coupled reactions that breaks the glucose molecule to obtain energy, is a central route, and plays an important role in the metabolism of living organism.

Since some time ago, temporal oscillations in glycolysis have been observed and studied in many kinds of biosystems, including yeast cell-free extracts [1,2] and extract of yeast cells [3,4,5,6,7,8] (for review on biochemical oscillations see [9]). In early studies, only simple oscillations were observed [1,2]. Later, with the constant, stochastic or periodic input of the substrate, quasiperiodicity and

---

* Corresponding author.

F. Masulli, R. Tagliaferri, and G.M. Verkhivker (Eds.): CIBB 2008, LNBI 5488, pp. 132–143, 2009.
© Springer-Verlag Berlin Heidelberg 2009

chaos were found [3,4,5]. Sustained complex oscillations in glycolysis, including period-doubled, quasiperiodic oscillations and chaos [6,7], have been achieved in a CSTR (continuous-flow, stirred tank reactor).

Linear or periodic perturbations of the kinetic of a metabolic reaction by means of the variation of some external parameters such as, for example, the rate constant of a metabolite or glucose input, is one of the most commonly used tools in basic and applied studies of glycolysis [10,11]. In general, living biological systems show a great tolerance to external perturbations. Thus, it is observed that the glycolytic system is strongly stable for a wide range of any chosen control parameter [12,13]. A perturbation introduced in any step of the reactions gives rise to a variation in the concentration of a specific substance, causing a modification in the subsequent steps, and, by means of feedback the system comes back to its previous configuration. These are transient glycolytic oscillations. However, it is possible to obtain sustained oscillations, complex oscillations and chaotic dynamic when the entry of influx glucose is constant or sinusoidal [3,4,5,10,11].

One of the best studied metabolisms from the experimental and modeling point of view is the glycolytic pathway of the yeast *Saccharomyces cerevisiae* [14,15,16,17]. In 1996 Richard et al. [18,19] experimentally studied the mechanism of glycolysis using that yeast. They reported the obtaining of stable oscillations in the cell from a chemical procedure during the glycolytic process. In a population of intact yeast cells Richard et al. [18] observed that all glycolytic intermediates oscillate with the same period, but with different phases, and that this period is in the range of 1 min. Some control parameters as the cellular density of the sample allow to change the amplitude of these oscillations. Moreover, the oscillations have to be synchronous in all the cells, because the results are not observable if that phase synchronization does not exist.

Accepting the fact that the amplitudes of oscillations produced in a step of the metabolic path must be lesser than the amplitudes of the oscillations in a previous step, Richard et al. postulate that some glycolytic steps do not operate as driver but stabilizer of the chemical reactions.

Since the publication of the papers of Richard et al. [18,19], there have been several attempts to achieve mathematical modeling of the glycolytic path followed by the cell from introduction of glucose [20,21]. Wolf et al. [22] proposed a model that, although it does not perfectly fit with the results of Richard et al., exhibits the same qualitative behavior as the experimental results. The model includes the main steps of anaerobic glycolysis, and the production of glycerol and ethanol. This glycolytic process is modeled by a set of nine differential equations that allows to study essential features of the glycolytic oscillations.

The purpose of this article is to study, from model proposed by Wolf et al., the response of the system to externally applied periodic·perturbations inside the oscillatory band that the autonomous system presents, to analyze the general behavior of the system and to explore the conditions that optimize the metabolic route.

2 The Model

The model of Wolf et al. [22] is a model of glycolysis in yeast that qualitatively describes the essence of the experimental observations of Richard et al. [18,19]. The model consists of lumped reactions and includes production of glycerol and fermentation to ethanol. Furthermore, it takes into account the exchange of acetaldehyde between the cells, the trapping of acetaldehyde (the synchronizing agent) by cyanide, and the presence of more than one cell. Therefore, this model can describe intercellular synchronization. The resulting model is a 9-variable model whose schematic representation is shown in Fig. 1.

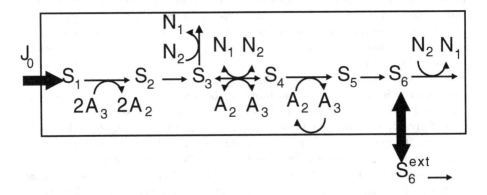

Fig. 1. Schematic representation of the glycolytic pathway. The glucose influx added, J_0, inside the cell determines the global behavior of the system. This model leads to the system of nine coupled differential equations described in Section 2.

Table 1. Metabolic variables

Variable	Metabolite
S_1	Glucose
S_2	Fructose-1,6-bisphosphate
S_3	Triosephosphates, glyceraldehyde-3-phosphate and dihydroxyacetone phosphate
S_4	3-phosphoglycerate
S_5	Pyruvate
S_6	Inner acetaldehyde
S_6^{ext}	Outer acetaldehyde
A_2	ADP
A_3	ATP
N_1	NAD$^+$
N_2	NADH

The variables consist of metabolic concentrations. These variables are listed in Table 1. In the following, we consider two conserved concentrations,

$$A = A_2 + A_3 = \text{constant},$$
$$N = N_1 + N_2 = \text{constant}.$$

The set of coupled nonlinear equations describing the time change of the metabolites in the one-cell version of the model of Wolf et al. is:

$$\frac{dS_1}{dt} = J_0 - k_1 S_1 A_3 \left[1 + \left(\frac{A_3}{K_i}\right)^n\right]^{-1} \tag{1}$$

$$\frac{dS_2}{dt} = k_1 S_1 A_3 \left[1 + \left(\frac{A_3}{K_i}\right)^n\right]^{-1} - k_2 S_2 \tag{2}$$

$$\frac{dS_3}{dt} = -\frac{k_{\text{GAPDH}} + k_{\text{PGK}} + S_3 (N - N_2)(A - A_3) - k_{\text{GAPDH}} - k_{\text{PGK}} - S_4 A_3 N_2}{k_{\text{GAPDH}} - N_2 + k_{\text{PGK}} + (A - A_3)} -$$
$$-k_8 S_3 N_2 + 2k_2 S_2 \tag{3}$$

$$\frac{dS_4}{dt} = \frac{k_{\text{GAPDH}} + k_{\text{PGK}} + S_3 (N - N_2)(A - A_3) - k_{\text{GAPDH}} - k_{\text{PGK}} - S_4 A_3 N_2}{k_{\text{GAPDH}} - N_2 + k_{\text{PGK}} + (A - A_3)} -$$
$$-k_4 S_4 (A - A_3) \tag{4}$$

$$\frac{dS_5}{dt} = k_4 S_4 (A - A_3) - k_5 S_5 \tag{5}$$

$$\frac{dS_6}{dt} = k_5 S_5 - k_6 S_6 N_2 - \kappa(S_6 - S_6^{\text{ext}}) \tag{6}$$

$$\frac{dS_6^{\text{ext}}}{dt} = \varphi\kappa(S_6 - S_6^{\text{ext}}) - k_9 S_6^{\text{ext}} \tag{7}$$

$$\frac{dA_3}{dt} = -2k_1 S_1 A_3 \left[1 + \left(\frac{A_3}{K_i}\right)^n\right]^{-1} + k_4 S_4 (A - A_3) - k_7 A_3 +$$
$$+\frac{k_{\text{GAPDH}} + k_{\text{PGK}} + S_3 (N - N_2)(A - A_3) - k_{\text{GAPDH}} - k_{\text{PGK}} - S_4 A_3 N_2}{k_{\text{GAPDH}} - N_2 + k_{\text{PGK}} + (A - A_3)} \tag{8}$$

$$\frac{dN_2}{dt} = \frac{k_{\text{GAPDH}} + k_{\text{PGK}} + S_3 (N - N_2)(A - A_3) - k_{\text{GAPDH}} - k_{\text{PGK}} - S_4 A_3 N_2}{k_{\text{GAPDH}} - N_2 + k_{\text{PGK}} + (A - A_3)} -$$
$$-k_6 S_6 N_2 - k_8 S_3 N_2 \tag{9}$$

The parameter values that are used in this model are shown in Table 2. These values are chosen so that the numerical results agree approximately with the experimental ones. For further details of the Wolf's model, the reader is referred to [22]. Public domain *Silicon Cell* version of the model can be found at http://www.jjj.bio.vu.nl/database/wolf

In the following, we have solved numerically the system of nine differential equations (Eqs. 1-9) by a sixth order Runge-Kutta-Felhberg method. This is a free software tool merged into the GNU plotting utilities, written by N. B. Tufillaro and R. S. Maier. The mean concentrations of the variables of the system (Table 3) are chosen as initial conditions in Runge-Kutta integration routine.

Our aim is to work with a specific control parameter: the influx of glucose, J_0. This choice arises from the possibility of controlling the system input to obtain the system output and analyze the general behavior.

Table 2. Parameter values

Parameter	Value	Parameter	Value
k_1	550 mM^{-1} min^{-1}	k_6	2000.0 mM^{-1}min^{-1}
K_i	1.0 mM	k_7	28.0 min^{-1}
k_2	9.8 min^{-1}	k_8	85.7 mM^{-1}min^{-1}
$k_{\mathrm{GAPDH+}}$	323.8 mM^{-1}min^{-1}	κ	375.0 min^{-1}
$k_{\mathrm{GAPDH-}}$	57823.1 mM^{-1}min^{-1}	φ	0.1
$k_{\mathrm{PGK+}}$	76411.1 mM^{-1}min^{-1}	A	4.0 mM
$k_{\mathrm{PGK-}}$	23.7 mM^{-1}min^{-1}	N	1.0 mM
k_4	80.0 mM^{-1}min^{-1}	n	4
k_5	9.7 min^{-1}	k_9	80 min^{-1}

Table 3. Initial values of metabolites

Metabolite	Initial Value (mM)
S_1	1.09
S_2	5.10
S_3	0.55
S_4	0.66
S_5	8.31
S_6	0.08
S_6^{ext}	0.02
A_3	2.19
N_2	0.41

The concentrations of A_3 versus the control parameter J_0 is shown in Fig. 2 as a bifurcation diagram. There is an oscillatory region between two steady regions. The system is oscillatory for $J_0 \in (42, 85)$ mM/min. Minimal and maximal (full circles inside oscillatory region) of A_3 oscillation amplitudes for varied glucose inflow concentration are also shown, as well as the time average of these amplitudes for each value of J_0 with empty circles (Fig. 2 (left)).

The Fig. 2 (right) is similar to Fig. 2 (left) but for larger values of J_0. We can observe the saturation of A_3 concentration for higher values of J_0.

The kinetic phase diagram of the model allows us to see two bifurcation points. The first critical point is obtained at $J_0 = J_0^{c1} = 42$ mM/min, and a soft phase transition can be observed between a steady behavior and an oscillatory one. A critical exponent α, obtained by linear regression from the expression $A_3 = (J_0 - J_0^{c1})^\alpha$, takes the value 0.493 ± 0.015, a result which is acceptable for that predicted for the Hopf bifurcation, that is, practically 0.5. This power-law is shown in Fig. 3. The second critical point is $J_0 = J_0^{c2} = 85$ mM/min, and it corresponds to a broken phase transition with a critical exponent $\alpha = -0.0018 \pm 0.0007$, that is, practically 0.

Figure 4 (left) shows a time series of S_2, S_5, A_3, and N_2 concentrations for $J_0 = 50$ mM/min. Difference of phase among the four metabolites are shown in Table 4. These values agree with those reported in the literature [19,22]. Finally,

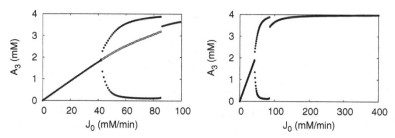

Fig. 2. Left. Bifurcation diagram for A_3 concentration as a function of J_0. The diagram shows an oscillatory region between two steady regions. The system becomes oscillatory if $J_0 \in (42, 85)$ mM/min. We also show in full circles inside oscillatory region maximal and minimal amplitudes, and the time average of these amplitudes for each value of J_0 with empty circles. **Right.** Similar to Left but for larger values of J_0. One can observe the saturation of A_3 for higher values of J_0.

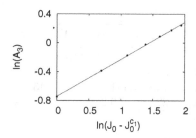

Fig. 3. Power-law. Filled circles exhibit the data obtained from amplitudes of period-1 (one maximum) oscillations of A_3 concentration as a function of J_0. The solid line corresponds to the power-law dependence, $A_3 = (J_0 - J_0^{c1})^\alpha$, with $J_0^{c1} = 42$ mM/min and $\alpha = 0.493$.

stable limit cycles for the 9-variable model in $S_1 - A_3$, $S_2 - A_3$, and $S_5 - A_3$ planes are shown in Fig. 4 (right) for $J_0 = 50$ mM/min.

3   The Perturbed Model

Once we have obtained the range of J_0 where the autonomous system exhibits an oscillatory behavior, we can attempt to force it with a determined frequency in order to study the response of the system to the perturbations. We assume that the influx of glucose into the system is modulated as

$$J_0^* = J_0 \left(1 + A \sin(2\pi f t)\right) \ , \tag{10}$$

where f is the perturbing frequency (different of the natural frequency of the non forced system, f_0) and A, the relative amplitude of the perturbation, both operating as control parameters. Obviously, the modulation of the external parameter J_0 causes simultaneous changes in the other processes of the glycolytic pathway.

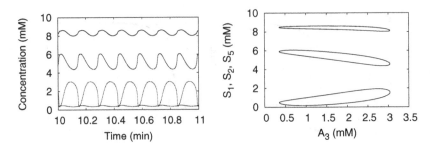

Fig. 4. Left. From top to bottom: time series of oscillations in S_5, S_2, A_3 and N_2 concentrations. In all of them we use the same glucose incoming flux $J_0 = 50$ mM/min. **Right.** Several stable limit cycles of the autonomous system with $J_0 = 50$ mM/min. From top to bottom, S_5, S_2, S_1 versus A_3.

Table 4. Difference of phase (in degrees) between some of the variables, using $J_0 = 50$ mM/min

	S_2	S_5	A_3	N_2
S_2	0	32.51	173.82	22.27
S_5	−32.51	0	141.31	−10.24
A_3	−173.82	−141.31	0	−151.55
N_2	−22.27	10.24	151.55	0

We have chosen a periodic state of the autonomous system corresponding to $J_0 = 50$ mM/min, keeping constant the other parameter values shown in Table 2. This state has a natural frequency $f_0 = 7.0954$ min^{-1} (the period of the oscillations is $T_0 = 0.141$ min, as it is shown in Wolf et al. [22]), which we have taken as a reference. Here our study involves the variation of the parameters A and f.

To simplify the presentation of results, ATP (A_3) concentration is chosen as the only output variable. For each pair of values of A and f, oscillations in A_3 concentration are obtained from the time series resulting from the solution of the kinetic Eqs. 1-9 by the method of Runge-Kutta. In general, the time series obtained last 20 min. We rule out the first 5 min to safely remove the initial transitory regime for all the cases and with the remaining points we calculate the time average of A_3, $\langle A_3 \rangle$, and its fluctuation, θ, so that $A_3 = \langle A_3 \rangle + \theta$. We have carefully analyzed the A_3 concentration fluctuaction, θ, in order to distinguish behaviors having a chaotic, quasiperiodic or periodic regime. Afterwards, a thorough study of the time series obtained is performed using typical tools of nonlinear dynamics and calculating, among other quantities, the Fourier transform and the Poincaré map.

First, the glucose influx is perturbed with $f = 6.0$ min^{-1} and A is varied. In Fig. 5 the time oscillations in the A_3 concentration are shown for five values of A. From top to bottom: $A = 0$ (system non perturbed, periodic state P1 (one maximum)), $A = 0.05, 0.10, 0.15$ (quasiperiodic states), and $A = 0.20$ (periodic

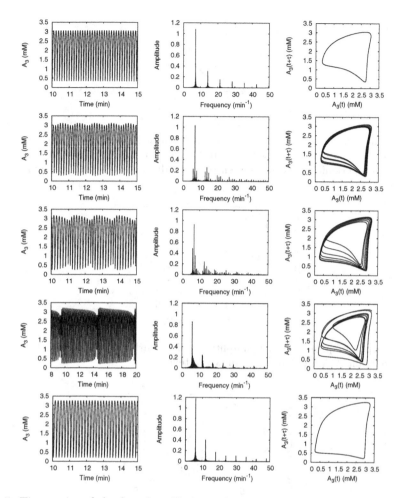

Fig. 5. Time series of the forced oscillations of concentration of A_3 obtained for different values of the perturbed amplitude, being $f = 6.0$ min^{-1} and $J_0 = 50$ mM/min. These series, obtained by simulations of the perturbed model, represent (from top to bottom): a periodic regime ($A = 0$, autonomous state), quasiperiodic behaviors ($A = 0.05, 0.10, 0.15$), and periodic state ($A = 0.20$). The Fourier spectra and Poincaré maps, with a time delay $\tau = 0.027$ min, are also attached to the time series.

state P1). The corresponding Fourier spectra and the Poincaré maps (for time delay $\tau = 0.027$ min) are shown as well. In all cases, the system shows a frecuency spectrum discrete. For $A = 0$ (natural oscillations), the fundamental frequency corresponds to the peak 7.0954 min^{-1}, the remaining peaks being its harmonics. For the quasiperiodic states two incommensurable frequencies appear. Denoting by f_1, f_2 and f_3 (from less to more intensity) the frequencies of the most prominent three peaks obtained by means of Fourier analysis, one can observe the quasiperiodic behavior of the signal with two incommesurable frequencies, because the third one is a linear combination of the other two ($f_3 - f_2 = f_2 - f_1$).

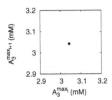
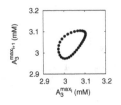
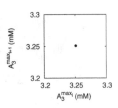

Fig. 6. Next-maximum maps of some simulated states shown in Fig. 5, being $f = 6.0$ min^{-1} and $J_0 = 50$ mM/min. **Left:** $A = 0$, **Center:** $A = 0.05$, **Right:** $A = 0.20$. We show the quasiperiodic behavior between two periodic states P1 (one maximum).

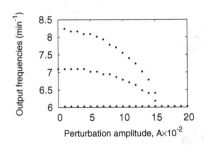

Fig. 7. Output frequencies versus amplitudes of perturbation for $J_0 = 50$ mM/min and $f = 6.0$ min^{-1}. More details are shown in text.

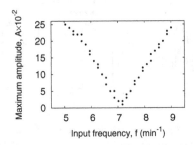

Fig. 8. Maximal amplitudes of the perturbation, A, as a function of frequencies, f, for $J_0 = 50$ mM/min. For each value of frequency there is a maximum perturbing amplitude from which the quasiperiodic behavior disappears.

For example, for $A = 0.10$, the three first peaks correspond to: $f_1 = 6.0273$ min^{-1}, $f_2 = 6.7903$ min^{-1}, and $f_3 = 7.5532$ min^{-1}. The remaining peaks are harmonics of these. Finally, for $A = 0.20$, there is only a fundamental frequency: 6.0273 min^{-1}.

Figure 6 shows three next-maximum maps of some time series of Fig. 5. If the behavior is periodic with a single maximum, the next-maximum map will show a single point. If the behavior is quasiperiodic, the points densely populate an invariant closed curve.

The quasiperiodic behavior keeps up to $A = 0.15$. For perturbing amplitudes greater than $A = 0.15$ a periodic state P1 appears with a frequency equal to

6.0273 min^{-1} (very close to the perturbing frequency $f = 6.0$ min^{-1}). This aspect can be observed in Fig. 7 where output frequencies versus perturbing amplitudes for $J_0 = 50$ mM/min and $f = 6.0$ min^{-1} are represented. It can also be observed that for $A = 0$ only an output frequency is obtained (f_0) and for $0 < A \leq 0.15$ the system presents a quasiperiodic behavior with two fundamental frequencies.

Next, in order to analyze whether the above results appear when the perturbing frequency varies, we have carried out a similar study for f ranging between 5 and 9 min^{-1}. The maximum perturbing amplitude A that the system can accept to keep the quasiperiodic state for a given perturbing frequency f is shown in Fig. 8. Out of this range, the behavior of the system is P1, with a frequency close to the perturbing one. We may notice that when we come closer to the value of the autonomous frequency of oscillation (f_0), either from lower frequencies than f_0, or from higher frequencies, the maximum amplitude of perturbation for which the system still exhibits quasiperiodic behavior keeps decreasing.

4 Conclusions

The model of Wolf et al. [22] allows to numerically solve some aspects of the problem of cellular glycolysis. Qualitatively it describes well the real behavior of the chemical reactions of this metabolic pathway. From this model one can demostrate that for certain values of the glucose influx sustained oscillations in the concentrations of the other substances in the reaction appear. When this influx is perturbed in a sinusoidal way, for a certain range of the perturbing frequency and relative amplitude, the state of the system becomes quasiperiodic, with two incommensurable frequencies. Above this range the system exhibits a periodic behavior with a frequency similar to the perturbing frequency. For the recognition of quasiperiodicity we have proceeded in the usual way, studying the time series obtained from integration of the differential equation system of the model, through the Poincaré map and Fourier transform, typical quantities of the nonlinear analysis.

We indicate that the periodic perturbation of the kinetics of a glycolytic route applied here and the results obtained are general ones, and can be applied to other models exhibiting kinetic oscillations. Thus, this analysis offers a way to modulate and to control the concentrations of different metabolites in the glycolytic pathway, and hence it allows to advance in the optimization of the whole process. Because knowledge of biochemical oscillations is still limited, it is necessary to continue with the research in the field of Biology.

Acknowledgments

This work is partially financed by the Junta de Andalucía (Project No. P06-TIC-02025).

References

1. Chance, B., Hess, B., Betz, A.: DPNH oscillations in a cell-free extract of *S. carlsbergensis*. Biochem. Biophys. Res. Commun. 16, 182–187 (1964)
2. Hess, B., Brand, K., Pye, K.: Continuous oscillations in a cell-free extract of *S. carlsbergensis*. Biochem. Biophys. Res. Commun. 23, 102–108 (1966)
3. Boiteux, A., Goldbeter, A., Hess, B.: Control of oscillating glycolysis of yeast by stochastic, periodic, and steady source of substrate: A model and experimental study. Proc. Nat. Acad. Sci. USA 72, 3829–3833 (1975)
4. Markus, M., Kuschmitz, D., Hess, B.: Chaotic dynamics in yeast glycolysis under periodic substrate input ux. FEBS Lett. 172, 235–238 (1984)
5. Markus, M., Müller, S., Hess, B.: Observation of entrainment, quasiperiodicity and chaos in glycolyzing yeast extracts under periodic glucose input. Ber. Bunsenges. Phys. Chem. 89, 651–654 (1985)
6. Nielsen, K., Sørensen, P., Hynne, F.: Chaos in glycolysis. J. Theor. Biol. 186, 303–306 (1997)
7. Nielsen, K., Sørensen, P., Hynne, F., Busse, H.: Sustained oscillations in glycolysis: an experimental and theoretical study of chaotic and complex periodic behavior and of quenching of simple oscillations. Biophys. Chem. 72, 49–62 (1998)
8. Danø, S., Sørensen, P., Hynne, F.: Sustained oscillations in living cells. Nature 402, 320–322 (1999)
9. Goldbeter, A.: Biochemical Oscillations and Cellular Rhythms. Cambridge University Press, Cambridge (1996)
10. Bier, M., Bakker, B., Westerhoff, H.: How yeast cells synchronize their glycolytic oscillations: A perturbation analytic treatment. Biophys. J. 78, 1087–1093 (2000)
11. de la Fuente, I., Martínez, L., Veguillas, J., Aguirregabiria, J.: Quasiperiodicity route to chaos in a biochemical system. Biophys. J. 71, 2375–2379 (1996)
12. Kitano, H.: Foundations of Systems Biology. MIT Press, Cambridge (2001)
13. Klipp, E., Herwig, R., Kowald, A., Wierling, C., Lehrach, H. (eds.): Systems Biology in Practice. Concepts, Implementation and Application. WILEY-VCH Verlag, Berlin (2006)
14. Rizzi, M., Baltes, M., Theobald, U., Reuss, M.: In vivo analysis of metabolic dynamics in *Saccharomyces cerevisiae*: II. mathematical model. Biotech. Bioeng. 55, 592–608 (1997)
15. Wolf, J., Heinrich, R.: Effect of cellular interaction on glycolytic oscillations in yeast: a theoretical investigation. Biochem. J. 345, 321–334 (2000)
16. Teusink, B., Passarge, J., Reijenga, C., Esgalhado, E., van der Weijden, C., et al.: Can yeast glycolysis be understood in terms of *in vitro* kinetics of the constituent enzymes? Testing biochemistry. Eur. J. Biochem. 267, 5313–5329 (2000)
17. Reijenga, K., Westerhoff, H., Kholodenko, B., Snoep, J.: Control analysis for autonomously oscillating biochemical networks. Biophys. J. 82, 99–108 (2002)
18. Richard, P., Teusink, B., Hemker, M., van Dam, K., Westerhoff, H.: Sustained oscillations in free-energy state and hexose phosphates in yeast. Yeast 12, 731–740 (1996)
19. Richard, P., Bakker, B., Teusink, B., van Dam, K., Westerhoff, H.: Acetaldehyde mediates the synchronization of sustained glycolytic oscillations in populations of yeast cells. Eur. J. Biochem. 235, 238–241 (1996)

20. Hynne, F., Danø, S., Sørensen, P.: Full-scale model of glycolysis in *Saccharomyces cerevisiae*. Biophys. Chem. 94, 121–163 (2001)
21. Reijenga, K., Snoep, J., Diderich, J., van Verseveld, H., Westerhoff, H., Teusink, B.: Control of glycolytic dynamics by hexose transport in *Saccharomyces cerevisiae*. Biophys. J. 80, 626–634 (2001)
22. Wolf, J., Passarge, J., Somsen, O., Snoep, J., Heinrich, R., Westerhoff, H.: Transduction of intracellular and intercellular dynamics in yeast glycolytic oscillations. Biophys. J. 78, 1145–1153 (2000)

Curating a Large-Scale Regulatory Network by Evaluating Its Consistency with Expression Datasets

Carito Guziolowski[1], Jeremy Gruel[2], Ovidiu Radulescu[3], and Anne Siegel[4]

[1] Centre INRIA Rennes Bretagne Atlantique, IRISA, Campus de Beaulieu,
35042 Rennes, France
carito.guziolowski@irisa.fr
http://www.irisa.fr/symbiose/carito_guziolowski
[2] INSERM U620, 2 Av Pr. L. Bernard, 35043 Rennes, France
[3] Université de Rennes 1, IRMAR Campus de Beaulieu, 35042 Rennes, France
[4] CNRS, UMR 6074, IRISA, Campus de Beaulieu, 35042 Rennes, France

Abstract. The analysis of large-scale regulatory models using data issued from genome-scale high-throughput experimental techniques is an actual challenge in the systems biology field. This kind of analysis faces three common problems: the size of the model, the uncertainty in the expression datasets, and the heterogeneity of the data. On that account, we propose a method that analyses large-scale networks with small – but reliable – expression datasets. Our method relates regulatory knowledge with heterogeneous expression datasets using a simple *consistency rule*. If a global consistency is found, we predict the changes in gene expression or protein activity of some components of the network. When the whole model is inconsistent, we highlight regions in the network where the regulatory knowledge is incomplete. Confronting our predictions with mRNA expression experiments allows us to determine the missing post-transcriptional interactions of our model. We tested this approach with the transcriptional network of *E. coli*.

1   Introduction

Reconciling gene expression data with large-scale regulatory network structures is a subject of particular interest due to the urgent need of curating regulatory models that are likely to be incomplete and may contain errors. Based on the large amount of genome-scale data yielded by high-throughput experimental techniques, many data-driven approaches for reconstructing regulatory network structures have been proposed [1,2,3]. Meanwhile, for some well studied organisms there are already large-scale models of regulations derived from literature curations or from computational predictions; as for example the RegulonDB database [4] for the bacteria *E. coli*. On account of this, large-scale models are presented as a compilation of interactions deduced from different methods or from experiments applied under different conditions. The accumulation of diverse sources in the construction of regulatory models may cause errors.

F. Masulli, R. Tagliaferri, and G.M. Verkhivker (Eds.): CIBB 2008, LNBI 5488, pp. 144–155, 2009.

A challenging solution is therefore to design automatic methods that integrate experimental datasets to known regulatory structures enabling biologists to conciliate heterogeneous data types, find inconsistencies, and refine and diagnose a regulatory model [5,6,7].

On the basis of these arguments, we propose an iterative method to identify and fix regions in a regulatory network model where proposed regulations are logically inconsistent with that of microarray expression data. Using our method we can: *(i)* detect the regions in the network inconsistent with the experimental data, and *(ii)* extend the initial set of expression data in order to generate computational predictions. We applied our method to the transcriptional regulatory network of *E.coli K12* (1763 products and 4491 interactions) extracted from the RegulonDB database.

By using fast and performant algorithms we are able to reason over the large system of constraints representing a large-scale regulatory network. Thus, we detect global inconsistencies involving contradictions among different regions in the network and go one step further than other methods [8] that detect only local contradictions involving only network modules formed by direct predecessors of a node in the graph. Regarding the *E. coli* model, we concluded that its transcriptional interactions did not adequately explain the initial experimental data composed by 45 differentially expressed genes under the exponential-stationary growth shift. On that account, we detected specific post-transcriptional interactions in order to obtain a globally explained model by the experimental data.

Once a global consistency is obtained, our method generates gene-expression computational predictions, which are the result of a consistency test performed previously to the entire network. As opposed to other methods [9,10], we do not simulate the response of our model to an artificial perturbation; we describe the (possibly huge) set of models that are consistent with the initial dataset, then we look for invariants in this set. In this way, our computational predictions correspond to the nodes whose expression change fluctuates in the same direction in all the models consistent with the initial dataset. Regarding the *E. coli* regulatory model, we calculated 502 gene-expression predictions that correspond to nearly 30% of the network products predicted to change considerably under the analysed condition. These predictions were validated with microarray outputs, obtaining an agreement of 80%. This percentage can be comparable to the one obtained by other methods working on *E. coli* data [11,12,13], and considerable, since we used only a transcriptional model without including metabolic regulations.

2   Approach

2.1   Global Approach

We analysed regulatory networks formed by the interactions among certain molecules. As molecules we refer to genes, sigma factors, active proteins, or protein complexes; and as interactions to Protein-DNA, Sigma-gene, and complex formation. Even so, any kind of interactions may be studied as long as we can map them as *influence* relations, i.e. *A influences B if increasing or decreasing A's concentration causes a change in B's concentration*. Molecules that hold influence relations form an *influence network*, the central object of this study (Fig. 1).

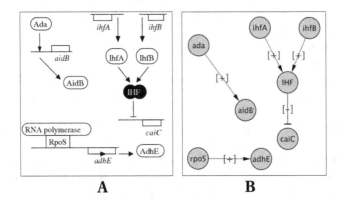

Fig. 1. A regulatory network (**A**) mapped into an influence network (**B**). Influences among molecules create an influence network; the arrows in the influence network represent a positive (+) or negative (−) influence.

The interactions of the influence network must be described qualitatively as +, −, and ?, where: +, represents a positive influence (e.g. activation of gene transcription, recognition of a gene promoter region, or formation of proteins complexes); −, a negative influence (e.g. inhibition of gene transcription, inactivation of a protein); and ?, a dual or complex regulation. Differential data issued from perturbation steady state experiments should be also provided in the form of qualitative (+, −) *concentration changes* of some network molecules. One type of concentration changes may be statistically significant mRNA-expression responses described qualitatively as: + up-regulation, − down-regulation.

An influence network is analysed using the following consistency rule: *"The variation of the concentration level of one molecule in the network must be explained by an influence received from at least one of its predecessors, different from itself, in the network"*. Checking the validity of this rule for an influence network alone, or for an influence network plus a set of reliable concentration changes, will be referred from now on as the *consistency check process*. The mathematical bases of the consistency rule are formally proven in [14,15]. In these studies the authors assumed that the concentration changes must correspond to changes between two stable conditions in the cell.

Let us intuitively explain the logic of the consistency rule in the simple influence network presented in Fig. 2. *A* and *B* may represent two proteins that activate the transcription of gene *C*. The consistency rule states that if *A* and *B* are both up-regulated (+) under certain condition, then *C* must be up-regulated, i.e. a + prediction will be assigned to *C* (Fig. 2A). Similarly, the concentration change of *C* will be predicted as − if both, *A* and *B*, were down-regulated. When

A is up-regulated ($+$) and B is down-regulated ($-$) the expression level of C cannot be *predicted*, as both expression levels (up/down regulated) are possible for C and do not contradict the consistency rule (Fig. 2B). Only if C was a protein complex formed by proteins A and B, we may *predict* the expression level of C depending on the change in expression of the limiting former protein. To state this, we have extended the theory in [14,15] in a context where *the weakest takes it all* concluding in a rule called the *protein complex behaviour rule.* This rule applies when the values of the concentrations of A and B are well separated (one of them is much smaller than the other). The details of this rule are provided in the Supplementary material.

A third situation may occur when all the molecules are observed, let us say, A is up-regulated, B is up-regulated, and C is down-regulated. The consistency rule states that C should be up-regulated; the experiment, however, shows the contrary. Thus, we arrive to a *contradiction* between the network and the experiment, also called *inconsistent* model (Fig. 2C). No predictions may be generated from an inconsistent model, yet, a region in the network is identified together with the expression data that created the conflict.

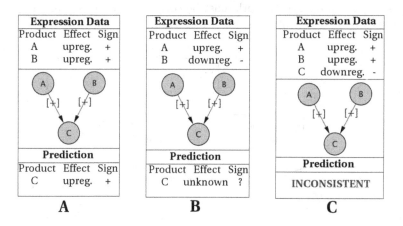

Fig. 2. Examples explaining the consistency check process. **A.** Expression predictions when a set of consistent expression data is provided. **B.** Consistent expression data may not generate a new prediction. **C.** Expression data provided resulted inconsistent with the influence network.

An influence network consistent with expression data is a network where all the expression data is explained by the (consensual or not) fluctuation of all the nodes in the network. In Fig. 2 we illustrated that depending on the expression data, we may obtain up to three different results: consistency and prediction, just consistency, or inconsistency. In order to evaluate the consistency of a large-scale network, we represent it as a system of qualitative constraints in $\{+, -, ?\}$, where each constraint relates a node with its direct predecessors. All constraints in the system should satisfy the consistency rule, and should not contradict any other. Computing the satisfiability of a large system of qualitative constraints is an NP-complete problem for even linear qualitative systems [16]. As classic methods of resolution do not allow to solve this kind of problems [17], we used an efficient representation based on binary decision diagrams for the set of solutions of a qualitative system proposed in [18]. The consistency of a large system of qualitative constraints is thus computed using *Pyquali*, a Python library devoted to solve huge systems of constraints in three values (+, −, ?) that is based on the mentioned representation. As a result it is possible to decide in a matter of seconds if a large-scale regulatory model is consistent with an initial dataset and to detect inconsistent regions in the model when needed.

In Fig. 3 we illustrate a flow-chart of the complete consistency check process. As input data we receive a qualitative influence network and a set of qualitative concentration changes. We build from this initial data a system of mathematical constraints, its consistency will be analysed afterwards using *Pyquali*. If the system is consistent we predict new concentration changes that after being compared with other experiments may generate new inputs to the initial set of concentration sets, however, it is also possible that thanks to this comparison new nodes and edges in the influence network will be added. If the system is inconsistent, the influence network must be corrected either by searching in the literature or by experimental results.

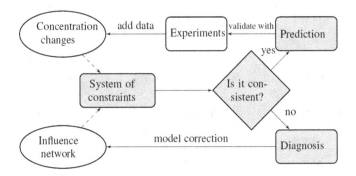

Fig. 3. Consistency check process. (1) We build a system of constraints from an influence network with a set of concentration changes, (2) we check the consistency of the system, (3) if it is consistent and an initial dataset of concentration changes was provided, we may predict new concentration changes of the molecules in the network that after compared with real measurements may question the original dataset and model. If it is not consistent, we report the inconsistent region in order to correct the network or initial dataset. The shaded blocks represent the automatically calculated steps.

2.2   Small Example

To understand how a consistent model may generate computational predictions, let us analyse in detail the consistency check process applied to a small region of the *E. coli* influence network. The influence network presented in Fig. 4 is analysed during the exponential-stationary growth shift. An initial dataset

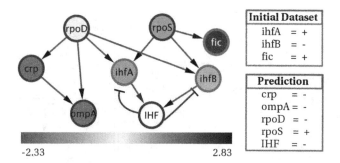

Fig. 4. Consistency check process for 8 products of the *E. coli* regulatory network under the exponential-stationary growth shift. All transcriptional influences that each product receives appear in the network. The grey-red intensity of each product reflects the experimentally observed change in mRNA expression (*log_2 ratio*) during the studied condition. Products with a green border refer to those present in the initial dataset obtained from the literature, whereas products with a blue border refer to our computational predictions. Experimentally observed mRNA-expression changes agree with the computational predictions except for RpoD, where the predicted changes correspond to variations on the *active protein* and cannot be observed on mRNA expression levels.

of concentration changes (obtained from the literature) was initially provided for *fic*, *ihfA*, and *ihfB* [19,20]. From this initial data the analysis proceeds as follows. As *fic* is up-regulated, and it only receives an influence from the sigma factor RpoS, then RpoS is fixed to be up-regulated. The protein complex IHF is fixed as down-regulated after applying the *protein complex behaviour rule* deduced from observing the concentration of its former proteins and the metabolically stable behaviour of IHF in its dimeric form (see the Supplementary material). The gene transcribing one of its former proteins, *ihfB*, appears down-regulated; hence, it must receive a negative influence from one of its three predecessors in the network: IHF's influence is positive as it is down-regulated but it inhibits *ihfB*'s expression; RpoS's influence is also positive as it was predicted to be up-regulated; only RpoD's value can be fixed to down-regulated (−), causing a negative influence over *ihfB*. The genes that receive a unique influence from RpoD, *crp* and *ompA*, are fixed as down-regulated because of the negative influence coming from RpoD. No other alternatives nor contradictions appear in the fixed values of IHF, RpoS, RpoD, *crp*, and *ompA*; consequently, they are the computational predictions.

3   Results

3.1   Construction of the Network

We constructed the influence network from the regulatory model of *E. coli* available in the RegulonDB database. The influences in our network represented transcriptional, complex formation, and Sigma-gene regulations. The products of the

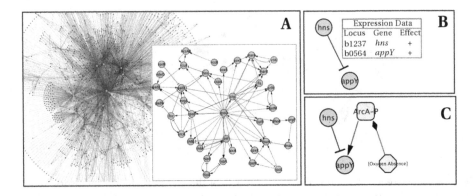

Fig. 5. A. *E. coli* influence network (1763 products and 4491 interactions). A small region of it is presented to the right (39 products and 94 interactions). The products forming this region control most of the components of the bigger network. **B.** The inhibition of gene *appY* by the *hns* product causes an inconsistency with the expression data related to the exponential-stationary growth shift. **C.** Correction of the inconsistency by adding a positive regulation from ArcA-P (phosphorylated protein ArcA) to *appY*; this regulation occurs in the absence of oxygen.

network were genes, active proteins, and protein complexes. The information to reconstruct this model was extracted from the database on June 2007. In Fig. 5A we show the complete and reduced network.

3.2   Consistency Test and Correction Using One Stress Condition

To start with the consistency check process, two initial data were provided: *(i)* *E.coli* influence network, and *(ii)* a set of 45 differentially expressed genes in the transition from exponential to stationary growth phase (see the complete list of observed genes in the Supplementary material). This set of phenotypes was collected from the literature based on initial information provided in RegulonDB. The first result obtained was an inconsistency between the model and the experimental data. The inconsistent region, highlighted by the method, is the one shown in Fig. 5B. It represents the inhibition of transcription of the *appY* gene by the H-NS protein; no other transcriptional regulation of the *appY* gene was reported in the RegulonDB database. These products (*appY*, *hns*, and therefore H-NS) are however, shown to increase their levels in the exponential-stationary growth shift of the cell [21,22]. Hence, the source of the conflict may be in the model.

Searching in the primary literature, we found that the *appY* gene is induced during entry into stationary phase, and that during oxygen-limiting conditions the stationary-phase induction is partially dependent on ArcA [23]. The protein ArcA is activated via phosphorylation by the ArcB sensor under conditions of reduced respiration [24]; the signal which leads to the activation of ArcA during entry into stationary phase may be the deprivation of oxygen caused by an increase in cell density [23]. Based on these studies we corrected our influence network adding new interactions (see Fig. 5C), obtaining a model consistent with the experimental data.

3.3   Computational Predictions and Validation

Once our network was consistent with the expression data provided for the exponential-stationary growth shift we generated the computational predictions. From the 45 initial expression phenotypes, 526 changes in other components of the network during the same experimental condition were predicted. We characterized them into 12 functional groups using the *DAVIS* software [25]. To validate our computational predictions, we obtained from the *Many Microbe Microarray Database* [2] a dataset of differentially expressed genes after 720 minutes of growth (stationary phase) in a rich medium [26]. This dataset was compared to our predictions (Fig. 6). The extended view of this comparison is included in the Supplementary material.

Expression profiling of this microarray dataset identified 926 genes that change significantly (2-fold) in transcription in response to the growth shift from exponential to stationary phase. The 526 products, computationally predicted, could be classified into four categories: 131 agreed with significant expression changes; 32 had a predicted expression change in a direction opposite to that of the experimental data; 329 had a predicted expression change that was not found to be statistically significant in the experimental data; and for 34 products there was no expression data available (some products of the network were protein complexes and could not be compared to mRNA expression). Thus, of the 163 (=131+32) significant differentially expressed genes that could be compared between the computational predictions and the experiment, 131 (or 80% consensus) agreed. Only the 31% (coverage) of our predictions could be compared with 2-fold expression changes; therefore we performed the same analysis choosing different thresholds (1.5-fold, 0 fold). In this way, new consensus and coverage percentages were calculated showing that the higher the threshold is, the better is the consensus and the worse is the coverage of the predicted data (Fig. 7).

The quality of our computational predictions is related to the experimental condition used to calculate them. Three important phases in the cell (exponential, early stationary-phase, late stationary-phase), appear in the exponential-stationary growth shift. During the shift among these phases, many molecules in the cell change their behaviour considerably. In our study we have chosen to compare the late stationary phase, when the number of cells does not change, with the late exponential growth phase, when the cell continues to grow exponentially. We believe that these two conditions represent instants in the cell where the genes and proteins do not significantly change their concentration. However, the 45 initial expression data may correspond to slightly different time points of the exponential phase and thus induce divergences in our computational predictions. In spite of this, a high percentage of our predictions was validated when compared with the microarray data.

A Functional groups:

A

Locus	Gene	L2R	P
b0911	rpsA	-1.33	-
b0884	infA	-4	-
b0852	rimK	-0.36	-
b0082	mraW	0.04	-
b4147	efp	-2.11	-
b3340	fusA	-1.87	+
b3321	rpsJ	-4.68	-
b3320	rplC	-4.73	-
b3319	rplD	-4.26	-
b3318	rplW	-5.05	-
b3317	rplB	-3.96	-
b3316	rpsS	-4.59	-
b3315	rplV	-4.33	-
b3314	rpsC	-4.05	-
b3313	rplP	-3.6	-
b3312	rpmC	-3.26	-
b3311	rpsQ	-4	-
b3298	rpsM	-3.19	-
b3297	rpsK	-2.9	-
b3296	rpsD	-2.92	-
b3294	rplQ	-2.94	-
b3231	rplM	-4.87	+
b3230	rpsI	-3.57	+
b2531	iscR	-0.12	-
b1718	infC	-0.84	-
b1717	rpmI	-1.63	-
b2026	hisI	-1.2	-
b2025	hisF	-1.95	-
b2024	hisA	-0.84	-
b2023	hisH	-1.02	-
b2022	hisB	-0.84	-
b4024	lysC	-2.02	-
b0004	thrC	-1.91	-
b0003	thrB	-1.32	-
b2838	lysA	-0.7	-
b2478	dapA	-1.75	-
b0002	thrA	-2.23	-
b0001	thrL	-1.12	-
b2763	cysI	-1.58	-
b2762	cysH	-2.82	-
b2752	cysD	-3.09	-
b2751	cysN	-2.34	-
b2421	cysM	-1.7	-

A,C,G,E,L

Locus	Gene	L2R	P
b2414	cysK	-4.82	-
b1275	cysB	-0.69	-
b0576	pheP	-0.09	-
b3460	livJ	-3.76	-
b3458	livK	-2.6	-
b3457	livH	-1.22	-
b3456	livM	-1.23	-
b3455	livG	-1.98	-
b3454	livF	-1.49	-
b2156	lysP	-1.84	-
b1453	ansP	0.04	+
b3656	yicI	0.05	+
b3338	chiA	0.33	-
b2132	bglX	-0.08	-
b3631	rfaG	-0.47	-
b3626	rfaJ	-1.54	-
b3429	glgA	-0.51	-
b3428	glgP	-0.27	-
b0182	lpxB	0.27	+
b3739	atpI	-2.46	-
b3738	atpB	-3.67	-
b3737	atpE	-3.37	-
b3736	atpF	-3.43	-
b3735	atpH	-3.31	-
b3734	atpA	-3.46	-
b3733	atpG	-2.57	-
b3732	atpD	-2.59	-
b3731	atpC	-1.9	-
b4025	pgi	-1.59	-
b3386	rpe	-0.8	-
b2464	talA	1.23	+
b2097	fbaB	1.8	+
b1723	pfkB	0.89	+
b3256	accC	-2.17	-
b3255	accB	-2.55	-
b2316	accD	-1.55	-
b0185	accA	-0.82	-
b3632	rfaQ	-0.92	-
b3630	rfaP	-0.67	-
b3629	rfaS	-0.81	-
b3625	rfaY	-1.1	-
b3624	rfaZ	-0.3	-
b1215	kdsA	-0.91	-

N,R

Locus	Gene	L2R	P
b3641	slmA	-0.82	-
b3438	gntR	0.23	-
b2405	xapR	0.19	-
b1564	relB	2.06	-
b1563	relE	2.64	-
b4312	fimB	-0.83	-
b3864	spf	-3.13	+
b2691	argQ	-0.97	-
b1977	asnT	0.83	-
b0143	pcnB	-2.46	-
b1084	rne	0.48	-
b0683	fur	0.31	-
b0064	araC	0.47	-
b0571	cusR	-0.54	-
b4401	arcA	1.5	+
b4398	creB	-0.09	-
b4393	trpR	-0.48	-
b4293	fecI	0.39	+
b4172	hfq	-1.03	-
b4133	cadC	0.02	-
b4118	melR	0.94	-
b4063	soxR	0.34	-
b4062	soxS	-0.46	-
b4043	lexA	0.48	-
b3961	oxyR	-0.49	-
b3934	cytR	-0.65	-
b0399	phoB	0.3	+
b3906	rhaR	-0.13	-
b3905	rhaS	-0.35	-

R

Locus	Gene	L2R	P
b3773	ilvY	0.34	-
b3753	rbsR	0.35	-
b3556	cspA	-1.41	-
b3555	yiaG	2.45	+
b3501	arsR	0.24	-
b3423	glpR	0.32	-
b0346	mhpR	1.1	-
b0345	lacI	0.21	-
b3357	crp	-2.33	-
b0338	cynR	0.61	-
b0330	prpR	0.55	-
b3237	argR	-0.76	-
b3181	greA	-0.62	+
b0313	agaR	-0.81	-
b0313	betI	-0.08	-
b3067	rpoD	0.61	-

Log2 Ratio shading

- L2R>+1.0
- 1>L2R>0.5
- +0.5>L2R>-0.5
- -0.5>L2R>-1.0
- -1.0>L2R

Comparison color coding (Prediction vs experiment)

- ▲△ or ▼▽
- ▲▽ or ▼△
- ▲□ or ▼□

B

Comparison		
Threshold	2-fold	1.5-fold
Exp. total	926	1757
Pred. total	502	502
▲△ or ▼▽	130	195
▲▽ or ▼△	32	62
▲□ or ▼□	340	211
No possible comparison		
Threshold	2-fold	1.5-fold
▲* or ▼*	34	34
*△ or *▽	458	1289

Fig. 6. Table of observed vs. predicted gene-expression responses in the *E.coli* network under the exponential-stationary growth shift condition. **A.** The locus numbers, gene names, and the log_2 ratio (L2R) of gene expression (exponential to stationary) are shown for some of the 526 predicted expression changes (+,−). Genes were divided by functional groups into: A, amino acids metabolism and biosynthesis; C, carbohydrates metabolism and biosynthesis; E, energy metabolism; G, glucose catabolism; L, lipid metabolism and biosynthesis; N, nucleic acids metabolism; R, regulatory function; S, cell structure; SI, signal peptides; T, transport; V, vitamin metabolism and biosynthesis; and U, unassigned. The L2R is shaded depending on the magnitude of the expression shift. Filled and open symbols indicate computational predictions and experimental data, respectively; squares indicate no change in gene expression; triangles indicate a change in expression, as well as the direction of change (up-regulated or down-regulated). **B.** Comparison between all predicted and observed expression changes. An * symbol indicates either that our model did not predict a gene expression or that no expression data related to a gene in our model was found.

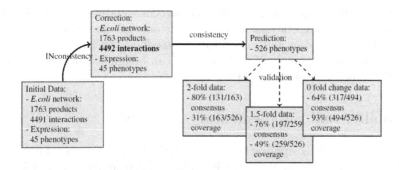

Fig. 7. Results of the consistency check process applied to the *E.coli* transcriptional network using 45 phenotypes related to the exponential-stationary growth shift. We validated our computational predictions using the observations in a microarray dataset filtered with three thresholds. Consensus refers to the percentage of validated model predictions; coverage indicates the percentage of compared predictions.

4   Discussion

We have introduced an iterative method that checks the consistency between a regulatory model and expression data. To validate our approach we chose the *E. coli* transcriptional regulatory network obtained from the RegulonDB database. By using an expression dataset of 45 *E. coli* genes significantly expressed under the exponential-stationary growth shift, we corrected the model and predicted the outputs of microarray experiments with an 80% of agreement. This percentage is comparable to the one obtained by other methods working on a more complex regulatory model for *E. coli* [11,12,13]. After the comparison of our computational predictions with microarray measurements, some disagreements were detected. A reasonable explanation for these divergences resides on our previously made assumption that some level of correlation exists between the transcription factor protein and the target gene expression without considering in detail the post-transcriptional effects; this problem was also reported in [8]. An example of this case of disagreement was illustrated in Fig. 4; nevertheless, we can still use this type of errors in our predictions to complete the regulatory model with post-transcriptional regulations. A second possible reason is the complexity of the exponential-stationary growth shift which induces heterogeneity in the observations obtained from different sources of the literature [27]. We illustrated that our automatized framework checks in very short time the global consistency of a large-scale system of constraints, into which regulatory and expression knowledge can be represented. Computationally, the solution of this type of problems is not trivial, nevertheless we analysed the complete large-scale regulatory network of *E.coli* in less than one minute. Biologically speaking, we are able to integrate large-scale regulatory data and confront it to experimental observations in order to provide a diagnosis of regions in the network where the regulatory knowledge is contradictory with the expression data. Moreover, we compute a set of computational predictions that represent the invariant regions in the network that explain our expression data consistently. These results reflect important global configurations in a regulatory network that can be practically used to diagnose models and in a future to modulate their global behaviour. What we have shown for the *E.coli* network, can be applied to other networks and to different expression datasets.

Availability and Supplementary Material

Sources and a working application of our method can be accessed on-line at: http://www.irisa.fr/symbiose/bioquali/. The supplementary material is available on-line at: http://www.irisa.fr/symbiose/projects/networks.

Acknowledgments

We would like to thank Michel Le Borgne and Philippe Veber for their invaluable work implementing Pyquali. The work of the first autor was financially supported by CONICYT (Comisión Nacional de Investigación Científica y Tecnológica), Chile. This work was supported by the ANR program ANR-06-BYOS-0004.

References

1. Joyce, A., Palsson, B.: The model organism as a system: integrating 'omics' data sets. Nat. Rev. Mol. Cell. Biol. 7, 198–210 (2006)
2. Faith, J., Hayete, B., Thaden, J., et al.: Large-scale mapping and validation of Escherichia coli transcriptional regulation from a compendium of expression profiles. PLoS Biol. 5, e8 (2007)
3. Bansal, M., Belcastro, V., Ambesi-Impiombato, A., Di Bernardo, D.: How to infer gene networks from expression profiles. Mol. Syst. Biol. 3, 78 (2007)
4. Salgado, H., Gama-Castro, S., Peralta-Gil, M., et al.: RegulonDB (version 5. 0): Escherichia coli K-12 transcriptional regulatory network, operon organization, and growth conditions. Nucleic Acids Res. 34, D394–D397 (2006)
5. Ideker, T., Galitski, T., Hood, L.: A new approach to decoding life: systems biology. Annu. Rev. Genomics Hum. Genet. 2, 343–372 (2001)
6. Palsson, B.: The challenges of in silico biology. Nature Biotechnology 18, 1147–1150 (2000)
7. Kitano, H.: Systems Biology: A Brief Overview. Science 295, 1662–1664 (2002)
8. Herrgard, J., Lee, B., Portnoy, V., Palsson, B.: Integrated analysis of regulatory and metabolic networks reveals novel regulatory mechanisms in Saccharomyces cerevisiae. Genome Res. 16, 627–635 (2003)
9. Ideker, T., Thorsson, V., Ranish, J.A., et al.: Integrated genomic and proteomic analyses of a systematically perturbed metabolic network. Science 292, 929–934 (2001)
10. Rawool, S.B., Venkatesh, K.V.: Steady state approach to model gene regulatory networks–simulation of microarray experiments. Biosystems 90, 636–655 (2007)
11. Covert, M., Knight, E., Reed, J., et al.: Integrating high-throughput and computational data elucidates bacterial networks. Nature 429, 92–96 (2004)

12. Covert, M., Palsson, B.: Transcriptional regulation in constraints-based metabolic models of Escherichia coli. J. biol. chem. 277, 28058–28064 (2002)
13. Edwards, J.S., Palsson, B.O.: The Escherichia coli MG1655 in silico metabolic genotype: its definition, characteristics, and capabilities. PNAS 97, 5528–5533 (2000)
14. Radulescu, O., Lagarrigue, S., Siegel, A., et al.: Topology and static response of interaction networks in molecular biology. J. R. Soc. Interface 3, 185–196 (2006)
15. Siegel, A., Radulescu, O., Le Borgne, M., et al.: Qualitative analysis of the relation between DNA microarray data and behavioral models of regulation networks. Biosystems 84, 153–174 (2006)
16. Dormoy, J.: Controlling qualitative resolution. In: 7th National Conference on Artificial Intelligence, pp. 319–323. Morgan Kaufman Press, Saint Paul (1988)
17. Travé-Massuyès, L., Dague, P.: Modèles et raisonnements qualitatifs. Hermes Sciences, Paris (2003)
18. Veber, P., Le Borgne, M., Siegel, A., et al.: Complex Qualitative Models in Biology: A new approach. Complexus, 3–4 (2005)
19. Hiratsu, K., Shinagawa, H., Makino, K.: Mode of promoter recognition by the Escherichia coli RNA polymerase holoenzyme containing the sigma S subunit: identification of the recognition sequence of the fic promoter. Mol. Microbiol. 18 (1995)
20. Aviv, M., Giladi, H., Schreiber, G., et al.: Expression of the genes coding for the Escherichia coli integration host factor are controlled by growth phase, rpoS, ppGpp and by autoregulation. Mol. Microbiol. 14, 1021–1031 (1994)
21. Dersch, P., Schmidt, K., Bremer, E.: Synthesis of the Escherichia coli K-12 nucleoid-associated DNA-binding protein H-NS is subjected to growth-phase control and autoregulation. Mol. Microbiol. 8, 875–889 (1993)
22. Atlung, T., Sund, S., Olesen, K., Brøndsted, L.: The histone-like protein H-NS acts as a transcriptional repressor for expression of the anaerobic and growth phase activator AppY of Escherichia coli. J. Bacteriol. 178, 3418–3425 (1996)
23. Brondsted, L., Atlung, T.: Effect of growth conditions on expression of the acid phosphatase (cyx-appA) operon and the appY gene, which encodes a transcriptional activator of Escherichia coli. J. Bacteriol. 178 (1996)
24. Iuchi, S., Chepuri, V., Fu, H., et al.: Requirement for terminal cytochromes in generation of the aerobic signal for the arc regulatory system in Escherichia coli: study utilizing deletions and lac fusions of cyo and cyd. J. Bacteriol. 172, 6020–6025 (1990)
25. Dennis, G., Sherman, B., Hosack, D., et al.: VID: Database for Annotation, Visualization, and Integrated Discovery. Genome Biol. 4 (2003)
26. Allen, T., Herrgård, M., Liu, M., et al.: Genome-scale analysis of the uses of the Escherichia coli genome: model-driven analysis of heterogeneous data sets. J. Bacteriol. 185, 6392–6399 (2003)
27. Gutierrez-Rios, R., Rosenblueth, D., Loza, J., et al.: Regulatory network of Escherichia coli: consistency between literature knowledge and microarray profiles. Genome Res. 13, 2435–2443 (2003)

Spatial Clustering of Molecular Dynamics Trajectories in Protein Unfolding Simulations

Pedro Gabriel Ferreira[1], Cândida G. Silva[2,3], Paulo J. Azevedo[4], and Rui M.M. Brito[2,3]

[1] Genome Bioinformatics Laboratory, Center for Genomic Regulation, Barcelona, Spain
pedro.ferreira@crg.es
[2] Chemistry Department, Faculty of Science and Technology, University of Coimbra, Coimbra, Portugal
[3] Center for Neuroscience and Cell Biology, University of Coimbra, Coimbra, Portugal
[4] Department of Informatics, CCTC, University of Minho, Braga, Portugal

Abstract. Molecular dynamics simulations is a valuable tool to study protein unfolding *in silico*. Analyzing the relative spatial position of the residues during the simulation may indicate which residues are essential in determining the protein structure. We present a method, inspired by a popular data mining technique called Frequent Itemset Mining, that clusters sets of amino acid residues with a synchronized trajectory during the unfolding process. The proposed approach has several advantages over traditional hierarchical clustering.

1 Introduction

Protein folding is the process by which the protein acquires its three-dimensional (3D) native structure. The 3D structure of a protein, ultimately determined by its linear sequence of amino acids, is essential for protein function. Recently, several human and animal pathologies, such as cystic fibrosis, Alzheimer's and mad cow disease, among others, have been identified as protein folding or unfolding disorders. Over the years, many experimental and computational approaches have been used to study these processes. The analysis of data obtained from molecular dynamics (MD) simulations of induced-unfolding processes may be an important tool to explore and understand the protein unfolding mechanisms.

Analyzing the behavior and the relationship among the residues during the simulation may provide important insights on the unfolding process. In particular, identifying groups of residues that show a synchronized behavior during the simulation can provide important clues on which residues play a critical role on protein folding and protein structure. Such residues may be involved in the formation of a core or nucleus that is preserved within the partially folded structures. In this work, by *synchronization* or *synchrony* between two or more

F. Masulli, R. Tagliaferri, and G.M. Verkhivker (Eds.): CIBB 2008, LNBI 5488, pp. 156–166, 2009.

residues we mean that they conserve their relative distance during the simulation. Traditional clustering techniques are not flexible enough to allow additional restrictions nor that an element (residue) may appear in more than one cluster. Therefore, they are not adequate to capture the behavior of residues that appear synchronized with more than one set of other residues.

Here, we propose a method to cluster sets of amino acid residues (AARs), that have a synchronized trajectory during the unfolding process of the protein Transthyretin (TTR) [1,2,3]. Since the clustering is determined by the spatial location of the alpha-carbon (C_α) atoms of each residue, we call it *Trajectory Spatial Clustering*. The method is inspired in a popular data mining technique called Frequent Itemset Mining (FIM) [4,5]. The method is devised in three main steps. First, all the pairs of AARs that during the simulation have a small distance variation are determined. These pairs will form the seed for larger clusters. In the second step, these clusters are successively extended with other AARs. If the elements of a cluster exceed a certain distance variation threshold (provided by the user), it means that the cluster is no longer synchronized and the extension process is stopped. Clusters that are contained in other clusters are considered redundant. In the last step, after all clusters have been determined, redundant cluster can be rejected and similar clusters merged.

2   Method

We start this section by formalizing the problem and then presenting our approach by describing the proposed algorithm. The goal of this work is to design a methodology that discovers sets of residues – clusters – that have a synchronized trajectory during the simulation. The residues in a cluster follow their own trajectory but conserve among each other their relative distance in 3D space. In order to measure the distance variation between two residues, we introduce a measure called *coefficient of distance variation*, denoted as *cdv*. Two AARs are considered to form a synchronized cluster if its *cdv* does not exceed a user defined threshold value, called cdv_{max}. An expected outcome is that this value increases with the cluster extension. To deal with this, we introduce a second parameter, cdv_{inc}, representing an increment in the cdv_{max} threshold for each new extension.

Residues that are contiguous in the linear sequence are expected to have a highly conserved movement due to physical restrictions. Thus, to avoid reporting such trivial relations between residues, a third parameter called *minimum sequence distance (msd)* is introduced. Imposing such restriction means that for residues to be considered in the same cluster, their distance in the linear sequence should be greater or equal to *msd*. The problem can be stated as follows:

Given the 3D coordinates of the C_α atom of each residue at each instant of the simulation, a maximum coefficient of distances variation (cdv_{max}), a cdv_{inc} increment for each new extension and a minimum sequence distance (msd) between residues, find all the clusters of residues that show a synchronized behavior.

The overall proposed approach can be summarized in the following three-step algorithm:

Step I - Rigid Link Determination. Find all pairs of residues with distance variation below a pre-defined threshold (cdv_{max}) and a sequence distance greater than msd. These pairs will form the seeds for the clusters.

Step II - Cluster Extension. In an agglomerative way, each cluster is successively extended with new residues. A residue is considered for extension if it forms a rigid link with one of the residues in the cluster. Extension stops when the cluster exceeds a given variation threshold ($cdv_{max} + cdv_{inc}$).

Step III - Cluster Filtering. This step is optional and consists in removing redundant clusters and merging similar ones. Clusters are then ranked according to their cdv value.

2.1   Rigid Link Determination

In the first step of the method, we find all the pairs of residues that in the 3D space have a coefficient of distance variation less than the initial cdv_{max} and that in the linear sequence are apart at least msd residues. These pairs of residues are called *rigid links* and correspond to clusters of size 2. The distance between the residues A and B, at a given instant i of the simulation, is given by the Euclidean distance between their C_α atoms, expressed by Equation 1.

$$dist_i(A, B) = \sqrt{(A_i.x - B_i.x)^2 + (A_i.y - B_i.y)^2 + (A_i.z - B_i.z)^2} \qquad (1)$$

For a simulation of N time points, the cdv of two residues A and B correspond to the root mean square deviation of the Euclidean distance between A and B and is given by Equation 2.

$$cdv(A, B) = \sqrt{\frac{\sum_{i=1}^{N} dist_i(A, B)}{N}} \qquad (2)$$

By performing a pairwise comparison of all the AARs the rigid links are found.

2.2   Cluster Extension

In step two, the initial set of clusters are extended with new residues. Transitivity is the basic idea behind the extension process and it can be described as follows: if a and b form a rigid link denoted as $rl(a, b)$, and b and c another rigid link $rl(b, c)$, then $\{a, b, c\}$ will potentially form a cluster with synchronized behavior.

However, for clusters with size greater than two, a way to measure the overall cluster synchrony is needed. Several approaches can be taken. One such approach would be to measure the variation of the cluster centroid. For each time frame the cluster centroid (central point) is calculated. Then, it is verified if during the simulation period the centroid positions vary significantly. This can be measured through the root mean square deviation of the Euclidean distance between two consecutive time points. If this variation is below a pre-defined threshold, the cluster is considered synchronized. We observed that this approach is incorrect since increasing the cluster size will always result in a decrease of centroid variation. Therefore, this measure is not able to capture correctly the cluster variation. A different approach to measure the overall cluster variation was considered. The global variation of a cluster C is given by the average value of cdv between all pairs of residues in cluster C. Algorithm 1 provides the way to calculate the global variation of the cluster. With this calculation each cluster can now be extended given that its global variation remains below a certain threshold.

input : *cluster C (cluster to extend)*; *msd*
1 **for** $i = 1$ **to** $|C| - 1$ **do**
2   **for** $j = i + 1$ **to** $|C|$ **do**
        /* Compare pairwise all AARs in C   */
3          $p = C[i]$;
4          $q = C[j]$;
5          $dSum = 0$;
6   **if** $abs(p - q) \geq mindist$ **then**
7              $dist = CDV(p, q)$;
8              $dSum = dSum + dist$;
9              $pairs = pairs + 1$;
10   **end**
11   **end**
12 **end**
13 **return** $gvar = dSum/pairs$;

Algorithm 1. GVar: Calculation of global variation of a cluster C

The cluster extension step relies on a bottom-up procedure, where larger clusters result from the extension of smaller ones. Given a synchronyzed cluster $C = \{c_1, \cdots, c_n\}$, $C \cup \{x\}$ is a candidate cluster if $\exists\, rl(c_n, x)$. This strategy, which is based in one of the most efficient strategies for FIM [6], allows to find all valid candidate sets while maintaining a reduced number of redundant candidates. This approach relies on the monotonic property, also called downward closure, largely used by itemset mining algorithms [5] to discover sets of items that co-occur frequently. Frequent itemsets appear in the database a number of times greater than a given threshold value called minimum support. From this property results that any subset of a frequent itemset must also be frequent. Analogously, we have that all sub-clusters of a synchronized cluster are also synchronized.

Extending the cluster with new residues will increase its variation. At a certain point the initial variation threshold will become too restrictive to allow for new extensions. Thus, the method has to account for a certain variation tolerance to each new extended cluster. This is expressed through the parameter cdv_{inc}. For each new extension of the cluster the value of cdv_{max} is increased by cdv_{inc}. From this results that a cluster is always tested for synchrony with the threshold relative to its size. Algorithm 2 describes the extension procedure (DFSExtend):

the cluster C is extended with all rigid links (line 1) where the first AAR of the rigid link is equal to the last AAR in C (line 3). C is extended with the second AAR in rl if the global variation of the new cluster is below the allowed threshold (line 6). In this case $newC$ is considered a synchronized cluster. If $newC$ cannot be further extended (verified by an outcome of 0 in DFSExtend - line 7) then it is added to the list of synchronized clusters $LSynC$. This test saves work in step III since it already eliminates some of the redundant clusters. If the option to report all synchronized clusters is activated, then all new clusters that pass the test in line 6 are added to list $LSynC$.

```
   input  : cluster C; Lrl(list rigid links); cdv_max; cdv_inc
 1 foreach rl in Lrl do
       /* AAR in rigid link rl                                      */
 2      a ↔ b = rl;
 3      if pop(C) = a then
           /* extend C with b                                       */
 4          newC = push(C, b);
 5          newCvar = GVar(newC);
 6          if newCvar ≤ (cdv_max + cdv_inc) then
               /* extend recursively new cluster                    */
 7              ex = DFSExtend(newC, Lrl, cdv_max + cdv_inc, cdv_inc);
 8              if ex == 0 then
 9                  push(LSynC, newC);
10              end
11          else
12              return 0;
13          end
14      end
15 end
```

Algorithm 2. DFSExtend: Recursive function for cluster extension

2.3   Cluster Filtering

Performing cluster search through the combination of an exhaustive enumeration and residue at a time extension results that some of the reported clusters are found to be sub-clusters of larger clusters, which leads to the notion of maximal cluster. A cluster is said to be *maximal* if it is not contained in any other cluster. Therefore non-maximal clusters can be rejected (not reported) since they provide redundant information. Cluster extension is performed through a depth first search. Reporting only the larger clusters of a branch of the search space tree eliminates many of the redundant clusters. For instance, in the following extension path $\{a, b\} \rightarrow \{a, b, c\} \rightarrow \{a, b, c, d\}$ only $\{a, b, c, d\}$ needs to be reported. Since some of the non-maximal clusters can be found at the end of a path, an additional test to verify cluster maximality is required. This consists in verifying if a synchronized cluster is contained or contains other synchronized cluster.

Since highly synchronized redudant clusters can also be of interest, the user may choose the option of reporting all synchronized clusters. Additionally, clusters with similar composition can also be reported. For instance, $\{R10, R21, R50\}$ and $\{R10, R22, R50\}$ can be found as synchronized clusters. To make easier the interpretation of the results, a post-processing step can be applied to merge such clusters. The previous clusters can be merged as $\{R10, R(21 - 22), R50\}$.

3   Application

In this section, we make use of the proposed algorithm to assist in the study of the unfolding mechanisms of the protein Transthyretin (TTR) (Figure 1), a human plasma protein involved in amyloid diseases such as Familial Amyloid Polyneuropathy, Senil Systemic Amyloidosis and Familial Amyloid Cardiomyopathy [1].

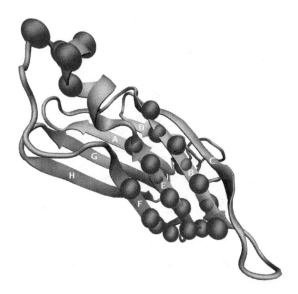

Fig. 1. Secondary structure ribbon representation of the WT-TTR monomer (PDB entry 1TTA). β-strands, α-helices, and turns and random structure are represented by arrows, coils and tubes, respectively. The eight β-strands are named A (residues 2-18), B (residues 28-36), C (residues 40-49), D (residues 54-55), E (residues 67-73), F (residues 91-82), G (residues 104-112) and H (residues 115-123). The residues identified in Table 1 are represented as beads.

Molecular dynamics protein unfolding simulations may be analyzed through the variation of chemical or geometric properties [1,2,3]. We focused on a particular dataset which consists of 127 time series, each representing the 3D coordinates of the C_α atom of each AAR along one MD unfolding simulation. Each time series is a collection of 8000 data frames, with the 3D coordinates of the C_α

Table 1. Output of the spatial clustering algorithm sorted by increasing value of average value of cdv for the cluster. The settings used for this analysis were: $cdv_{max}=8$, $cdv_{inc} = 0.6$ and $msd = 2$.

Cluster	Avg cdv	Cluster	Avg cdv	Cluster	Avg cdv
34 69 94	6.79	33 70 92	7.85	31 71 92	8.15
77 80 83	7.05	35 68 94	7.88	73 76 79	8.18
33 69 94	7.08	75 78 81	7.92	19 112 115	8.24
76 79 82	7.24	34 67 96	7.98	32 45 58	8.29
76 80 83	7.54	32 72 92	8.01	29 73 76	8.40
32 69 94	7.57	10 13 105	8.07	75 79 82	8.44
32 71 92	7.59	32 70 92	8.08	33 72 92	8.45
35 69 94	7.60	30 72 92	8.11	30 71 92	8.52
34 68 94	7.74	35 67 96	8.13	79 85 88	8.57
31 72 92	7.83	33 71 92	8.14	30 73 76 79	9.21

3D View of the Rigid Links

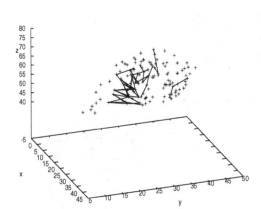

Fig. 2. Rigid Links between the c_α of the AARs of the clusters in Table 1 mapped into the native state of the protein

atom of each AAR per picosecond (ps) of simulation, for a total of 8 nanoseconds (ns) of simulation.

A prototype in C++ that implements the algorithm described in section 2 was developed. Different combinations of values of the parameters cdv_{max}, cdv_{inc} and msd were tested with computation times around 2 seconds. Here, we show and analyze the results obtained with the following parameter values: $cdv_{max} = 8$; $cdv_{inc} = 0.6$ and $msd = 2$. 148 rigid links are found and 30 synchronized clusters are reported. With $cdv_{inc} = 0$, only 15 clusters are found; with $cdv_{max} = 20$, $cdv_{inc} = 0.0$ and $msd = 2$, we obtain 3775 rigid links and 41277 clusters. Table 1 presents the 30 clusters ordered by the average cdv value. Note that these clusters were reported before step III and can be further tested for maximality and merged.

From the analysis of Table 1, three sets of AARs appear highly represented: 29 to 35, 67 to 73, 79 to 83 and 92, 94 and 96. In Figure 2, it is graphically represented the AARs preponderance on the reported synchronized clusters. Each point depicts the position of the C_α of each AAR in the native structure of the protein. Lines represent the rigid links in the cluster. From this figure we can see that only a small fraction of the AARs in the protein are covered by the clusters. When represented in the secondary structure of the protein (Figure 1), the highly represented AARs are placed in the following structures: β-strand B (AARs 28 to 36), β-strand E (AARs 67 to 73), α-helix (AARs 75 to 82) and β-strand F (AARs 91 to 97). More interestingly, we observed that the clusters tend to relate AARs in β-strands B, E and F mixed together, while the residues from the α-helix seem to relate more with each other. Another fact worth noticing is that among the residues from β-strands B, E and F two large "super-clusters" can be identified: (i) AARs 32 to 35 with AARs 69 and 72, and (ii) AAR 30 to 33 with AAR 71 or AAR 72, and AAR 92. In the experimental work on TTR by Liu and colleagues [7], it is suggested that the disruption of the structure of β-strands B, C, E and F is a cooperative process. By analyzing the trajectory of the different AARs during an unfolding simulation and their relation with each other, our results seem to point in the same direction. Thus, the method proposed here seems to prove useful in the analysis and comprehension of MD unfolding simulation data.

We selected the merged cluster $\{(32, 33, 34, 35)\ 69\ 94\}$ to visualize the synchrony of the elements in a cluster during the simulation. The 3D coordinates of the AARs are plotted separately in Figure 3. We can clearly see that the six AARs in this cluster have an almost perfect synchronization during the entire simulation.

In order to contrast the method proposed with a more traditional approach, Figure 4 presents the dendrogram obtained from the hierarchical clustering of the AARs based on the cdv. To obtain these results we first built a similarity matrix based on the cdv values for all the pairs of AARs. Then, a hierarchical clustering algorithm using the "average" agglomerative method was applied. As can be seen from Figure 4, the hierarchical clustering approach is mainly able to detect trivial relations, i.e relations between AARs that are contiguous in the linear sequence and therefore expected to have high synchrony. Nevertheless, a few surprising relations appear (for instance between residues 39 and 83) that may be worthwhile to look at in detail.

4   Discussion

In this paper, we propose a method for the analysis of data from MD simulations of the unfolding process of the protein Transthyretin (TTR). This method is intended to find groups of amino acid residues (AARs) that show a synchronized behavior during the entire simulation. Such AARs may be essential for the protein structure and will help to better understand the unfolding process. The applied strategy for the discovery of the synchronized clusters closely resembles the strategy applied by FIM algorithms, with the difference that the criterium to

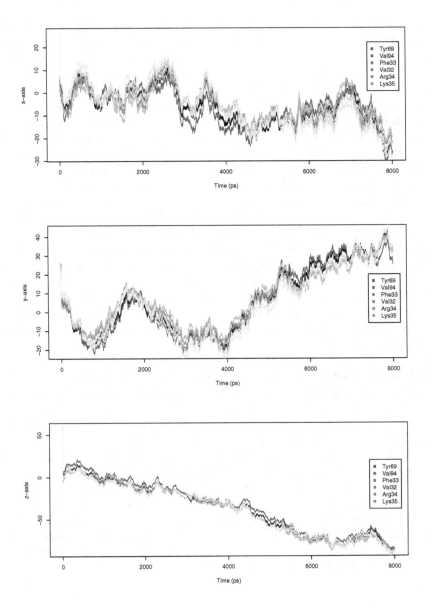

Fig. 3. 3D coordinates of the AARs in the cluster $\{(32, 33, 34, 35)\ 69\ 94\}$

report the sets of items is not their frequency but the variation of their spatial location. Although it uses a simple approach, this method has revealed to be quite efficient regarding the characteristics of the target data. When compared to the hierarchical clustering approach, our method has the following advantages: (i) by imposing a minimum distance in the linear sequence for the AARs in the cluster, it avoids reporting trivial matches between contiguous (in linear

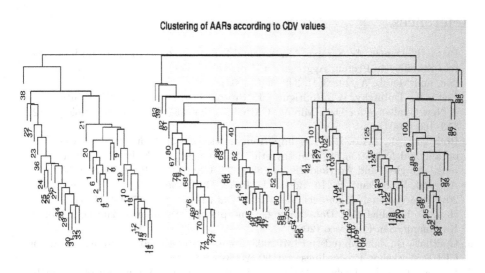

Fig. 4. Hierarchical Clustering of the AARs using the *cdv* as the similarity measure

sequence) AARs; (ii) it allows that an AAR may be present in more than one cluster and that they are located further apart in the linear sequence and in the 3D space; (iii) separating the AARs into clusters has the drawback of requiring a threshold value, but the advantage of providing a more intuitive interpretation of the results and allowing the extension of the analysis to multiple simulations.

Regarding this last aspect, it is worth mentioning that, as in many other data mining applications, the definition of the correct parameters value is an iterative and interactive process. An apriori indication for the range of values where the most relevant solutions are expected to be found can be provided, but the exact values will always depend on the characteristics of the data.

If data from several simulations is available, this approach can be easily extended to find clusters that are conserved across several simulations. This can be done by adapting a frequent itemset mining algorithm [5] to discover itemsets that are conserved in different simulations.

Acknowledgments

The authors acknowledge the support of the "Fundação para a Ciência e Tecnologia", Portugal, and the program FEDER, through grant PTDC/BIA-PRO/ 72838/2006 (to PJA and RMMB) and the Fellowships SFRH/BPD/42003/2007 (to PGF) and SFRH/BD/16888/2004 (to CGS). We thank the Center for Computational Physics, Departamento de Física, Universidade de Coimbra, for the computer resources provided for the MD simulations.

References

1. Brito, R., Damas, A., Saraiva, M.: Amyloid Formation by Transthyretin: From Protein Stability to Protein Aggregation. Current Medicinal Chemistry - Immun. Endoc. and Metab. Agents 3(4), 349–360 (2003)
2. Brito, R., Dubitzky, W., Rodrigues, J.: Protein Folding and Unfolding Situations: A New Challenge for Data Mining. OMICS: A Journal of Integrative Biology 8(2), 153–166 (2004)
3. Rodrigues, J., Brito, R.: How important is the role of compact denatured states on amyloid formation by Transthyretin? In: Grateau, G., Kyle, R.A., Skinner, M. (eds.) Amyloid and Amyloidosis, pp. 323–325. CRC Press, Boca Raton (2004)
4. Agrawal, R., Srikant, R.: In: Bocca, J., Jarke, M., Zaniolo, C. (eds.) Proceedings of 20th International Conference Very Large Data Bases, pp. 487–499 (1994)
5. Han, J., Kambler, M.: Data Mining, Concepts and Techniques, 2nd edn. Morgan Kaufmann, San Francisco (2006)
6. Goethals, B., Zaki, M. (eds.): FIMI 2003, Frequent Itemset Mining Implementations. CEUR Workshop Proceedings, vol. 90 (2003)
7. Liu, K., Cho, H., Lashuel, H., Kelly, J., Wemmer, D.: A Glimpse of a Possible Amyloidogenic Intermediate of Transthyretin. Nature Structural Biology 7(9), 754–757 (2000)

Clustering Bacteria Species Using Neural Gas: Preliminary Study

Claudio Mirabello[1], Riccardo Rizzo[2], Alfonso Urso[2], and Salvatore Gaglio[1,2]

[1] Dipartimento di Ingegneria Informatica
University of Palermo, Viale delle Scienze, 90128 Palermo, Italy
[2] ICAR-CNR
Italian National Research Council, Viale delle Scienze, 90128 Palermo, Italy
{ricrizzo,urso}@pa.icar.cnr.it

Abstract. In this work a method for clustering and visualization of bacteria taxonomy is presented. A modified version of the Batch Median Neural Gas (BNG) algorithm is proposed. The BNG algorithm is able to manage non vectorial data given as a dissimilarity matrix. We tested the modified BNG on the dissimilarity matrix obtained from sequences alignment and computing distances using bacteria genotype information regarding the16S rRNA *housekeeping* gene, which represents a stable part of bacteria genome. The dataset used for the experiments is obtained from the Ribosomal Database Project II, and it is made of 5159 sequences of 16S rRNA genes. Preliminary results of the experiments show a promising ability of the proposed algorithm to recognize clusters of the actual bacteria taxonomy.

1 Introduction

Clustering is a technique applied to understand or highlight mechanisms at work in the object representation domain, mechanisms that cause strong similarities among patterns. The aim of the clustering is to find correlations, to confirm relationships among objects.

Many data clustering approaches are based on a feature space where objects are represented, however biological datasets usually contain large objects (long nucleotides sequences or images), and a vector space representation of such objects can be difficult and typically results in a high dimensional space where the Euclidean distance is a low contrast metric. Moreover this approach requires the choice of a set of numerical features that can faithfully describes all the meaningfull characteristics of the objects This can be easy for some objects, obtaining an effective description, or difficult for images or DNA sequences. For DNA sequences it is possible to obtain a dissimilarity matrix because there are meaningful and effective methods to measure their similarity [1]. If objects are represented using a dissimilarity matrix a possible strategy for a clustering algorithm is to select one of the objects as a cluster center and to define the clusters using the distances in the dissimilarity matrix.

F. Masulli, R. Tagliaferri, and G.M. Verkhivker (Eds.): CIBB 2008, LNBI 5488, pp. 167–176, 2009.

A set of useful neural algorithms that connect clustering to visualization is Self Organizing Maps [3] that have a duofold application: mapping and visualization and allows to better understand relation among data. Self Organizing Maps generated a breed of self organizing algorithms: Neural gas [7],[6], Growing Neural Gas [5], and a set of hierarchical and growing neural algorithms [4].

Neural Gas network [7] is a clustering neural network that produces a set of cluster centers spread of the data manifold. The original neural gas algorithm is not suitable for data available only as a dissimilarity matrix, but the algorithm was modified in [8] in order to manage this kind of data, and renamed Batch Median Neural Gas (BNG).

In this paper we modified the original BNG algorithm to add weighted links between units, these links are filtered using a threshold and highlight the pattern clusters. In the next section the BNG original algorithm is reported together with the link mechanism; the third section reports some considerations about the proposed algorithm and results from the experiments on a test data set; the fourth section shows the biological dataset used and the clusters that we expected to obtain, then the results are reported and the conclusions are given.

2   Batch Median Neural Gas Algorithm

Batch neural gas (BNG) is a self organizing neural algorithm that uses neurons "overimposed" to patterns in order to define the cluster during training. One of the basic properties of the self organizing neural algorithm is that during learning the weight adjustment is propagated from one unit to the other using neighborhood relationships that are defined using links among units.

In BNG algorithm neighborhood relationships are not obtained from links among neural units but using the neuron distances. These distances are weighted by $k_{ij} = h_\lambda(k(x_j, w))$ and used to calculate the cost function in eq.1; the neural units are free to move during learning (as particles in a gas).

In [8] the batch neural gas is derived from the following cost function

$$E_{NG} \sim \sum_{i=1}^{n} \sum_{j=1}^{p} h_\lambda(k(x_j, w)) \, d(x_j, w_i) \tag{1}$$

where k_{ij} are treated as hidden variables, and the values k_{ij} are a permutation of $\{0, 1, 2, \ldots, n-1\}$ for each point x_j.

In BNG the units are superimposed to the pattern, so that the k_{ij} are obtained from the distance among patterns and the learning algorithm goal is to find the right pattern.

The batch algorithm optimizes E_{NG} in two steps:

– E_{NG} is minimized with respect to k_{ij} assuming fixed w_j
– E_{NG} is minimized with w_j assuming fixed k_{ij}.

the choice of the neural units is made following the equation below:

$$w_i = x_l \text{ where } l = \arg\min_{l'} \sum_{j=1}^{p} h_\lambda(k_{ij})d(x_j, x_{l'}) \tag{2}$$

the convergence proof is in the original paper [8].

After the training of the Batch Neural Gas using the original algorithm the neural units are linked using a simple mechanism. The basic idea is to use the hebbian rule so that: if two units are excited together by an input pattern, i.e. are the best matching unit w_i and the second best matching unit w_k, a link should be established, and it is maintained if it is renewed frequently. The implemented algorithm is the following:

1. while there are more patterns in X

   (a) take a pattern $x \in X$
   (b) find $w_i \mid \forall w_j \in W \ j \neq i \ d(x, w_i) < d(x, w_j)$
   (c) find $w_k \mid \forall w_j \in W - \{w_i\} \ d(x, w_k) < d(x, w_j)$
   (d) increase the strength of the link between k and i: $c_{ki} + +$

2. *normalization step*: for each connection c_{ij} set $c_{ij} \leftarrow \frac{c_{ij}}{|X|}$

where symbols have the same meaning above.

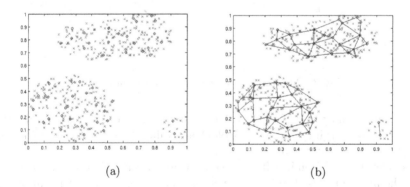

(a)   (b)

Fig. 1. 1(a) The result of the training of the BNG network using a simple artificial dataset; the neural units, small circles, are mixed with the patterns; 1(b) The result obtained using the proposed linking method

Fig. 1(a) and Fig. 1(b) show the meaning of the link method using a simple set of 2D patterns represented with a distance matrix. The BNG network is trained using the algorithm in [8] and the result is in Fig. 1(a) where neural units are superimposed to the corresponding patterns. It is difficult to highlight the clusters: each blob is populated by a set of neural units. If the units are linked together using the proposed algorithm the clusters emerge from the resulting graph. If the units are linked each other it is easy to identify the clusters using the connected parts of the graph Fig. 1(b).

3   Algorithm Details and Preliminary Results

Some experiments using the well known iris dataset can explain the behavior of the clustering algorithm and of the linking procedure. The iris dataset is a collection of four dimensional patterns labelled and pre-classified, so it is a ideal test bed for a clustering algorithm. We calculated the dissimilarity distance matrix of the patterns and made ten different run of the BNG algorithm.

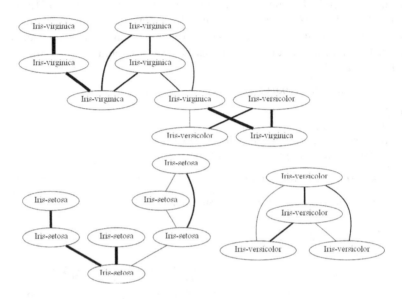

Fig. 2. One result obtained with the iris dataset

Figs 2 and 3 show two of the obtained results: it is possible to see that the clusters (setosa, virginica, and versicolor) are (almost) correctly identified in fig. 2 and are wrong in fig. 3.

This difference is dependent from the initialization of the algorithm and is only due to the procedure used to build the connection among units. It is important to highlight that the distribution of the patterns among the neural units changes only slightly among different runs, this was observed by the clusters structure and is confirmed by the low variance of the measures compactness and cluster proximity that were calculated for all the runs of the algorithm (the average value of compactness is 2.91 and variance is 0.02, the proximity average value is 0.2 and the variance is $5 * 10^{-5}$). It is the linking procedure that is critical and can give different results.

One way to solve this problem is to use a threshold and to consider links that have a strength above the threshold. In this way it is possible to filter the link set and see the true clusters.

In the rest of the paper the links are plotted using a width proportional to $\frac{C_{ij}}{|X|}$ so that the width of the link represent the refresh frequency of the single link. In order to filter the number of links plotted a threshold is used.

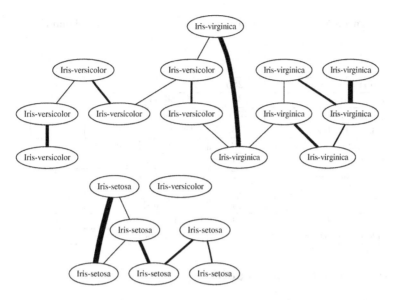

Fig. 3. A wrong result obtained with the iris dataset

4   Sequence Dataset

Microbial identification is crucial for the study of infectious diseases. In recent years a new identification approach based on bacteria genotype has been proposed and is currently under development [2]. In this new approach phylogenetic relationships of bacteria could be determined by comparing a stable part of the genetic code. The part of the genetic code commonly used for taxonomic purposes for bacteria is the 16S rRNA "housekeeping" gene. The 16S rRNA gene sequence analysis can be used to obtain a classification for rare or poorly described bacteria, to classify organisms with an unusual phenotype in a well defined taxon, to find misclassification that can lead to the discovery and description of new pathogens.

The dataset used for the experiments is obtained from the Ribosomal Database Project II [9], that provides ribosome related data services, including on line data analysis, rRNA derived phylogenetic trees, and aligned and annotated rRNA sequences. The dataset is made of 5159 sequences of 16S rRNA "housekeeping" genes. For these sequences we calculate the dissimilarity matrix using Jukes and Cantor method [1].

The bacteria taxonomy is shown in table 1; for each phylum the number of sequences belonging to the phylum is reported. From the analysis of the table it is possible to notice that there are three big clusters that contain 88% of the total number of sequences, and many small clusters, 11 phyla contain less than ten sequences.

Table 1. The number of sequences available for each phylum

Phylum	sequences	Phylum	sequences
Proteobacteria	1920	Verrucomicrobia	10
Actinobacteria	1285	Chlorobi	9
Firmicutes	1338	Chloroflexi	8
Bacteroidetes	355	Planctomycetes	7
Spirochaetes	53	Nitrospira	5
Deinococcus-Thermus	40	Thermodesulfobacteria	4
Fusobacteria	34	Fibrobacteres	2
Thermotogae	27	Lentisphaerae	2
Aquificae	18	Chrysiogenetes	1
Chlamydiae	14	Dictyoglomi	1
Cyanobacteria	14	Thermomicrobia	1
Deferribacteres	10	Gemmatimonadetes	1

5   Sequence Dataset Results

The goal of the proposed algorithm is to group the sequences into meaningful cluster that confirm or reproduce the phylum classification made by experts. We expect to find bacteria from a phylum gathered by a neuron or a set of linked neurons.

Considering the phylum classification shown in Table 1, a good result should be achieved by obtaining three clusters representing the three larger phylum, i.e. Proteobacteria, Actinobacteria and Firmicutes. That means to separate the 88% of the patterns correctly, while the other patterns should be considered as "noise". A set of experiments were made using different numbers of neural units (25, 50, 100 and 200), and many different values of other parameters, i.e. the λ value and the link threshold value.

The first observation is that the training algorithm moves the network toward a stable configuration (see Fig. 4) and in a local minima of the quantization error, as Fig. 5 shows. The quantization error is defined as:

$$E = \sum_{i=1}^{|X|} d(x_i, w_b) \tag{3}$$

where the w_b is the neural unit nearest to the pattern x_i.

After the training, the 25 neurons network, with $\lambda_{initial} = 50$, $\beta = 0.8$, and 50 epochs of training, produces as result a graph composed by three subgraphs, strongly connected inside and weakly connected each other. So, it is possible to correctly separate the three main phylum by applying an ad hoc link threshold. Fig. 6 is obtained using a threshold $\tau = 0.015$. Each circle contains the name of the superimposed sequence with the sequence number, and the list of sequences associated to the neuron organized by phylum. For example the neuron top left is superimposed to the sequence 3945 of Actinobacteria phylum and also gather 260 sequences of Actinobacteria phylum.

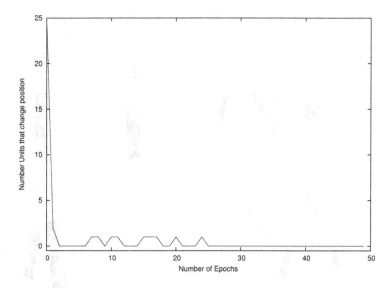

Fig. 4. The number of neurons moved during the training

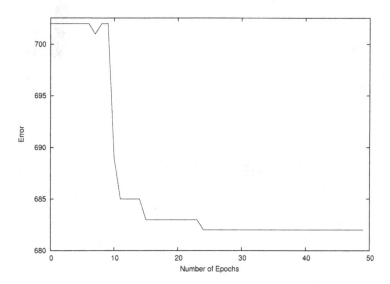

Fig. 5. The graph of the approximation error during the training

We found that raising the number of neurons causes less difference between the intra-cluster and inter-cluster link width, so that selecting the correct line threshold for clusterization of the phylum becomes more and more difficult. Also, if 50 or 100 neuron are trained, the fourth cluster (Bacteroidetes), becomes represented by a single neuron that links strongly with all the other clusters, so that obtaining a correct separation becomes nearly impossible. As result, while

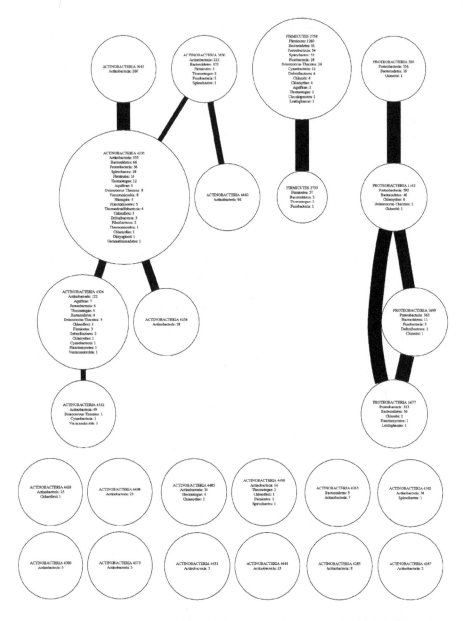

Fig. 6. The cluster structure obtained using the proposed algorithm. The three basic clusters are detected, but it is possible to see that the minor clusters are mixed to he big one and there are some isolated clusters of Actinobacteria sequences. Each circle contains the name of the superimposed sequence with the sequence number, and the list of sequences associated to the neuron organized by phylum. For example the neuron top left is superimposed to the sequence 3945 of Actinobacteria phylum and also gather 260 sequences of Actinobacteria phylum.

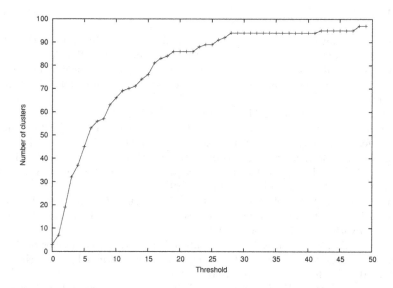

Fig. 7. The number of clusters versus the threshold value

raising the threshold, the clusters are pulverized in many different subclusters with no particular meaning. Similar problems can be found using 200 neurons: the four big clusters can be separated but a large set of neurons are associated to few sequences generating many meaningless clusters.

The link threshold plays a fundamental role in the clusterization of the neurons regardless the number of neurons or training parameter values. We tried to find a way to automatically select the threshold by finding a plateau for the number of connected components in the graph while increasing the threshold, but the number of connected parts raise constantly with the threshold as Fig. 7 shows for the 100 neurons network.

6 Conclusions

The Batch Median Neural Gas is an effective and reasonable fast clustering algorithm that can be used for large set of biological data and in this paper it was used to classify sequences from the Ribosomal Database Project II. Neurons group together the sequences in a meaningful way but it is possible to obtain a better result if a link structure is built after the training, obtaining a graph where connected components highlight the obtained clusters. A link weight is calculated and can be used to filter the graph but the threshold is critical and further investigations are needed to obtain a full automatic clustering.

References

1. Jukes, T.H., Cantor, R.R.: Evolution of protein molecules. In: Munro, H.H. (ed.) Mammalian Protein Metabolism, pp. 21–132. Academic Press, New York (1969)
2. Garrity, G.M., Julia, B.A., Lilburn, T.: The revised road map to the manual. In: Garrity, G.M. (ed.) Bergey's manual of systematic bacteriology, pp. 159–187. Springer, New York (2004)
3. Kohonen, T.: Self-Organizing Maps. Springer, New York (1997)
4. Pakkanen, J., Iivarinen, J., Oja, E.: The Evolving Tree - a novel self-organizing network for data analysis. Neural Processing Letters 20(3), 199–211 (2004)
5. Fritzke, B.: A growing neural gas network learns topologies. In: Tesauro, G., Touretzky, D.S., Leen, T.K. (eds.) Advances in Neural Information Processing Systems, vol. 7, pp. 625–632. MIT Press, Cambridge (1995)
6. Martinetz, T.M., Berkovich, S.G., Schulten, K.J.: Neural-Gas Network for Vector Quantization and its Application to Time-Series Prediction. IEEE Transactions on Neural Networks 4(4), 5558–5569 (1993)
7. Martinetz, T.M., Schulten, K.J.: A "neural-gas" network learns topologies. In: Kohonen, T., Makisara, K., Simula, O., Kangas, J. (eds.) Artificial Neural Networks, pp. 397–402. North-Holland, Amsterdam (1991)
8. Cottrell, M., Hammer, B., Hasenfuss, A., Villmann, T.: Batch and median neural gas. Neural Networks 19(6), 762–771 (2006)
9. Ribosomal Database Project II, http://rdp.cme.msu.edu/

A New Linear Initialization in SOM for Biomolecular Data

Antonino Fiannaca[1,3], Giuseppe Di Fatta[2], Alfonso Urso[1],
Riccardo Rizzo[1], and Salvatore Gaglio[1,3]

[1] ICAR-CNR, Consiglio Nazionale delle Ricerche, Palermo, Italy
[2] School of Systems Engineering, University of Reading, UK
[3] Dipartimento di Ingegneria Informatica, Universitá di Palermo, Italy

Abstract. In the past decade, the amount of data in biological field has become larger and larger; Bio-techniques for analysis of biological data have been developed and new tools have been introduced. Several computational methods are based on unsupervised neural network algorithms that are widely used for multiple purposes including clustering and visualization, i.e. the Self Organizing Maps (SOM). Unfortunately, even though this method is unsupervised, the performances in terms of quality of result and learning speed are strongly dependent from the neuron weights initialization. In this paper we present a new initialization technique based on a totally connected undirected graph, that report relations among some intersting features of data input. Result of experimental tests, where the proposed algorithm is compared to the original initialization techniques, shows that our technique assures faster learning and better performance in terms of quantization error.

Keywords: Self Organizing Maps, Unsupervised Learning, Linear Initialization.

1 Introduction

In the last years, most of biochemical databases are growing rapidly in volume, number and complexity. Most of biological data is then available for the extraction of knowledge, in order to discovery new concept relationships hidden in these databases. For this reason the Knowledge Discovery in Databases (KDD) is widely used to obtain and manipulate informations into available data The most commonly used form to represent matematically a collection of elements is a vectorial space featured by a collection of real vectors. For investigations over unsupervised field, a data mining process can easily lead to use of a machine learning algorithm based on Self Organizing Maps (SOM). This family of unsupervised neural networks is use to support the exploration and the visualization of multidimensional patterns for several data types and applications. For example in the biological field, where the amount of available data is growing day by day, Self-Organizing Maps have been used extensively. In chemistry [4,5] and biology applications the SOM has been used successfully with analysis and

F. Masulli, R. Tagliaferri, and G.M. Verkhivker (Eds.): CIBB 2008, LNBI 5488, pp. 177–187, 2009.
© Springer-Verlag Berlin Heidelberg 2009

classification purposes and as a tool for clustering and visualization([6]), in order to obtain an *"executive summary"* of a massive gene expression data set.

Unfortunately, it is also well known that SOM results are heavily influenced by the initialization of the neuron weight vectors, because a proper initialization can reduce both the time required for the training phase and the network performances. Moreover, a good initialization technique could avoid the local minima during the organization of neurons and, specially, could reveal emergent elements characteristics in dataset. For the above reasons, it is clear that an efficient initialization algorithm plays a crucial role in computation of large datasets. In this work, a new technique based on the fundamental idea of linear initialization will be introduced.

The initialization techniques can be grouped in two main classes: *random initialization techniques* in which the neural weights are not linked to the features of the input pattern set, and *dataset-based techniques* in which the analysis of the training data exploits some features of the input data.

In the first class, random initialization, the neural weights are chosen randomly inside an hypercube that contains the input dataset[1]; In the second class there are two sub-casses: sample and linear initialization; in the former[2], weight values are a random selection of input patterns, while in the latter[1,3] weight vectors are defined along the subspace spanned by the principal eigenvectors of the input dataset.

The paper has the following structure: the next section reports a brief overview of the SOM learning algorithm used in the paper; the section 3 shows in detail the proposed initialization algorithm; the section 4 reports the evaluation criteria and the experimental results. Finally some conclusions are reported in section 5.

2   SOM Learning Algorithms

As seen in section 1, Self-Organizing Maps have been used for various of problems, according to specific purpose. In the literature are reported several optimization algorithms consistent with users needs and requirements, and for most of data type. Most of the differences among implementations occur in learning process. Above all, there are two classes of learning algorithm: the on-line and the batch learning; advantages and drawbacks of these implementations have been discussed in[7].

Notice that the algorithm introduced in this paper is indipendend of the selected learning process; for this reason, it can be applied to most of the SOM implementations. However, during experimental tests, the incremental learning algorithm [8] trained for epochs is used; this version of the learning algorithm has been reported in [9].

3   The Initialization Algorithm for SOM

As mentioned in section 1, there are three different techniques to initialize a map. In this paper, a new technique based on the fundamental idea of linear initialization will be introduced;

Notice that this initialization procedure arises from the analogy between SOM and principal curve analysis algorithm highlighted by [12th-cercare], where a principal curve is just a non-linear generalization of a principal component (from Principal Component Analysis). Here we uses singular eigenvalues obtained by *Singular Value Decomposition (SVD)* [10], rather than the most used Principal Components Analysis (PCA); this replacement is allowed by dualism between PCA and SVD [11].

Moreover, the use of SVD offers some advantages combined with SOM [12], and results faster than PCA with large or sparse datasets [13].

The SVD is a factorial analysis technique able to reduce the multidimensional vector space representing the input patterns. In the proposed analysis process, the multidimensional input pattern set can be represented by a reduced number of *latent factors* that are able to describe the information contained in the original set, without loss of information. Considering m elements and n features, the application of the SVD of matrix A defines a mapping between the m elements and n features into a vector space S, in which it is possible to represent each element and each feature. The SVD is defined by the product

$$A = U \Sigma V^T \tag{1}$$

where $U(m,n)$ and $V(n,n)$ are orthonormal matrices and, $\Sigma(n,n)$ is a diagonal matrix whose elements, $(\sigma_1, \sigma_2, ..., \sigma_n)$ with $\sigma_1 \geq \sigma_2 \geq ... \geq \sigma_n \geq 0$, are called the *singular values* of A. The columns of U are called *left singular vectors* of A and the columns of V are called *right singular vectors* of A.

The vector space reduction is performed using the *Truncated Singular Value Decomposition* (TSVD) [14]. Given a positive integer k with $k < n$, the TSVD is defined by

$$\widetilde{A} = \widetilde{U} \widetilde{\Sigma} \widetilde{V}^T \tag{2}$$

where the matrix $\widetilde{U}$ is obtained from the matrix U by suppressing the last $n - k$ columns, the matrix $\widetilde{V}$ is obtained from the matrix V by suppressing the last $n - r$ columns, and the matrix $\widetilde{\Sigma}$ is obtained from the matrix Σ by suppressing the last $n - k$ rows and the last $n - k$ columns. Notice that, in our study, the SVD is not used to reduce the vectorial space, but to project both patterns and features in a new space, where both of them have the same dimensions; in this way, we can project features and patterns over the same SOM.

The main idea of the proposed algorithm is the projection, over the Kohonen map, of some highly representative features of the input dataset, preserving theirs mutual relationships. The data model used for this aim is a totally connected undirect graph $G(V, E)$ (here called "*InitGraph*"). The vertices V of *InitGraph* represent the distinctive features and the edges E are the mutual distance among all features; more in details, each vertex v represents a highly representative feature and the length of each edge e is proportional to the mutual distance between a pair of selected features. Regarding the lenght of the edges, the proposed algorithm respects the geometry of the graph using "elastic edges" that would balance all pairwise relationships using a mean square error.

The *InitGraph* plays a fundamental role during initialization process, in fact it performs a bridge between multidimensional space of dataset and two-dimensional space of Kohonen map, preserving most of the mutual relations among selected patterns.

The initialization is made in the following three steps:

1. *Selection* of the k highly representative features of the input dataset; they are selected using the largest absolute values of features along some projection axes;
2. *Arrangement* of selected features into a two-dimensional space using the graph model introduced with *InitGraph*;
3. *Fitting* the *InitGraph* projection into the two-dimensional lattice Kohonen layer.

The first step is the retrieval of highly representative features, whose identification is done by SVD projection. Following projection, patterns and features will be spread along some new principal axis. In the new vector space, the elements of dataset will be distributed along the direction of features that better represent them. The aim is to obtain a sub-set of features holding information about most of the patterns. These features lie in principal directions and are defined in the matrix of left singular vectors (given by SVD), where a mapping from the feature space to a newly generated abstract vector space is defined. The first k singular vectors describe a k-dimensional subspace of the abstract SVD vector space. The negative components represent features that are anti-correlated with the positive components. In other words, for patterns in which positively weighted features tend to appear, the negatively weighted features do not tend to appear. As reported in [15], it is possible to extract highly representative features for input patterns, from the analysis of left singular value components. The pseudo code for *step*1 is reported in Table 1.

The second step is devoted to build a model able to translate relation among selected patterns into a two dimensional structure, using *InitGraph*; this model must observe the geographic distance among areas where selected patterns should be located. The pseudo code of *step*2 is shown in Table 2.

The third step is the adaptation of the initialization graph on the Kohonen map. Obviously the graph will be properly scaled before being overlapped on the map. After that, highly representative features will be located over some neurons, which set theirs weights equal to the corresponding representative features. The remaining neurons will be initialized depending on the previous ones. A pseudo learning epoch will assign the weights of remaining neurons, where the patterns used for training are the selected features and the gaussian radius is equal to the width of the grid. Moreover the position of vertices over the SOM lattice is done in order to arrange patterns in a smart fashion; in fact all vertices are located far from the borders of neurons' grid, avoiding this way the border effect during first steps of learning process. The pseudo code of *step*3 is shown in Table 3.

In order to better understand this algorithm, a very simple example is reported. We suppose to initialize the SOM with an artificial dataset of 200 elements represented with 14 features: it describes 4 clusters, each of them having

50 elements. Using the SVD, we project the dataset into a new space fourteen-dimensional. As shown in figure 1, at step 1 we take in account the first two singular values ($k= 2$) and obtain a two-dimensional space where elements and features are projected. In this space the elements of dataset are best represented by only four features ($p = 4$), therefore these four features carry out information about most of the elements.

As shown in figure 2(a), at step 2 the algorithm builds a totally connected graph (*initGraph*) with the selected features as vertices. Distances among vertices respect the mutual distances among selected features into two dimensional spaces (figure 1).

Finally, at step 3, the initialization graph is projected over a 30×30 Kohonen map and four neurons are overlapped by four vertices/features. In this way, the weight of each neuron overlapped is equal to the vectors of corresponding feature. The other neurons are initialized with a pseudo learning epoch. Figure 2(b) shows a U-Matrix representation of the SOM, when the initialization phase is completed; here a pseudo learning epoch has been executed with only four patterns (most significant features) and with a gaussian radius equals to the width of the grid.

Table 1. Pseudo code of initialization algorithm. Step 1.

1. Decompose dataset (elements Vs. features) with SVD;
2. Set $k =$ number of principal components to take in account;
3. Let $p =$ number of rows and columns that contain the k biggest values ($p \leq 2 \times k$);
4. Select first k left singular vectors (k principal axes);
5. For each column of k left singular vectors matrix,
    (a) find both biggest and smallest value, and mark the two related features; (All marked features are the p most significant features for input dataset)
6. Generate the pairwise matrix S ($p \times p$) of all marked features, using cosine rule distance between each pair of weights vectors;

4   Experimental Results

In this section the comparison among the proposed, the sample and the random initialization method is evaluated. In order to reach this goal, three SOMs are implemented with the same learning algorithm, but with different initialization algorithms, over two real world datasets. The result of training process will be evaluated using the evolution of the Quantization Error (QE) in order to compare the effectiveness of the different initialization methods in terms of map resolution. The evolution of QE during training process was monitored over 50 experiments for each initial configuration and means of achieved values will be analyzed. The significance level of analyzed means are evaluated by the t-Test, that checks if the means of compared groups are statistically different from each others.

Table 2. Pseudo code of initialization algorithm. Step 2.

1. Set tol = tolerance adopted for elastic edges;
2. Generate a totally connected undirect graph $G(V, E)$ where vertices V are the first three marked elements of S;
3. Set lengths of edges $E(i, j)$ equal to value (distance) between features i, j in pairwise dissimilarity matrix (with respect to geometry of graph);
4. Calculate the *Mean Square Error* among all edges $E(i, j)$ and theirs theoretical distances (the pairwise dissimilarity between patterns i, j in pairwise matrix S) according to $MSE(\|E(i, j) - S(i, j)\|)$;
5. While the tolerance adopted for elastic edges is not satisfied ($MSE \geq tol$),
   (a) Add a random little value ϵ_1 to each edge $E(i, j)$, in order to get each edge closer to its theoretical measure (according to pairwise matrix);
   (b) Decrease parameter tol with a random little value ϵ_2 ($tol = tol - \epsilon_2$);
6. For $c = 4$ to p (for each remaining $p - 3$ rows or columns of the matrix S),
   (a) Add a new vertex v, corresponding to element c of S, and $c(c - 1)/2$ new edges;
   (b) Calculate measures of new edges using S matrix (with respect to geometry of graph);
   (c) Calculate the *Mean Square Error* among all edges $E(i, j)$ and theirs theoretical distances, as in step 4;
   (d) While the tolerance adopted for elastic edges is not satisfied ($MSE \geq tol$),
       i. Add a random little value ϵ_1 to each edge $E(i, j)$, in order to get each edge closer to its theoretical measure (according to pairwise matrix);
       ii. Decrease parameter tol with a random little value ϵ_2 ($tol = tol - \epsilon_2$);
7. Resulting graph is the "initialization graph".

Table 3. Pseudo code of initialization algorithm. Step 3.

1. Let m = height and n = width of Kohonen map, thus number of neurons is $m \times n$;
2. Build a bounding box R for the *initGraph*;
3. Scale the box R in order to obtain a rectangle $m \times n$;
4. For $s = 1$ to p (for each vertex of *initGraph*),
   (a) Find the neuron overlapped with s-vertex and assign to its weight vector the s-feature vector;
5. In order to initialize the other ($m \times n - p$) neurons, execute a SOM learning epoch where p patterns are the most significant features, using a neighbourhood function with a wide gaussian function.

4.1   Evaluation of the Proposed Algorithm

To allow comparison among SOM initialization algorithms, a stop condition for learning process, depending on maximum number of learning epochs, has been adopted. To choose this number, several tries have been run over dataset. The number selected, 20 epochs, was sufficient to have a stable configuration of neuron weights for all SOMs and for both datasets. All the maps have a 50×50 square lattice. The training phase for each epoch is done with a learning rate function and a neighbourhood radius function decreasing exponentially. Proposed initialization technique uses a neighbourhood radius equal to 50 units during the third step (pseudo learning epoch), whereas during the learning process of the SOM the neighbourhood radius decreases from 32 up to 2 units; these parameter values are those that give the best results for several datasets widely used in literature.

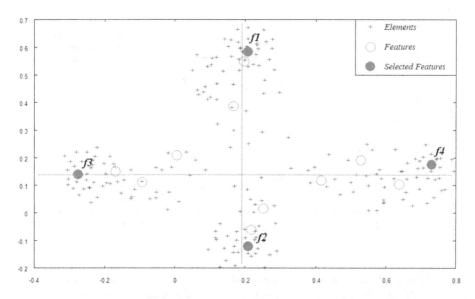

Fig. 1. Selection of the most representative features in the space generated by the first two singular values (*Step 1*)

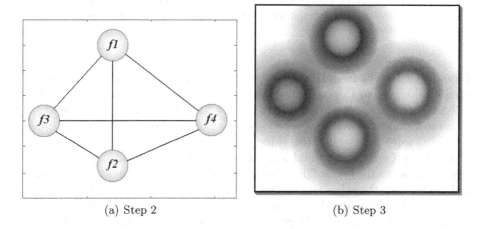

| (a) Step 2 | (b) Step 3 |

Fig. 2. (a) Arrangement of selected patterns into a 2D space using *initGraph* (*Step 2*). (b) Fitting the *initGraph* projection into the Kohonen Map (*Step 3*).

4.2   Validation of the Proposed Initialization Process

The validation of the algorithm has been carried out using two real world dataset. The first one has been used for nuclear feature extraction for breast tumor diagnosis, and it is referred to the "Wisconsin Diagnostic Breast Cancer" [16], here called *wdbc*, made of 569 instances and 30 real-valued input features. The second dataset, is obtained by the publicly available DTP AIDS Antiviral Screen dataset [17] of the National Cancer Institute. The screen measures the protection

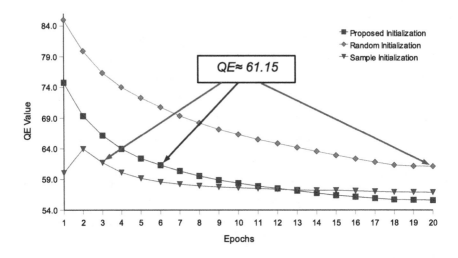

Fig. 3. Quantization error versus number of epochs for breast-cancer dataset. The random initialization algorithm reaches the stop condition at the epoch $p = 20$ with a $QE \approx 61.15$. Almost the same QE value was reached by proposed initialization algorithm at epoch $p = 6$ and by sample initialization algorithm at epoch $p = 3$. Even though the sample initialization seems better than proposed one, the former falls down in a local minimum configuration, in fact at epoch $p = 20$ reaches a greater value than proposed algorithm.

of human CEM cells from HIV-1 infection. In particular, we used the 325 compounds belonging to the confirmed active class. The used dataset, here called *NCI-325*, are represented according to [12] by 20 real-valued input features.

Figure 3 shows the average evolution of quantization error during the training process of *wdbc* dataset for the three initialization algorithms. In this chart, the curve representing the SOM learning process initialized with the proposed algorithm reaches the lowest QE value among the maps: this means our technique work better then the other ones in terms of resolution of the map. Moreover the random initialization algorithm reaches the stop condition at the epoch $p = 20$ with a $QE \approx 61.15$, whereas the proposed one reaches almost the same QE value at epoch $p = 6$. Notice that the evolution of SOM learning process, initialized with the sample method, starts with the lower value, with respect to the other maps, but it ends with a greater QE value: this means that the SOM learning process with the sample initialization maybe is fallen out in a local minimum for the network. Since the curves in the figure seem very close and have almost the same shape, and it should be possible that they come from the same distribution, the t-Test has been calculated. The test rejects the null hypothesis with a level of significance = 0.17%.

Figure 4 shows the average evolution of quantization error during the training process of *NCI-325* dataset for proposed, sample and random initialization algorithms. Once again the proposed algorithm works better than the other ones.

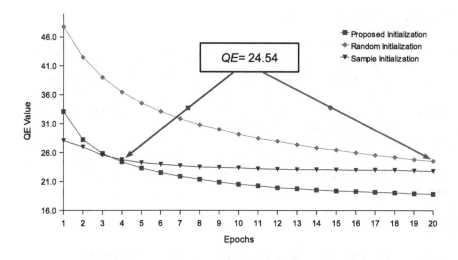

Fig. 4. Quantization error versus number of epochs for NCI-325 dataset. The Random initialization algorithm reaches the stop condition at the epoch $p = 20$ with a $QE \approx 24.54$. Almost the same QE value was reached by both proposed and sample initialization algorithms at epoch $p = 4$.

Table 4. Average execution time measured in sec. of SOM learning process for both datasets and for all initialization algorithms.

Average Execution Time of training processes			
	Proposed	**Sample**	**Random**
NCI-325 Initialization	0.020	–	–
NCI-325 Learning process	13.630	13.828	13.830
WDBC Initialization	0.056	–	–
WDBC Learning process	151.548	152.093	151.836

Notice that the evolution of SOM using sample initialization exhibits nearly the same behaviour shown in the previous chart.

Table 4 reports average execution times for SOM initialized with the three different techniques. All quoted times are derived from tests on a machine having a 3.00GHz Pentium IV processor, 1014Mb of RAM, Windows Vista Business 32bit operating system. Our technique spends 0.020 sec and 0.056 sec. to execute respectively the NCI-325 and the WDBC initialization process. We suppose the cost of the other initialization techniques equal to 0 sec. The cost percentages, for NCI-325 and WDBC training (initialization + learning), are respectively about 0.146% and 0.037%. This is a very small value with respect to the advantages shown in previous figures.

5   Conclusions

In this paper a new initialization algorithm for Self Organizing Maps with vectorial data input is proposed. Unlike previous initialization methods, the proposed one introduces a totally connected undirect graph (here called *initGraph*) that connects some key patterns; the edges of *initGraph* take into account the distance among all key patterns. Finally, a projection of this graph over the neuron grid reports the distance informations into the map.

This method has been compared with both sample and random initialization techniques. Results of experimental tests, carried out on real biological datasets, demonstrate the good performances obtained by our technique in terms of resolution of the map and execution time.

References

1. Kohonen, T.: Self-Organizing Maps, 3rd edn. Springer, Berlin (2001)
2. Varsta, M.: Self organizing maps in sequence processing. Dissertation, Department of Electrical and Communications Engineering, Helsinki University of Technology (2002)
3. Vesanto, J., Alhoniemi, E.: Clustering of the Self-Organizing Map. J. IEEE-NN 11(3), 586–600 (2000)
4. Zupan, J., Gasteiger, J.: Neural Networks for Chemists: An Introduction, pp. 121–122. VCH, Weinheim (1993)
5. Manallack, D.T., Livingstone, D.J.: Neural networks in drug discovery: have they lived up to their promise? European Journal of Medicinal Chemistry 34, 95–208 (1999)
6. Nikkilä, J., Törönen, P., Kaski, S., Venna, J., Castrén, E., Wong, G.: Analysis and visualization of gene expression data using Self-Organizing Maps Neural Networks. Special issue on New Developments on Self-Organizing Maps 15(8-9), 953–966 (2002)
7. Fort, J., Letrémy, P., Cottrell, M.: Advantages and drawbacks of the Batch Kohonen algorithm. In: ESANN 2002, pp. 223–230 (2002)
8. Van Hulle, M.: Faithful Representations and Topographic Maps: From Distortion-to Information-Based Self-Organization. John Wiley, New York (2000)
9. Fiannaca, A., Di Fatta, G., Gaglio, S., Rizzo, R., Urso, A.: Improved SOM Learning using Simulated Annealing. In: de Sá, J.M., Alexandre, L.A., Duch, W., Mandic, D.P. (eds.) ICANN 2007. LNCS, vol. 4668, pp. 279–288. Springer, Heidelberg (2007)
10. Golub, G.H., Van Loan, C.F.: Matrix Computations, 3rd edn. Johns Hopkins University Press, Baltimore (1996)
11. Wall, M.E., Rechtsteiner, A., Rocha, L.M.: Singular value decomposition and principal component analysis. In: Berrar, D.P., Dubitzky, W., Granzow, M. (eds.) A Practical Approach to Microarray Data Analysis, pp. 91–109. Kluwer, Norwell (2003)
12. Di Fatta, G., Fiannaca, A., Rizzo, R., Urso, A., Berthold, M., Gaglio, S.: Context-Aware Visual Exploration of Molecular Datab. In: ICDM Workshops 2006. Sixth IEEE International Conference on Data Mining, Hong Kong, pp. 136–141 (2006)
13. Papadimitriou, C.H., Raghavan, P., Tamaki, H., Vempala, S.: Latent semantic indexing: A probabilistic analysis. In: Proc. 17th ACM Symp. on the Principles of Database Systems, pp. 159–168 (1998)

14. Cullum, J.K., Willoughby, R.A.: Real rectangular matrices. In: Lancozos algorithms for large symmetric eigenvalue computations, Theory, vol. 1. Birkhäuser, Boston (1985)
15. Holt, F.B.: Subspace Representations of Unstructured Text. In: IEEE ICDM Workshop on Text Mining (TextDM 2001), San Jose, California, USA (2001)
16. Mangasarian, O.L., Street, W.N., Wolberg, W.H.: Breast cancer diagnosis and prognosis via linear programming. Operations Research 43(4), 570–577 (1995)
17. National Cancer Institute, Self-organized map (som) of compounds tested in the nci anti-hiv screen, http://cactus.nci.nih.gov/services/somqsar/

3D Volume Reconstruction and Biometric Analysis of Fetal Brain from MR Images

Paola Campadelli[1], Elena Casiraghi[1], Gabriele Lombardi[1], and Graziano Serrao[2]

[1] Università degli Studi di Milano,
Dipartimento di Scienze dell'Informazione,
Via Comelico 39-41, 20135 Milano, Italy
campadelli@dsi.unimi.it, casiraghi@dsi.unimi.it,
lombardi@dsi.unimi.it
[2] Università degli Studi di Milano,
Dipartimento di Morfologia Umana,
Via Mangiagalli 31, 20133 Milano, Italy
graziano.serrao@unimi.it

Abstract. Magnetic resonance imaging (MRI) is becoming increasingly popular as a second-level technique, performed after ultrasonography (US) scanning, for detecting morphologic brain abnormalities. For this reason, several medical researchers in the past few years have investigated the field of fetal brain diagnosis from MR images, both to create models of the normal fetal brain development and to define diagnostic rules, based on biometric analysis; all these studies require the segmentation of cerebral structures from MRI slices, where their sections are clearly visible. A problem of this approach is due to the fact that fetuses often move during the sequence acquisition, so that it is difficult to obtain a slice where the structures of interest are properly represented. Moreover, in the clinical routine segmentation is performed manually, introducing a high inter and intra-observer variability that greatly decreases the accuracy and significance of the result. To solve these problems in this paper we propose an algorithm that builds a 3D representation of the fetal brain; from this representation the desired section of the cerebral structures can be extracted. Next, we describe our preliminary studies to automatically segment ventricles and internal liquors (from slices where they are entirely visible), and to extract biometric measures describing their shape. In spite of the poor resolution of fetal brain MR images, encouraging preliminary results have been obtained.

Keywords: Magnetic Resonance Imaging, image de-noising, 3D fetal brain reconstruction, image segmentation, biometric analisys.

1   Introduction

Over the last 30 years US scanning techniques have become the standard instrument to evaluate and follow fetal development. These images are analyzed to extract biometric measures that are helpful for fetal gestational age estimation [5], for prenatal diagnosis [12], and for diagnosis of possible brain malformations [6]. But a limitation of US is its inability to precisely represent the cerebral structures, making it difficult to ascertain

F. Masulli, R. Tagliaferri, and G.M. Verkhivker (Eds.): CIBB 2008, LNBI 5488, pp. 188–197, 2009.

malformations of the brain development. Indeed, although this technique gives a full view of the fetus, US evaluation of the fetal brain is limited by a series of factors such as limited resolution, ossification (which obscures visualization of posterior fossa structures), nonspecific appearance of some abnormalities, and subtle abnormalities which cannot always be visualized via ultrasound.

To overcome these limitations, presently magnetic resonance imaging is becoming increasingly popular as a second-level technique for detecting morphologic abnormalities, thanks to its more precise representation of the structural anatomy of the brain; as a rule, fetal US is performed as the first routine evaluation of the fetus, and if the brain looks abnormal on US, a more detailed analysis is expected from MRI, to precisely characterize each type of malformation of the fetal brain. Indeed, as reported in [11], from mid-pregnancy to term, there are no technical limitations to this imaging approach, and the diagnostic rules that apply to the newborn child could be helpful in detecting fetal abnormalities too. For this reason, in the past few years increasing interest has been devoted to MR image analysis, and several medical researchers have worked to create magnetic resonance templates describing the normal fetal brain development [3,9,15], while some others have concentrated on the principles of brain malformations to derive the salient (i.e. highly discriminative) biometric features, noticeable from MR images of fetuses after mid-gestation [11].

In this work, we are interested in biometric analysis of ventricles and liquors in brain images of fetuses at the 19th-33th week. This study could be relevant to provide quantitative indexes describing the shape of these cerebral structures, to estimate the extent to which they are allowed to further grow and, in general, to gain better knowledge about the radial process of cellular migration, generally starting during the 19th gestational week.

In the clinical setting, quantitative data for fetal biometry are obtained after segmentation of the cerebral structures (e.g. ventricles and liquors) from slices where their sections are entirely visible (see the top row of figure 1).

The problem of this approach is that fetuses often move during the sequence acquisition, so that it is difficult to obtain a slice where the ventricles are properly represented. Furthermore, segmentation is performed manually by experts, and quantitative biometric measures are obtained by measuring the length of segments manually traced on the MR slice. The weakness of this routine is that no standard rule is established to define which segments should be traced and which measurements should be employed. Conversely, each diagnostician exploits his 'personal' experience (see the top row of figure 1), thus introducing much inter-observer and intra-observer variability both in shape description and in diagnosis. It is clear that the high variability involved when employing such a biometric analysis technique necessarily reduces the accuracy and significance of the result.

To address these problems in this paper we at first propose an algorithm that processes the image sequence to build a 3D representation of the fetal brain; from the reconstructed volume the desired section of the cerebral structures can be extracted. Next, we describe our preliminary studies to automatically segment ventricles and liquors from fetal brain MR slices, and to compute biometric measures describing their shape.

2   Materials and Methods

The MR volumes used to develop and test our method have been acquired at the Buzzi Hospital in Milan; they are thirty volumes, performed at gestational ages from 19 to 33 weeks. For each patient, a sequence of about ten DICOM images (12 bits/pixel) is acquired, although some patient's brain is not entirely acquired, and the first slices belonging to different fetuses are not located at the same level. The sequences are acquired as balance (that is T2 weighted) sequences with the following parameters: repetition time (RT) and echo time (TE) set at their shortest values, number of exercitations=1, 28×28 cm field of view, 256×256 acquisition matrix; slice thickness of 3 mm, voxel size of 1.46×1.46 mm, reconstructed voxel size of 1.09 mm.

In section 3 we recall some related work; in section 4 we propose our method, to reconstruct the 3D gray level volume of the fetal brain; in section 5 we describe our preliminary studies both to segment liquors and ventricles, and to extract landmark points and segments that could be used to describe the segmented shapes; section 6 reports preliminary results and future work.

3   Related Work

To the best of our knowledge, automatic analysis of fetal brain MR images has been reported in [4], where the purpose is the biometric analysis of the posterior fossa from an image region strictly containing it, manually signed by an expert. In the work presented by Schierlitz *et al.* [14] the authors build the 3D volumes of ten fetal brains, by manually signing pixels on the boundary of structures of interest, and then applying a classic 3D volume reconstruction procedure. The aim of their work is to demonstrate that 3D modeling of the brain is useful both to provide a more precise diagnosis and to plan treatment.

Other works have been presented on MR images of human brain, whose main purposes are the segmentation of cerebral structures [8,13], lesion detection [10], brain functional localization [7], and cortical surface shape analysis [17]. These approaches employ different techniques, such as deformable models, image registration techniques relying on atlases, edge detection, histogram analysis, stochastic relaxation methods, wavelet analysis.

In general, the major difficulties encountered when working with MRI are related to intensity inhomogeneities, due to technical limits in the image acquisition process that cause tissues of identical composition to be imaged with different intensities. For this reason, several works about intensity inhomogeneity correction have been proposed in the past, and reviewed in [16]; unfortunately, due to the small image area occupied by the fetal brain (the fetal braincase occupies no more than 30×30 pixels in fetuses at the 19-33th week), none of the proposed correction technique is applicable, for they would necessarily affect small details that are significant for brain characterization. The same holds for segmentation, registration, and detection algorithms.

Anyhow, fetal MR imaging remains poor, in comparison with postnatal imaging, for physical reasons; the fetal brain is small and the fetus is buried within the womb and the amniotic cavity (surrounded by the abdominal air-containing viscera), the mother breath causes the movement of the diaphragm and abdominal wall, and the fetus himself is moving.

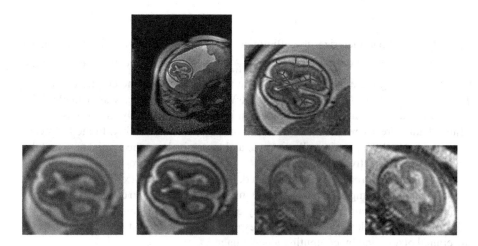

Fig. 1. Top row: Manually chosen slice where the cerebral structures are well visible, and the segments manually located to perform biometric analysis (for a better visualization, the image has been preprocessed as described in section 4). Bottom row: Original slices of two patients and the slices after inhomogeneities correction and contrast enhancement.

4   3D Volume Reconstruction of Fetal Brain

In this section we describe the algorithm employed to process the image sequence, $I(\mathbf{x}, z)$, where $\mathbf{x} = (x, y)$, to build a 3D representation of the fetal brain. To this aim, alignment of consecutive slices is necessary since they are often shifted and rotated with respect to each other, due to image artifacts and to fetal movements. After the alignment, a classical gray level interpolation method is employed to reconstruct the 3D gray level volume; this volume can then be sectioned along the plane containing the desired slice to be segmented to extract biometric measures, as described in section 5.

The alignment algorithm begins by a user-selected axial slice $I(\mathbf{x}, Beg)$ where the fetal braincase is at first segmented. The shape of the fetal braincase in the 2D slice can be approximated by an ellipse; therefore, a solution to the segmentation problem could be obtained by manually placing five points, sufficient to derive an accurate analytical description of an ellipse. Nevertheless, in order to reduce the manual interaction, so as to decrease the inter and intra-observer variation, we use a model fitting technique, whose initialization requires the user to locate only the two extrema of the ellipse's major axis. More precisely, we describe the ellipse as a geometrical model with the following parameter set $\theta = \{\mathbf{A}, \mathbf{B}, \phi, \mathbf{C}\}$, where A and B are, respectively, the major and minor axis sizes, ϕ is the rotation angle, and $C = (x_c, y_c)$ is the ellipse center. The initialization of all the parameters but the minor axis size, B, is computed directly from the extrema of the major axis, whilst B is initialized to $B = 0.88A$; the $\frac{B}{A}$ ratio has been estimated on 100 fetal brain slices.

Given this initialization of the parameter set, θ, an optimization process is employed to minimize the following energy function:

$$E(\theta) = \int_0^1 -(\nabla_I(Ell(t;\theta)) \cdot \hat{\mathbf{r}}) \, f(\nabla_I(Ell(t;\theta)) \cdot \hat{\mathbf{r}}) dt \qquad (1)$$

where $t \in [0, 1]$ parameterizes the elliptic curve $Ell(.)$, defined by θ, ∇_I is the image gradient, $\cdot$ denotes the dot product, $\hat{\mathbf{r}} = \frac{Ell(t;\theta)-C}{\|Ell(t;\theta)-C\|}$ is the unit vector oriented as the radius connecting the ellipse center C to the point $Ell(t;\theta)$ on the elliptic curve, and $f(x) = 1$ if $x > 0$, $f(x) = 0$ otherwise. We used the simplex search method for minimization; note that, in the optimization process the manually given axis is also adjusted, thus greatly reducing the inter and intra-observer variability. In the following we will refer to the segmented elliptic area, and to its pixels, as the cranial area and the cranial pixels, respectively.

Once the cranial area has been segmented, the image $I(\mathbf{x}) = I(\mathbf{x}, Beg)$ is processed to simultaneously compensate for intensity inhomogeneities and to enhance the contrast in this area. To this aim we initially apply a histogram expansion transformation, by selecting the thresholds, T_L and T_H, corresponding to the darkest and lightest 10% of the cranial pixels, and by computing a new image

$$I'(\mathbf{x}) = \frac{1}{T_H - T_L}(I(\mathbf{x}) - T_L)$$

Next, to recover from inhomogeneities, we perform unsharp masking by applying the negative of the Laplacian filter (computed with $\sigma = 0.3$) to $ln(I'(\mathbf{x}))$. The resulting image, $I''(\mathbf{x})$, is then processed by exponential transformation to enhance the contrast in its brighter regions, and it is normalized to have zero mean and unitary standard deviation in the cranial area. More precisely, we compute a new image

$$\tilde{I}(\mathbf{x}) = \frac{exp(I''(\mathbf{x})) - \nu}{\sigma}$$

whereby ν and σ are the mean and the standard deviation of the cranial pixel values in $exp(I''(\mathbf{x}))$. Finally, $\tilde{I}(\mathbf{x})$ is scaled onto the integer range $[0,..,255]$ and it is used as the new $I(\mathbf{x}, Beg)$. In the bottom row of figure 1 we show the effect of applying this fast, simple, and effective procedure.

The information provided by the segmented braincase ellipse in $I(\mathbf{x}, Beg)$ is exploited to start an iterative process, that searches for the best-fitting cranial ellipse in consecutive slices, running both toward the top and the bottom; more precisely, given the cranial ellipse $Ell(\theta(i))$, detected in the i-th slice $I(\mathbf{x}, i)$, a tracking procedure is employed to find $Ell(\theta(i+1))$ in the following slice $I(\mathbf{x}, i+1)$.

At first, both $I(\mathbf{x}, i)$ and $I(\mathbf{x}, i+1)$ are processed to compensate intensity inhomogeneities and to enhance the contrast, as previously described. Then, the best-fitting $Ell(\theta(i+1))$ is found in $I(\mathbf{x}, i+1)$ by minimizing the following energy function, by means of the simplex search method:

$$E\left(Ell(\theta(i)), Ell(\theta(i+1))\right) = \sum_{\mathbf{p} \in Ell(\theta)} \left(\sum_{\mathbf{j} \in W\left(T_{\theta(i+1)}^{\theta(i)}(\mathbf{p})\right)} (I(\mathbf{p}, i) - I(\mathbf{j}, i+1))^2 \right) \qquad (2)$$

where $T_{\theta(i+1)}^{\theta(i)} : \Re^2 \mapsto \Re^2$ is the mapping function between point $\mathbf{p} \in Ell(\theta(i))$ and point $\mathbf{j} \in Ell(\theta(i+1))$, and $W(\mathbf{j})$ is the 8-neighborhood of pixel $\mathbf{j}$. In few words, an energy term is computed for each pixel $\mathbf{p} \in Ell(\theta(i))$, by computing the sum of square differences between its gray level value and that of the pixels, $\mathbf{j}$ in $W(\mathbf{j})$. Note that the gray level comparison is not directly performed among corresponding pixels $\mathbf{p}$ in $Ell(\theta(i))$ and $\mathbf{j}$ in $Ell(\theta(i+1))$, but we employ the neighborhood of $\mathbf{j}$, $W(\mathbf{j})$, to compensate for image noise.

Once the optimum $\theta(i+1)$ is found, $Ell(\theta(i+1))$ is taken as the cranial ellipse in the current slice $I(\mathbf{x}, i+1)$, and the iterative procedure is repeated until either the last slice has been processed, or the detected elliptic area is less then $\frac{Area_{Beg}}{5}$, where $Area_{Beg}$ is the area of the braincase ellipse segmented in $I(\mathbf{x}, Beg)$. Figure 2 shows the aligned slices of two patients.

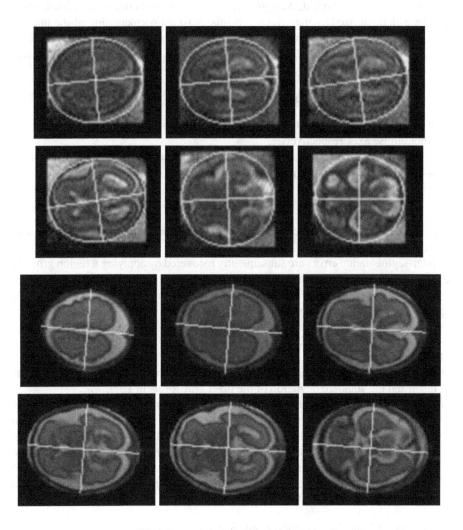

Fig. 2. Top rows : the 2nd, 3th, 4th, 5th, 7th, 9th slices of a patient at the 19th gestational week after alignment; in the image we show the axis of the tracked ellipse. Bottom rows: the 3rd, 5th, 7th, 8th, 9th, 12th tracked slices of a patient at the 27th gestational week.

5   Biometric Analysis of Fetal Brain Slices

In this section we describe our algorithm to perform segmentation (and biometric feature extraction) of liquors and ventricles on a manually chosen section, $I(\mathbf{x})$, $\mathbf{x} = (x, y)$, where these cerebral structures are entirely visible; if this section is not already represented in one of the slices of the patient's sequence, it can be extracted from the 3D fetal brain volume, reconstructed as described in section 4. At this stage, we suppose that the cranial ellipse in $I(\mathbf{x})$, $Ell(\theta)$ ($\theta = \{\mathbf{A}, \mathbf{B}, \phi, \mathbf{C}\}$), has been already segmented, the inhomogeneities have been corrected, and the contrast has been enhanced (see section 4).

At first, we apply a gray level clustering algorithm [2] to find two clusters: the cluster composed by pixels with darkest gray level values corresponds to ventricles, while liquors are characterized by the brightest pixels. A rough segmentation of the internal liquor, $IntLiq_{Bin}$, is then obtained by selecting those connected regions whose centroid C_R is in the ellipse $Ell_{Small}(\chi)$, $\chi = \{\frac{2}{3}\mathbf{A}, \frac{2}{3}\mathbf{B}, \phi, \mathbf{C}\}$, contained in $Ell(\theta)$, that is

$$\frac{(\hat{\mathbf{a}} \cdot (C_R - C))^2}{A^2} + \frac{(\hat{\mathbf{b}} \cdot (C_R - C))^2}{B^2} < \frac{2}{3}$$

where $\cdot$ denotes the dot product, $\hat{\mathbf{a}}$ and $\hat{\mathbf{b}}$ are unit vectors oriented along the major and the minor ellipse axis, respectively.

Over-segmentation errors in $IntLiq_{Bin}$ are then deleted by statistical thresholding, that is by deleting those pixels, $\mathbf{p}$, whose gray level $I(\mathbf{p})$ is such that

$$||I(\mathbf{p}) - \nu_{IntLiq}|| > 3\sigma_{IntLiq}$$

where ν and σ are the gray level mean and standard deviation of the pixels in the segmented internal liquor.

Under-segmentation errors are subsequently recovered by applying a region growing technique that accounts of both gray level and edge information. It takes as seed points the pixels on the perimeter of $IntLiq_{Bin}$, and considers the 4-connected neighborhood of each seed. Each analyzed pixel, $p = (x, y)$, is included into $IntLiq_{Bin}$, and it is used as a new seed, if:

(i) it has not been considered yet;
(ii) its gray level $I(p)$ is such that $||I(p) - \nu|| < c\,\sigma$, where $c = 2.0$, ν and σ are the mean and the standard deviation of the pixel gray levels in $IntLiq_{Bin}$;
(iii) it is not an edge point in the binary edge map, $Edge$, that is computed by applying the first order derivative of a gaussian function (with $\sigma = 0.5$) evaluated in eight directions. To keep only the significant edge pixels, the result is thresholded with hysteresis, by using 0.15 and 0.05 of the maximum gradient value as the high and the low threshold, respectively. The region growing stops either when it finds no more pixels that can be added to $IntLiq_{Bin}$, or when a maximum number of 10 iterations has been reached.

The internal liquor segmentation obtained so far is shown in the first image of figure 3. The resulting segmentation can be fragmented in two or more regions, due

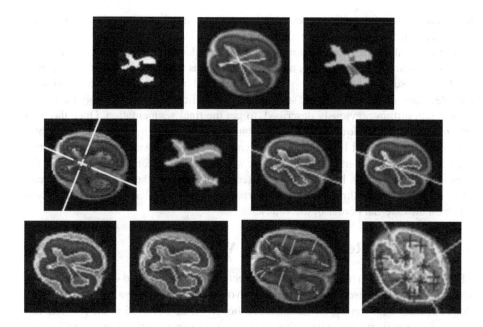

Fig. 3. Top row: the segmented internal liquor, $IntLiq_{Bin}$. The four segments drawn in the internal liquor, and the segments added to obtain a closed shape. Central row: the eight points that are the furthest from the cranial ellipse axes (squares), and the four points at the intersection between the liquor shape perimeter and the cranial ellipse axes (dots). The segmented internal liquor and its skeleton, and the upper and lower parts of the internal liquor and their symmetrical projections with respect to the major axis. Bottom row: the segmented internal and external liquor; the final ventricle segmentation; the segments used to estimate the ventricles' thickness; an example of landmark points detected on another patient slice.

to the presence of dark liquor areas caused by image artifacts. The next step extracts landmark points and segments that allow both to eventually close the segmented shape, and to perform its biometric analysis.

More precisely, given the major and the minor cranial ellipse axes, the internal liquor is divided into four sections; for each section ($sec = 1, .., 4$), we detect the two points, P_{sec}^{Maj} and P_{sec}^{Min}, that are the furthest from the major and the minor axis, respectively. Joining the corresponding points in alternate sections (e.g. P_1^{Maj} to P_3^{Maj}, P_2^{Maj} to P_4^{Maj}, and so on), we create four segments that describe the internal liquor shape, and also help to obtain a final segmentation, eventually closing the internal liquor shape (see figure 3). These points can be exploited to derive quantitative biometric measures, such as their reciprocal distances, the distance from the axes, as well as the mean and standard deviation of such values. Furthermore, other landmark points, and similar quantitative biometric measures, can be located on the perimeter of the liquor shape, such as the four points at the intersection with the ellipse axes (see figure 3). Other quantitative biometric measures could be derived also by the skeleton of the internal liquor shape, while symmetry measures can be computed from the shape of the whole liquor region, by imposing one of the ellipse axes as symmetry axis, and then computing the specular similarity of the two half shapes it defines (see figure 3).

The external liquor segmentation is obtained by selecting the biggest connected region composed of those pixels, $\mathbf{p}$, not yet segmented as internal liquor, and whose gray level $I(\mathbf{p})$ is such that $||I(\mathbf{p}) - \nu|| < \sigma$, where ν and σ are gray level mean and standard deviation of the pixels in the segmented internal liquor. Next, we apply the region growing algorithm previously described, to get the final segmentation. Once the final internal and external liquor segmentation has been obtained, the ventricle segmentation is simply obtained by selecting the area between them. An example of obtained segmentation is shown in figure 3, together with segments, automatically drawn in the ventricles, whose mean length could be used as an index of the ventricles thickness. They start from the landmark points, are perpendicular to the internal liquor border, and finish on the external border of the ventricles.

6   Preliminary Results and Future Work

The 3D reconstruction method described in section 4 has been tested on the thirty image sequences in our dataset; the computed 3D volumes have been evaluated visually by diagnosticians who were pretty satisfied. Although errors occurred in the first and last slices of four patients, we believe that these results are encouraging (see figure 2).

The biometric analysis algorithm, described in section 5 has been tested on 15 slices of different patients, chosen by diagnosticians since they contained a proper representation of ventricles and liquors. Due to the high inter and intra-observer variability the comparison among the computed results and those manually drawn by experts would not provide a meaningful assessment of the system performance. Therefore, the obtained results were visually evaluated by diagnosticians themselves; they were satisfied both with the segmentation results, and with the extracted biometric structures and measures.

Although the promising preliminary results and the consensus obtained in the clinical environment, this research work needs several improvements.

At first, we need to collect more data of both pathological and non-pathological patients; these data can be used to exhaustively test, and eventually improve, both the 3D reconstruction algorithm and the biometric analysis. Furthermore, medicians will exploit the data to find the slicing plane that allows to extract, from the 3D reconstructed volume, the 2D slice where ventricles and liquors are entirely visible. Finally, we still need to understand which biometric measures are useful to discriminate among normal and pathological cases.

References

1. Campadelli, P., Casiraghi, E.: Liver Segmentation from CT Scans: a Survey and a New Algorithm. Artificial Intelligence in Medicine (to appear, 2008)
2. Arbib, M.A., Uchiyama, T.: Color image segmentation using competitive learning. IEEE Transactions on Pattern Analisys and Machine Intelligence 16(12), 1197–1206 (1994)
3. Chong, B.W., et al.: A magnetic resonance template for normal neuronal migration in the fetus. Neurosurgery 39(1), 110–116 (1996)

4. Claude, I., et al.: Fetal Brain MRI: Segmentation and Biometric Analysis of the Posterior Fossa. IEEE Transactions on Biomedical Engineering 51(4), 617–626 (2004)
5. Cuddihy, S.L., et al.: Cerebellar vermis diameter at cranial sonography for assessing gestational age in low-birth-weight infants. Pediatratric Radiology 29(8), 589–594 (1999)
6. Ghidini, A., et al.: Dilated subarachnoid cisterna ambiens: A potential sonographic sign predicting cerebellar hypoplasia. Journal of Ultrasound in Medicine 15, 413–415 (1996)
7. Gholipour, A., et al.: Brain Functional Localization: A Survey of Image Registration Techniques. IEEE Transactions on Medical Imaging 26(4), 427–451 (2007)
8. Han, X., Fischl, B.: Atlas Renormalization for Improved Brain MR Image Segmentation Across Scanner Platforms. IEEE Transactions on Medical Imaging 26(4), 479–486 (2007)
9. Huang, H., et al.: White and gray matter development in human fetal, newborn and pediatric brains. Neuroimage 33(1), 27–38 (2006)
10. Johnston, B., et al.: Segmentation of Multide Sclerosis Lesions in Intensity Corrected Multispectral MRI. IEEE Transactions on Medical Imaging 15(2), 152–169 (1996)
11. RayBaud, C., et al.: MR imaging of fetal brain malformation. Child's Nervous System 19, 455–470 (2003)
12. Sanders, M., et al.: Gestational age assessment in preterm neonates weighing less than 1500 grams. Pediatratrics 88, 542–546 (1991)
13. Schwarz, D., et al.: A Deformable Registration Method for Automated Morphometry of MRI Brain Images in Neuropsychiatric Research. IEEE Transactions on Medical Imaging 26(4), 452–461 (2007)
14. Schierlitz, L., et al.: Three-dimensional magnetic resonance imaging of fetal brains. The Lancet 357, 1177–1178 (2001)
15. Triulzi, F., et al.: Magnetic resonance imaging of fetal cerebellar development. The Cerebellum 5(3), 199–205 (2005)
16. Vovk, U., et al.: A Review of Methods for Correction of Intensity Inhomogeneity in MRI. IEEE Transactions on Medical Imaging 26(3), 405–421 (2007)
17. Xia, Y., et al.: Automatic Segmentation of the Caudate Nucleus From Human Brain MR Images. IEEE Transactions on Medical Imaging 26(4), 509–517 (2007)
18. Yu, P., et al.: Cortical Surface Shape Analysis Based on Spherical Wavelets. IEEE Transactions on Medical Imaging 26(4), 154–169 (2007)

Searching for Glycomics Role in Stem Cell Development

Anil Sorathiya[1], Tadas Jucikas[2], Stephanie Piecewicz[3], Shiladitya Sengupta[3],
and Pietro Liò[1]

[1] University of Cambridge, Computer Laboratory, 15 JJ Thomson Avenue,
Cambridge CB3 0FD, UK
{as833,pl219}@cam.ac.uk
[2] University of Cambridge, DAMTP, Wilberforce Road, Cambridge CB3 0WA, UK
tj238@cam.ac.uk
[3] Massachusetts Institute of Technology, 65 Landsdowne Street, Cambridge, MA
02139, USA
{shiladit,smp}@mit.edu

Abstract. We present a methodology to analyze zebrafish knock-out experiment replicated time series microarray data. The knock-out experiment aimed to elucidate the transcriptomal regulators underlying the glycocalyx-regulation of vasculogenesis by performing global gene expression analysis of NDST mutants and wild-type siblings at three distinct time points during development. Cluster analysis and the construction of a genetic interaction network allows to identify groups of genes acting in the process of early stage vasculogenesis. We report the following findings: we found a large number of gene clusters, particularly glycans, during the three developmental steps of the zebrafish organism. In each step, genes connectivity changes according to two different powerlaws. The clusters are highlighted in such a way, that it is possible to see the dynamics of the interactions through the time points recorded in the microarray experiment. Vegf-related genes seem not to be involved at transcriptomics level, suggesting alternative regulative pathways do exist to modulate transcriptomal signatures in developing zebrafish. Our results show that there are several glycan-related genes which may be involved in early processes such as vasculogenesis.

Keywords: Vasculogenesis, zebrafish, complex network, microarray data.

1 Introduction

There is currently a tremendous growth in the amount of life science high through-put data and bioinformatics is challenged to provide the methodologies complementing life science experimental research and to keep the pace with the growing availability of a variety of molecular biology high through-put data of both normal workings of organisms and processes which cause disease. The analysis of these datasets, which has become an increasingly important part of

F. Masulli, R. Tagliaferri, and G.M. Verkhivker (Eds.): CIBB 2008, LNBI 5488, pp. 198–209, 2009.
© Springer-Verlag Berlin Heidelberg 2009

biology, poses a number of interesting statistical problems, largely driven by the complex inter-relationships between measurements.

To explore this methodology microarray data of Zebrafish endothelial stem cell N-deacetylase/ N-sulfotransferase enzyme (NDST) gene knock-out experiment is analysed. The knock-out organism has impaired blood vessel formation which is recently highly studied in the context of stem cells and cancer medicine [20]. Endothelial precursor cells, angioblasts, migrate and differentiate during vasculogenesis in response to local signals such as growth factors and extracellular matrix to form new blood vessels [2]. Three replicates are produced for both wild-type and knock-out phenotypes in three time points: 6, 20 and 24 hours from the start of embryonic development.

Glycans formed from complex arrays of sugars thus constitute a language for multicellular biology that some have designated as third language of life and morphogenesis. The past fifty years have witnessed a substantive and growing field of structural analysis and sequencing of glycans as well as chemical and enzymatic synthesis of these molecules. The time is at hand for the emergence of a theoretical glycobiology and search for the third language. The central dogma of molecular biology describes DNA, RNA and proteins as the key molecules in biological information flow. It tends to ignore other essential biomolecules including glycans (multimers, oligomers, or polymers of sugars) that are the most plentiful molecules on the planet in terms of mass. Glycosylation is the most common and abundant post-translational modification of proteins and lipids. Every cell is covered with glycans and most of the secreted proteins in higher organisms are glycosylated.

This study proposes a network clustering analysis approach to group genes active while compensating for NDST gene absence. The initial network is constructed from time step microarray data set. The main steps of the approach are: statistical microarray data analysis, identification of central differentially expressed genes, genetic interaction network construction, genetic network clustering, combination of clusters in different time steps, cluster analysis and genetic interaction network visualization.

2   Methodology

Affy and **limma** packages of R statistical tool were used for data loading and statistical analysis. These packages are used to normalise and remove background noise from gene expression data. We have used "RMA" method to correct background and "quantiles" method to normalise the data. Script set is running outside programs for clustering and produce a number of files at each algorithm step. The code can be obtained from the corresponding author upon request [21].

2.1   Exploratory Data Analysis Using FANOVA

Functional analysis of variance (FANOVA) is a functional data analysis (FDA) form of the classical ANOVA [17] [18] [19] which has a strong link to multivariate

analysis, such as principal component analysis, and to multivariate linear modeling and regularization methods that assume a particular class of smooth functions
for the estimators. We give a brief description of the technique we implemented.
Let $Y_{ij}(t)$ denote the signal j in the i-th experiment. Observations are modeled
by a fixed effect ANOVA model

$$y_{ij}(t) = u(t) + a_i(t) + e_{ij}(t) \tag{1}$$

where $i = 1, ..., p$, $l = 1, ..., n_i$; $\sum_{i=1}^{p} n_i = n$, $u(\cdot)$ is the grand mean, $a_i(\cdot)$ is
the deviation of the mean in experiment i from the grand mean, and $e_{ij}(\cdot)$ are
i.i.d. zero-mean normal random variables with variance σ^2. Here we use standard
functional ANOVA setting as in [19]. Following the standard ANOVA treatment
[19], for each t step, the function

$$\mathbf{F(t)} = \frac{\sum_i n_i [y_{i\cdot}(t) - \overline{y}_{\cdot\cdot}(t)]^2 / (p-1)}{\sum_{i,l} [y_{il}(t) - \overline{y}_{i\cdot}(t)]^2 / (n-p)} \tag{2}$$

is distributed as non central

$$\mathbf{F_{p-1,n-p}} = \sum_i \frac{n_i [\overline{y}_{i\cdot}(t) - \overline{y}_{\cdot\cdot}(t)]^2}{\sigma^2}. \tag{3}$$

This methodology is quite general and it is ideal for data exploration of the
meaningful variables for the gene expression data. In our case, the F statistics
gives us information about the link between measurement response and disease
phenotype and provide the distribution of p-values which can be evaluated using
complex networks analysis.

2.2   Network Analysis

When analyzing complex networks, the class of topology needs to be compared
with regular lattices, small worlds or random networks by making use of several
measures that describe the network's topology. One such quantity is the connectivity. In order to visualize the influence of connectivity on network topology,
let us introduce the *connectivity* distribution $P(k)$, which is proportional to the
number of sites with connectivity k. The probability distribution of a regular
lattice is extremely peaked around the lattice's *coordination number* (the number of vertices directly connected in the lattice), being a δ-function; in a random
graph it follows a Poisson distribution, centered around the mean connectivity
$\langle k \rangle$. Both of those choices are characterized by a well defined value of the mean
connectivity $\langle k \rangle$, and small variance $\langle k^2 \rangle - \langle k \rangle^2$. As shown by Watts and Strogatz
[8], the simple rewiring of a small fraction of links in an otherwise regular lattice
results in a sudden lowering of the diameter of the graph, without affecting the
average connectivity or the degree of clustering. This *small world* effect mani-
fests itself in a dramatic shortage of the distance between any two nodes, almost
without affecting the local perception of the network of contacts. A meaningful
measure is the degree of *clustering* of the network, which can be defined as the

number of neighbors of a given node which share an edge. In a regular lattice the clustering coefficient is high, while in a random graph it is vanishingly low. Finally, one is interested in the average length of the minimum path that connects any two vertices, called the *diameter* of the graph. This is large in a regular lattice, and low in a random graph. The study of biological networks has shown major differences from regular and random graphs because the degree distribution often follows a power-law (i.e. $P(k) \simeq k^{-\gamma}$, with $\gamma \simeq 2$). This distribution is characterized by a relatively large number of highly connected nodes (called *hubs*). Moreover, such distributions have a diverging second moment $\langle k^2 \rangle$ for $\gamma \leq 3$ and a diverging average connectivity $\langle k \rangle$ for $\gamma \leq 2$. An important class of biological networks show scale-free connectivity properties [9]. A simple model to generate networks with this property is based on the *preferential attachment* and can be exemplified in the following way. We start with a ring of K nodes, and we add the other $N - K$ nodes by choosing, for each of them, K connected nodes $j_n, n = 1, 2, \ldots, K$, with probability $k(j_n)/\sum_{l=1}^{N} k(l)$ (preferential attachment). The new node is added to the *inputs* of the chosen nodes. We also choose another node at random and add it to the list of input nodes of the new node. For example, this process simulates the growing of a sequence network in which a new node (a paralogous gene) is duplicated from another one. We have extracted from the data a list of gene names and expression levels. This list was prepared by selecting differentially expressed genes in NDST knock out experiment at 6 hour against wild-type experiment at 6h. The genes with expression at least one log-fold high were selected for further analysis. The idea behind it is that upon knocking out NDST gene these differentially expressed genes are likely the ones compensating or reacting to the lack of vasculogenesis process. The list ended up being 609 genes long. A graph of 609x609 elements was created where each node represented a gene and each edge represented the extent of their interaction. Each element of this matrix was filled with interaction value, which was calculated from expression data and showed how much the expression pattern differs between two genes. The interaction value is based on the formula:

$$\theta_{ij} = e^{\gamma \frac{\sum_{\forall c \in \Psi} |\epsilon_i^c - \epsilon_j^c|}{\sum_{\forall c \in \Psi} |\hat{\epsilon}_i^c - \hat{\epsilon}_j^c|}} \tag{4}$$

Here θ_{ij} represents the interaction value between genes i and j, γ is a tuning parameter which helps to control the sensitivity of clustering, Ψ is a set of conditions from which expression values need to be included in the calculation, ϵ means expression value of a particular gene in a particular condition and $\hat{\epsilon}$ is the same expression but in the wild type case. The θ_{ij} is accepted if it is bigger than a certain threshold, thus either producing or removing the edge of the complete graph. Then we introduce a clustering procedure.

2.3 Clustering

We use a clustering algorithm that does not need information on the number of clusters which is often unknown in large-scale comparisons. Although there is now a wide range of clustering algorithms, only a restricted number can successfully handle a network with the complete and weighted graph properties. Among them, we cite the recent method proposed by [10] that is based on simulated annealing to obtain clustering by direct maximization of the modularity. The modularity has been introduced by [11] and it is a measure of the difference between the number of links inside a given module and the expected value for a randomized graph of the same size and degree distribution. The modularity Q of a partition of a network is defined as $Q = \sum_s \left[\frac{l_s}{L} - \left(\frac{d_s}{2L} \right) \right]$ where the sum is over all modules of the partition. l_s and d_s describe the number of links and the total degree of the nodes inside module s and L the total number of links of the network [12]. In a recent work on resolution limits in community detection [12] the authors give evidence that modularity optimization may fail to identify modules smaller than a certain scale, depending on the total number of links in the network and on the number of connections between the clusters. Because of its properties, at the end, we implemented the Markov Clustering Algorithm (MCL, [13]). Its input is a stochastic matrix where each element is the probability of a transition between adjacent nodes. To increase the computational speed of the community detection, a nuisance parameter was used to assign an exponential weight to the transition probabilities. The weights between m_i and m_j were given by $e^{(\gamma \omega_{ij})}$ where ω_{ij} is -in the case of the 'variable graph'- the frequency of variables m_i and m_j being simultaneously altered in the same patients. γ was always set to 3. Additionally, we found the procedure to perform better when nodes are given a weight proportional to their node degree. Therefore, the sum of a given's node weights was added to the respective diagonal entry of the adjacency matrix. We used the strategy of Gfeller et al., [15] which allow detecting unstable nodes and compare results obtained with different granularity (inflation) parameters. In this algorithm, the starting matrix is modified to produce a novel matrix with a certain amount of noise added. The noise is homogeneously distributed between $-\sigma w_{ij}$ and σw_{ij} where w_{ij} is the edge weight and σ a fixed noise parameter, $0 \leq \sigma \leq 1$. The noise is added randomly to edges and the MCL clustering is performed on many noisy realization of the matrix. At each 'noisy' repetition, the algorithm records all the nodes belonging to the same cluster. After the prefixed number of repetitions has been concluded we end up with a matrix storing P_{ij} values corresponding to the fraction of times nodes i and j have been grouped together. Unstable nodes are identified as those having edges with less than a fixed values θ. We then calculate several distinct measures informing on the clustering and its stability such as the following clustering entropy:

$$\mathbf{S} = -1/L \sum_{ij} [P_{ij} log_2 P_{ij} + (1 - P_{ij}) log_2 (1 - P_{ij})] \tag{5}$$

where the sum is over all edges and the entropy is normalized by the total number of edges, L [16]. This might be used to detect the best clustering obtained after a long series of clusterings with different granularity parameters each time.

The entropy can be used to study the stability of gene networks/gene communities obtained from the clustering procedures. Due to the repeated noisy realizations of the original matrix, nodes may be attached to different communities after the clustering procedure. However, if the investigated system is very stable, nodes tend to cluster with the same communities regardless of the added noise. The stability of the different communities can be investigated by analyzing the entropy as a function of the clustering parameter r as the network breaks down into increasingly separated clusters as r increases.

Properties of networks can be assessed by various statistical measures such as the connectivity distribution $P(k)$, which is proportional to the number of sites with connectivity k. Other measures include graph density, diameter or transitivity all of which can be assessed before as well as during the clustering procedure. We want to point out that these measures may provide interesting insights into differences of network topologies between different disease states. This topic will be addressed in a different manuscript.

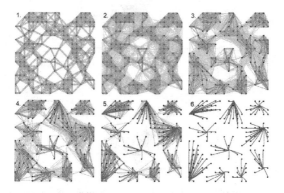

Fig. 1. Steps performed by MCL algorithm on an example graph. The input and the output are shown in 1 and 6 respectively. 2, 3, 4, 5 illustrates the algorithm steps of expansion and inflation. The black nodes represent the nodes with high influx of random walks [13].

Visualization. Visone social network visualization tool was used to draw the networks [5] (http://visone.info/). This tool enables to organize, color nodes of interest and calculate network properties like degree, radiality, authority, hub, etc. The original Manufacturer ID was used to label nodes.

3   Results and Discussion

In this work we aim at analysing the network structure of pattern of gene activation along time in wild type with respect to the perturbed state. Figure 2 shows the result of the FANOVA. For the two shifts 6h-20h and 20h-24h, the pvalues are distributed according to two powerlaws. The first plot (top) of figure 2 contains all the data disregarding of the time variable. In the middle plot, pvalues follow an initial exponent of 2, then 0.83; in the bottom plot the initial region

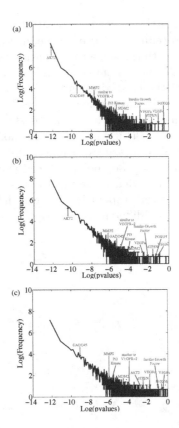

Fig. 2. The pvalues distributions which describe powerlaws. We show the position of some of the genes analysed in this study, on the distribution (a) represent the overall pvalues disregarding of time, (b) plot reports on 6h-20h and (c) shows the distribution for 20h represents.

has exponent 1.2 and the following region has exponent 1.05, thus becoming linearly related. These distributions suggest that most genes are partitioned in two groups with different connectivity pattern. The initial region of the 6h-20h is richer in large hubs than the other regions; the second region may be related to the existence of a large number of genes which are in networks dedicated to other intracellular/extracellular tasks and indirectly activated, probably as a result of loose interactions, by the group of genes in the first region.

Along the slopes we showed some of the glycan-related genes we are currently investigating (see also table 1). We have not found evidences of the involvement of the Vegf gene family. We suppose that this gene family acts at a level different from the transcriptomics level. In comparison, the distribution of pvalues from *Saccharomyces pombe* cell-cycle genes show a single slope, i.e. a single power law 3 suggesting a homogeneous distribution of gene connectivity in the cell cycle [16].

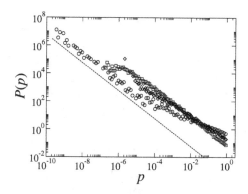

Fig. 3. The pvalues distribution of cell cycle gene expressions

Table 1. Differentially Expressed Genes using Fanova. Affymetrix ID marked as * is the gene related to glycan activity. Column 1 is Affymetrix ID, column 2 represents description or activity of gene in the cell, column 3 and 4 are pvalue calculated by FANOVA between 6h-20h and 20h-24h respectively and column 5 is ratio of 6h-20h and 20h-24h pvalues.

Afftmetrix ID	Description/Activity	6h-20h pvalue	20h-24h pvalue	pvalue change(ratio)
$Dr.3089.1.A1\_at$	membrane attack complex ,membrane, integral to membrane	0.9958033942	0.0020465606	486.5741
$Dr.11595.1.S1\_at*$	phosphatidylinositol glycan, class Q : GPI anchor biosynthetic process	0.0039521237	1.14E-005	346.6775
$Dr.17421.1.A1\_at$	DNA binding, regulation of transcription	0.9967489529	0.0028974576	344.0081
$Dr.20010.3.S1\_at$	translation initiation factor activity	0.991249865	0.0058106616	170.5916
$Dr.5308.1.A1\_at*$	phosphatidylinositol glycan, class P	0.4979120101	0.0310960065	16.0121
$Dr.5905.1.S1\_at*$	Similar to phosphatidylinositol glycan	8.67E-006	2.91E-005	0.2979
$Dr.17191.1.A1\_at*$	phosphatidylinositol glycan, class S (Homo sapiens)	2.74E-005	0.0008640627	0.0317
$Dr.2655.1.S1\_at*$	proteoglycan 4: catalytic activity, protease inhibitor activity, metal ion binding	0.0048006655	0.9925221503	0.0048
$Dr.19887.1.S1\_at$	DNA binding,transcription factor activity,sequence-specific DNA binding	0.0046563515	0.9974464812	0.0047
$Dr.813.1.S1\_at$	catalytic activity,transferase activity	0.0043537062	0.9954326037	0.0044
$Dr.12470.1.A1\_at$	protein serine/threonine kinase activity, ATP binding	0.0034396708	0.9912141197	0.0035
$Dr.822.1.S3\_at$	chemokine (C-X-C motif) ligand 12a (stromal cell-derived factor 1)	0.0015087484	0.9869259812	0.0015
$Dr.24983.1.S1\_at$	DNA binding , DNA-directed RNA polymerase activity	0.0013869684	0.9919293013	0.0014
$Dr.24527.1.S1\_at$	translation initiation factor activity binding, protein binding	0.0012446362	0.9987848827	0.0012
$Dr.7209.1.S1\_at$	nucleoside diphosphate kinase activity, ATP binding , kinase activity, transferase activity	0.0011987484	0.995562472	0.0012
$Dr.17204.1.A1\_at*$	transferase activity, transferase activity, transferring glycosyl groups	0.0004151669	0.4449264197	9.33E-004
$Dr.3019.1.A1\_at$	Opticin: protein binding	5.75E-005	0.9871951222	5.82E-005
$Dr.10640.1.S1\_at$	growth factor activity	5.34E-008	0.9994233223	5.34E-008
$Dr.15530.1.A1\_at$	parkinson disease (autosomal recessive, early onset) 7	4.98E-008	0.9961612658	5.00E-008

The clustering procedure using MCL confirmed the presence of components of the network. The weights of the edges in the network are calculated by Formula 4. Interaction value is estimated by taking a ratio of expression differences between two genes from perturbation samples and the same two genes from wild-type samples. The network constructed by the method is shown in Figure 4. Since

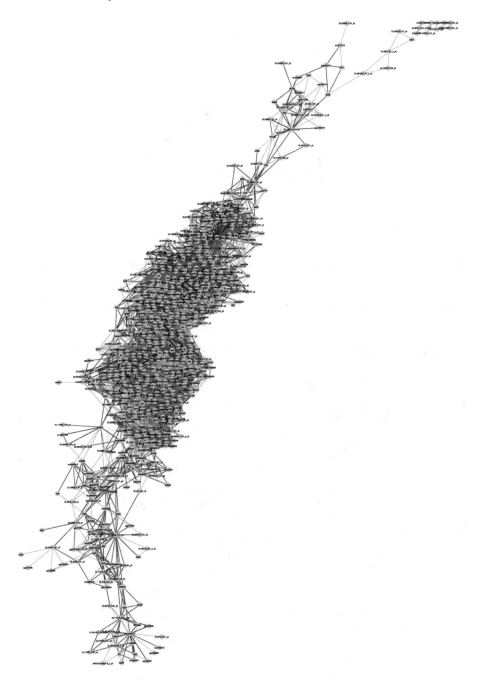

Fig. 4. Large network component from clustering procedure. Color code is: gray edges are present only in the 6h-20h developmental step; black edges are present in both 6h-20h and 20h-24h developmental steps; and red (wider dark gray in greyscale version of this image) edges are present only in the 20h-24h developmental step. Nodes are also colored according to their degree - blue nodes have high degree.

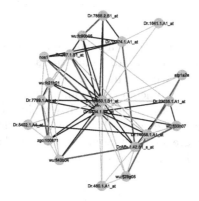

Fig. 5. The cluster discovered by the MCL-derived method. Gray color of edges represents the connections found only in 6h-20h, black edges represent connections found in both 6h-20h and 20h-24, and red edges represent connections found only in 20h-24h. The node coloring is based on network degree property - blue nodes have higher degree.

the number of genes used for network construction is rather big (609 genes) it is quite hard to see the details of the network. However, this network contains the most significant putative gene interaction information and shows which edges are present in which time steps. The development of the network through time is extremely useful information, showing the robustness and dynamics through the developmental stages. The gray edges of the network are present only in the 6h-20h developmental step, the black edges are present in both 6h-20h and 20h-24h developmental steps and red (wider dark gray in greyscale version of Figure 4) edges are present only in 20h-24h developmental step.

These data sets provide useful insights into two important aspects of stem cell behaviour: does the knock-out of NDST affect other glycan related genes? Does the involvement of FOXO transcription factors different from the results observed in the other types of experiments? The results show that the knock-out of NDST affects the FOXO transcription pathway, which was also observed in the FOXO transcription factors related experimental settings; the FOXO transcription factors response to reduced insulin and nutrient levels can act to inhibit growth (see [22] [26] [24]). The response to cellular stresses, such as nutrient deprivation or increased levels of reactive oxygen in species(such as Drosophila), dFOXO is activated and inhibits growth through the action of target genes such as d4E-BP [25]. In addition, FOXO regulates many genes associated with vascular destabilization, remodeling and endothelial cell apoptosis, which is responsible for regulation of endothelial cell survival and blood vessel stability [26].

Further insights may be achieved by fine tuning the clustering parameters to several smaller clusters. Figure 5 shows the most important cluster, with key genes related to glycan metabolism.

4   Conclusions

Our preliminary analysis on differentially expressed genes in zebrafish provides a layer of easily accessible information to be incorporated into further experimental studies or other analysis. We have investigated the patterns of activation with respect to the knock-out experiment of a glycan metabolism-related enzyme. We show that the network analysis and entropy-driven clustering allows to identify groups of genes, particularly glycans which may act in the early stages of vasculogenesis. Work in progress consist in understanding the glycan-related gene network underlying the three stages of vasculogenesis.

Acknowledgment

SS is supported by a AHA Scientist Development Grant. PL thanks the British Council which is a registered charity 209131 (England and Wales) SC037733 (Scotland) for fundings. Authors are very grateful to Prof. Joseph V Bonventre for access to resources in his laboratory and Prof. Ram Sasisekharam for his insights and critical inputs.

References

1. Hanahan, D., Weinberg, R.A.: The hallmarks of cancer. Cell 100(1), 57–70 (2000)
2. Fuster, M.M., et al.: Genetic alteration of endothelial heparan sulfate selectively inhibits tumor angiogenesis. Journal of Experimental Medicine 204, i16 (2007)
3. Ihaka, R., Gentleman, R.: A Language for Data Analysis and Graphics. Journal of Computational and Graphical Statistics 5(3), 299–314 (1996)
4. Enright, A.J., Van Dongen, S., et al.: An efficient algorithm for large-scale detection of protein families. Nucleic Acids Res. 30(7), 1575–1584 (2002)
5. Brandes, U., Wagner, D.: Visone - Analysis and Visualization of Social Networks. In: Junger, M., Mutzel, P. (eds.) Graph Drawing Software, pp. 321–340. Springer, Heidelberg (2003)
6. Veiga, D.F., Vicente, F.F., et al.: Gene networks as a tool to understand transcriptional regulation. Genet Mol. Res. 5(1), 254–268 (2006)
7. Caretta-Cartozo, C., De Los Rios, P., et al.: Bottleneck Genes and Community Structure in the Cell Cycle Network of S. pombe. PLoS Comput. Biol. 3(6), e103 (2007)
8. Watts, D.J., Strogatz, S.H.: Collective dynamics of 'small-world' networks. Nature 393, 440–442 (1998)
9. Park, J., Barabasi, A.-L.: Distribution of node characteristics in complex networks. PNAS (2007) 0705081104v1-0
10. Guimera, R., Sales-Pardo, M., Amaral, L.A.: Modularity from fluctuations in random graphs and complex networks. Phys. Rev. E Stat. Nonlin. Soft Matter Phys. 70(2 Pt 2), 025101 (2004)
11. Newman, M.E., Girvan, M.: Finding and evaluating community structure in networks. Phys. Rev. E Stat. Nonlin. Soft Matter Phys. 69, 026113 (2004)
12. Fortunato, S., Barthelemy, M.: Resolution limit in community detection. Proc. Natl. Acad. Sci. U S A. 104, 36–41 (2007)

13. van Dongen, S.: Graph Clustering by Flow Simulation. PhD thesis, University of Utrecht (May 2000)
14. http://www.arbylon.net/projects/
15. Gfeller, D., Chappelier, J.C., De Los Rios, P.: Finding instabilities in the community structure of complex networks. Phys. Rev. E Stat. Nonlin. Soft Matter Phys. 72, 056135 (2005)
16. Caretta-Cartozo, C., De Los Rios, P., Piazza, F., Lio, P.: Bottleneck genes and community structure in the cell cycle network of S. pombe. PLoS Comput Biol. 3(6), e103 (2007)
17. Ramsay, J.O., Dalzell, C.J.: Some tools for functional data analysis. Journal of the Royal Statistical Society, Series B 53, 539–572 (1991)
18. Ramsay, J.O., Munhall, K.G., Gracco, V.L., Ostry, D.J.: Functional data analysis of lip motion. Journal of the Acoustical Society of America 99, 3718–3727 (1996)
19. Ramsay, J.O., Silverman, B.W.: Functional Data Analysis. Springer, New York (1997)
20. Abramsson, A., Kurup, S., Busse, M., Yamada, S., Lindblom, P., Schallmeiner, E., Stenzel, D., Sauvaget, D., Ledin, J., Ringvall, M., Landegren, U., Kjellén, L., Bondjers, G., Li, J., Lindahl, U., Spillmann, D., Betsholtz, C., Gerhardt, H.: Defective N-sulfation of heparan sulfate proteoglycans limits PDGF-BB binding and pericyte recruitment in vascular development. Genes and Development 21, 316–331 (2007)
21. Ritchie, M.E., Diyagama, D., Neilson, J., van Laar, R., Dobrovic, A., Holloway, A., Smyth, G.K.: Empirical array quality weights for microarray data. BMC Bioinformatics 7, 261 (2006)
22. Neufeld, T.P.: Shrinkage control: regulation of insulin-mediated growth by FOXO transcription factors. Journal of Biology 2, 18 (2003)
23. Lam, E.W.-F., Francis, R.E., Petkovic, M.: FOXO transcription factors: key regulators of cell fate. Biochemical Society Transactions 34, 5 (2006)
24. Jünger, M.A., Rintelen, F., Stocker, H., Wasserman, J.D., Végh, M., Radimerski, v., Greenberg, M.E., Hafen, E.: The Drosophila Forkhead transcription factor FOXO mediates the reduction in cell number associated with reduced insulin signaling. Journal of Biology 2, 20 (2003)
25. Harvey, K.F., Mattila, J., Sofer, A., Bennett, F.C., Ramsey, M.R., Ellisen, L.W., Puig, O., Hariharan, I.K.: FOXO-regulated transcription restricts overgrowth of Tsc mutant organs. The Journal of Cell Biology, 691–696 (2008)
26. Daly, C., et al.: Angiopoietin-1 modulates endothelial cell function and gene expression via the transcription factor FKHR (FOXO1). Genes Dev. 18, 1060–1071 (2004)

A New Protein Representation Based on Fragment Contacts: Towards an Improvement of Contact Maps Predictions

Pietro Di Lena[1], Luciano Margara[1], Marco Vassura[1], Piero Fariselli[2], and Rita Casadio[2]

[1] Department of Computer Science, University of Bologna, Italy
{dilena,margara,vassura}@cs.unibo.it
[2] Biocomputing Group, Department of Biology, University of Bologna, Italy
piero@biocomp.unibo.it, casadio@alma.unibo.it

Abstract. Prediction of intramolecular residue-to-residue contacts is a difficult problem. In this paper we consider a related subproblem: the fragment-to-fragment contact prediction problem. A *fragment contact map* is a coarse-grained approximation of a residue contact map. Although fragment contact maps are approximations, they still contain most information about residue contacts and, most importantly, they are more tolerant to errors than contact maps (at least random error). We compare with an extensive set of experimental tests the error tolerance of fragment contact maps with respect to the error tolerance of residue contact maps. Moreover, as a simple concrete application, we show how to build a fragment contact predictor from a residue contact predictor and we show that in some cases this technique can improve contact predictions.

Keywords: Protein contacts, protein structures, coarse-grained contact maps, protein fragments.

1 Introduction

According to the Anfinsen's hypothesis, under the environmental conditions at which folding occurs, the native structure of a protein is uniquely determined by its amino acid's sequence. Despite the considerable research efforts devoted in the last decades to this subject, the problem to predict the three-dimensional structure of a protein from its one-dimensional amino acidic sequence is still one of the most challenging open problems in Bioinformatics [5].

Currently, protein structures are typically solved experimentally by time-consuming X-ray crystallography or NMR spectroscopy. Computational methods for the folding problem are generally divided in two main classes: *ab initio protein modeling* and *comparative protein modeling*. Ab initio protein modeling methods attempt to recover three-dimensional protein models from scratch, usually relying on physical laws. Comparative protein modeling methods, on the contrary,

F. Masulli, R. Tagliaferri, and G.M. Verkhivker (Eds.): CIBB 2008, LNBI 5488, pp. 210–221, 2009.

rely on already solved structures as starting points for the model. Comparative protein modeling is actually effective since it appears that the number of possible protein structures is quite limited compared to the number of possible protein sequences. Moreover, comparative modeling methods are based on the widely accepted (and experimentally confirmed) assumption that two proteins which have similar amino acid sequence will tend to have very similar structures. Since protein evolutionary process tends to preserve more the structure than the sequence, a target protein can be usually modeled with a good accuracy on a distantly related protein template by performing sequence alignment. However, when sequence similarity is very poor, methods based on sequence alignment are no more effective. In this context, one possible approach is to try to identify related sub problems such as prediction of secondary structures, solvent accessibility or, in particular, prediction of intramolecular contacts.

The residue-to-residue contact representation of a protein structure is usually called *contact map*. A contact map is an approximate (two-dimensional) representation of the (three-dimensional) structure of a protein. In detail, a protein contact map is defined as a binary symmetric matrix which contains 1 in the entry i, j if and only if the i-th and the j-th residues of the protein chain are in *contact*, i.e. their distance is below some given threshold. Despite contact maps are approximations, they generally contain enough information to infer the structures with a good level of accuracy. The reconstruction problem from contact maps has been already addressed and it seems to be promising [6,7,8,9,10,11]. In particular, it is generally possible to infer a three-dimensional structure with a good level of accuracy also from partially noised or incomplete (just 25% of the entries of the map) contact maps [7,8,9]. In general, even a small number of correctly predicted contacts for some target protein can be useful to select among all possible template models the most appealing ones. Every two years, the performances of current intramolecular contact prediction methods are assessed in the CASP experiments [2]. Despite several CASP sessions, the current accuracy performed in contact prediction is still unsatisfactory for available reconstruction methods to recover fold models with an acceptable level of accuracy.

The motivation of this work is to describe a new general approach which can be used to improve contact prediction. Here we consider *fragment contact maps*, which are coarse-grained approximations of residue contact maps obtained by considering fragment-to-fragment contacts instead of residue-to-residue contacts. Prediction of coarse-grained contact maps as an intermediate step to fold prediction have been already considered [6,12]. Our approach differs from the cited one: the authors of [6,12] consider coarse-grained contact maps obtained from secondary structure element contacts while we consider coarse-grained contact maps obtained by considering contacts between all possible protein fragments of some fixed length. In this paper we mainly investigate error tolerance fragment contact maps, showing that they are much more tolerant to errors than contact maps (at least random error). We compare with an extensive set of experimental tests the error tolerance of fragment contact maps with respect to the error tolerance of residue contact maps. Moreover, as a simple application, we show how to build a fragment contact predictor from a residue contact predictor. We test our technique on the predictions of the best performing group at last CASP7 [2] and on CORNET [3] predictions for the same test set obtaining some improvement in the last case.

The paper is organized as follows. In the Section 2 we describe the evaluation criteria which are adopted to measure the quality of contact predictions. In Section 3 we describe our approach and we show experimental results. Section 4 is devoted to concluding remarks.

2   Evaluation Criteria for Contact Prediction

In this section we describe the evaluation criteria for contact predictions adopted in CASP [2]. We will use such criteria to evaluate our tests in the next section.

Two residues are in *contact* if the distance between their respective C_β atoms is less than or equal to 8 Å (C_α for glycines). We obtain an equivalent definition of contact if we consider C_α atoms only. The most important evaluation measure for contact prediction is the *accuracy* which is defined as

$$(number\ of\ correctly\ predicted\ contacts)\ /\ (predicted\ contacts).$$

Usually contact predictors just provide a score to residue pairs: higher is the score, higher is the probability (according to the predictor) that the pair is in contact. In this case the list of predictions is sorted in decreasing score order and the accuracy is computed by taking the first $2L$, L, $L/2$, $L/5$ and $L/10$ pairs, where L is the length of the protein. It makes sense to distinguish between short range contacts (i.e. contacts between residues which are close in the sequence) and long range contacts (i.e. contacts between residues which are distant in the sequence). Long range contacts are more sparse than short range contacts and also more difficult to predict. Moreover, long range contacts are more informative than short range contacts about the protein structure. For this reason, pairs or residues are split in three classes according to their *sequence separation*[1] (i.e. the number of residues between them): short range contacts for sequence separation ≥ 6, medium range contacts for sequence separation ≥ 12 and long range contacts for sequence separation ≥ 24. The accuracy is generally evaluated for these three sets and for number of contacts $2L$, L, $L/2$, $L/5$ and $L/10$. In CASP experiments the most important evaluation parameter is the accuracy for $L/5$ number of contacts and sequence separation ≥ 24.

It is difficult to evaluate in terms of accuracy what is the state-of-art of intramolecular contact prediction since the evaluations depend essentially on the targets difficulty. From this point of view there is no contact predictor which is absolutely best performing than others but there are contact predictors which are more specialized than others on some protein target. In a "superficial" way, we can say that currently good performing contact predictors have on average performances on the order of $\sim 20\%$ of accuracy for $L/5$ number of contacts and sequence separation 24 [1,2].

---

[1] In last CASP7 the three separation ranges have been defined in a slightly different way, except for long range contacts.

3   Fragment Contact Maps

As we reviewed in the previous sections, the problem to predict protein intramolecular contacts is a complex and unsolved problem. In this section we describe a new general approach which can be used to improve contact prediction.

In first instance, we consider here a slightly relaxed problem than residue contact prediction: the problem to predict whether two given protein fragments are in contact, i.e. if there is a contact between at least one residue in the first fragment and one residue in the second fragment. For simplicity we consider here only fragments of the same length. We can represent all possible contacts between fragments of a protein by using again a binary symmetric matrix which we call *fragment contact map*. A fragment contact map can be easily recovered from a contact map. Formally, given a protein P and its contact map $M \in \{0,1\}^{n \times n}$ the *fragment contact map of order w* of P is the binary symmetric matrix $F \in \{0,1\}^{(n-w+1) \times (n-w+1)}$ such that $F_{i,j} = 1$ if and only if for some $i' \in [i, i+w-1]$ and for some $j' \in [j, j+w-1]$ we have $M_{i',j'} = 1$. By definition, the matrix F contains complete information about all contacts between all P's fragments of length exactly w. As an example, in Fig. 1, it is shown a contact map and the related fragment contact maps of orders $4, 7, 10$.

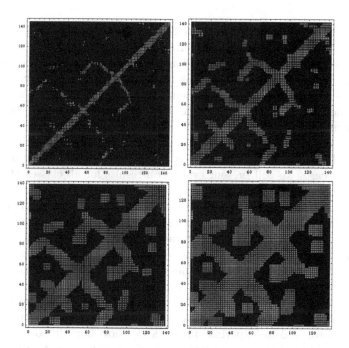

Fig. 1. Example of protein contact map of threshold 8 (pdb code 1a3a chain A) in the left upper corner. The related fragment contact maps of widths 4, 7 and 10 are shown in the right upper, left down and right down corners, respectively.

It is clear that a fragment contact map is a coarse-grained approximation of a contact map where the order of the approximation depends uniquely on the fragments length. Although a fragment contact map is an approximation, it still contains relevant information about protein contacts. We don't want to deal here explicitly with the problem to predict fragment contacts, but we want to show what are the advantages in considering *fragment pair contact prediction* as an intermediate step for *residue pair contact prediction*. We will discuss here mainly two properties related to fragment contact maps:

1. *The problem to predict a fragment contact map (of any given order) is not more difficult than the problem to predict the related contact map.* The reason is clear: if we are able to correctly predict a residue pair contact then we are able to correctly predict all contacts between all fragments which contain such residues.

2. *Fragment contact maps are much more tolerant to errors than contact maps.* This is a consequence of the *geometry* of fragment contact maps: note that any single 1 in the contact map produces a $w \times w$ square of 1s in the related fragment map of order w (to be more precise, this is true only for contacts with residues which are not in the head or in the tail of the protein sequence). In some sense, a fragment contact map contains redundant information about residue contacts and such redundancy offers some robustness against local errors.

In the rest of the paper we will discuss by experimental tests point 2. Our tests will be performed on a test set of 1092 proteins obtained by selecting from PDB those sequences whose structures have been solved with X-ray experiments with resolution <2.5 Å (without missing internal residues) and which have been subsequently filtered by removing sequence redundancy and by removing those proteins which have less than L contacts for sequence separation ≥ 24. To measure the accuracy of our tests we will use the same evaluation criteria adopted in CASP as described in Section 2. Accordingly, our main evaluation parameter will be the accuracy for sequence separation ≥ 24 and number of contacts $L/5$. In most cases we will present evaluation results for all classes of sequence separation and number of contacts (except $2L$).

The rest of this section is organized as follows. In the Section 3.1 we show a general method to generate residue contact predictions from fragment contact maps. Moreover, we show how the accuracy of the prediction is affected by the order of the map (assuming the fragment contact map contains no errors at all). In Section 3.2, we test the error robustness of fragment contact maps of different orders. Moreover, we compare the error tolerance of fragment contact maps with the error tolerance of contact maps. To conclude, in Section 3.3 we show, as a simple application, how to build a fragment contact predictor from a residue contact predictor and by experimental tests we show how, in some cases, this fragment-based technique can improve the performances of contact predictors.

3.1   From Fragment Contact Maps to Contact Predictions

In this section we show how to recover residue-to-residue contact predictions from fragment contact maps.

Recall the definition of fragment contact map. Let $F \in \{0,1\}^{(n-w+1) \times (n-w+1)}$ be the fragment contact map of order w of some protein P. Let $M \in \{0,1\}^{n \times n}$ be the contact map of P. By definition, if $M_{i,j} = 1$ then $F_{i',j'} = 1$ for all $i' \in [\max(1, i-w+1), \min(i, n-w+1)]$ and $j' \in [\max(0, j-w+1), \min(j, n-w+1)]$. In other words, every contact in M is replicated in F at most w^2 times. We can use this information to recover a probability for entries of the contact map M. More formally, given the (predicted) fragment contact map F of order w, for every $i, j \in [w, n - w + 1]$ we define the probability that residues i, j are in contact by

$$P_{i,j} = \frac{\sum\limits_{l=i}^{i-w+1} \sum\limits_{m=j}^{j-w+1} F_{l,m}}{w^2} \tag{1}$$

Note that, in order to simplify our analysis, we choose to recover probabilities just for those entries which are *not too close* to the border of the contact map, that is those entries which determine exactly a square of dimension $w \times w$ in the related fragment contact map (this is not a limit of the method but just a simplification since for residues close to the border of the map we should average a prediction on a smaller number of entries). Moreover, note that we can use formula (1) also when we have to deal with *predicted fragment contact maps*, i.e. maps whose entries are real numbers in the interval $[0, 1]$.

Now that we have a formula to recover contact predictions from fragments contacts it is natural to ask how the length of the fragments affects the accuracy of these predictions in presence of no error at all. In Figure 2 are shown the contact maps predicted from fragment contact maps of Figure 1. Darker colors correspond to lower probabilities. Note the black borders in the predicted contact maps. These black regions correspond to the entries for which we don't provide a prediction (according to what discussed above). The evaluation accuracy (for seq. sep. ≥ 24) for the predictions from fragment contact maps in Figure 1 decreases from 82% (order 4) to 75% (order 10). Not surprising, the accuracy decreases with the increase of the length of the fragments (order), which corresponds (see Figure 1) to an increase in the roughness of the approximation.

3.2   Error Tolerance: Fragment Contact Maps vs. Contact Maps

In this section we test the error tolerance of fragment contact maps with respect to the error tolerance of contact maps. We adopt random error as error model.

In our tests we perturbed fragment contact maps with some amount of random error (from 35% to 45%) and we then recovered the contact predictions from the perturbed maps by using formula (1). We perturbed native contact maps with the same amount of error in order to compare the error tolerance of the two representations. In order to make a comparison possible, we compute the

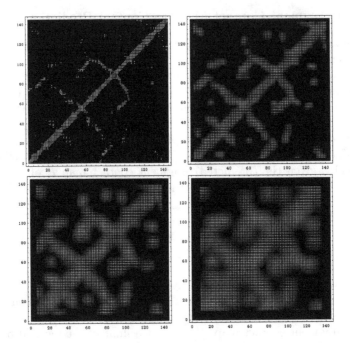

Fig. 2. Predicted contact maps for the protein of Figure 1. The ordering of the maps in the picture is the same as Figure 1. Entries with higher probability to be a contact are represented with lighter colors.

accuracy for perturbed native contact maps in a way which is very similar to what we do with fragment contact maps (it would be unfair to use for these maps the accuracy measure: *number of correctly predicted contacts/predicted contacts*). Given a *radius r*, for every entry i, j of the perturbed contact map we compute a score by taking the average value of the entries in the square of radius r centered in i, j (also in this case we do not consider entries which are too close to the border of the map). We just considered radii from 2 to 5 which roughly correspond to orders from 3 to 6.

In the first column of Figure 4, there are some examples of perturbed maps. The first and second map are the native contact map of Figure 1 perturbed with 35% and 45% of random error, respectively. The third and fourth maps are the related fragment contact map of order 7 perturbed with 35% and 45% random error, respectively. It is necessary to spend some words on how the random error is applied: the $x\%$ of random error means that the $x\%$ of the entries of the map are flipped. The error is not applied to all diagonals: by definition, for a fragment contact map of order w, all diagonals from 1 (main diagonal) to $w + 1$ must contain ones, then the error is applied starting from diagonal $w + 2$. By looking at the pictures in Figure 4, note that with 35% of error part of the original map is still visible in the perturbed fragment contact map while the structure of the native contact map is no more visible. With 41% of random error also

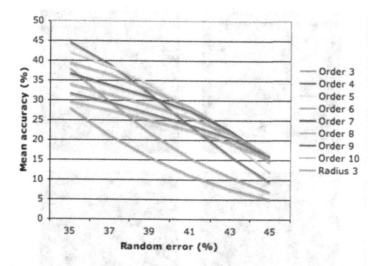

Fig. 3. Mean accuracy in function of the random error (sequence separation ≥ 24 and $L/5$ contacts). The picture shows the mean accuracy for predictions recovered from fragment contact maps of orders from 3 to 10 and from predictions recovered from native maps by using probability squares of radius 3 as described in Section 3.2.

for the fragment contact map the original shape seems to be almost completely lost. In the second column of Figure 4 it is possible to see the predicted contact maps from the related perturbed maps in the first column (the prediction for the native contact map have been computed with radius 3, as described above). In order to make the pictures more readable, all entries in the predicted maps whose values are less than 0.4 have been recoded to 0 (i.e. black color).

The error tolerance of fragment contact maps is largely a consequence of their *contact information redundancy*. In Figure 3, it is possible to see the mean accuracy (for sequence separation ≥ 24 and number of contacts $L/5$) of our test set in function of the percentage of random error (from 35% to 45% with step 2). To evaluate the accuracy of perturbed contact maps we used radius 3. For every percentage of error and for every map, 10 perturbed maps have been generated. First of all, it is evident that, fragment contact maps are more tolerant to errors than contact maps (at least for random error and for the fragment lengths we considered). From our tests it seems that, when the random error is relatively small, we can obtain better accuracies from fragment contact maps of lower orders while, when the error becomes bigger, maps of higher orders are more robust. In general, it seems that up to 43% of random error fragment contact maps of order 6, 7 are enough robust to offer performances which are comparable with the state-of-art in contact prediction. Moreover, it seems that no fragment contact map seems to be "reliable" when perturbed with 45% of random error or when the order is too large (above 8).

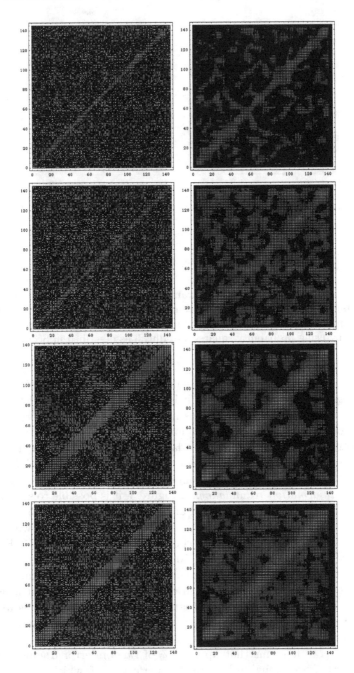

Fig. 4. Some maps from Fig. 1 perturbed with random error (first column) and related predictions (second column). In the first column: the first and second maps are the native map perturbed with 35% and 45% of error, respectively; the third and fourth maps are the fragment density map of order 7 perturbed with 35% and 45% of error, respectively. In the second column: predictions from the related maps in the first column. Higher probabilities have lighter colors.

3.3   An Application of Fragment Contact Maps

We can convert a residue-to-residue contact predictor into a fragment-to-fragment contact predictor in a very simple way: we define the probability of two fragments to be in contact as the highest score among all scores the (residue-to-residue) predictor assigns to residue pairs in the two fragments (where, obviously, a pair is defined by a residue in the first fragment and a residue in the second fragment). From a fragment contact map predicted in such way is then possible to use formula (1) to obtain a new contact map prediction. In some cases, going from residue contact predictions to fragment contact predictions and back can help to improve contact predictions. Before to show an example, it is useful to understand what happens when we pass through fragment contact predictions. It is not difficult to check that passing through fragment contact predictions has the effect to apply a *re-scoring* of the original predictions which never decreases the original score assigned to a pair. On the contrary, the scores are increased for all pairs which are in the neighborhood of some high-scored pair. This procedure has some sense for the fact that usually contacts are not isolated and, for a statistical reason, predictors are more likely to predict contacts which are not isolated.

As an example, we applied this re-scoring procedure to the predictions of the best performing group in last CASP7, the Karplus lab group, identified in CASP7 as group RR010 [2,4] and on the predictions obtained by CORNET [3] on the same target set. The CASP7 targets were split in two classes: Template Based Modeling (TBM) targets and Free Modeling (FM) targets. The residue-to-residue contact predictions were evaluated only on the 19 FM targets. Four FM targets (T0287, T0309, T0314, T0353) were considered the most difficult to predict since at the time of the experiments they lacked any homologs. We re-scored RR010's predictions by passing through fragment contact maps of order 3 for RR010 (RR010+F3) and 4 for CORNET (CORNET+F4). We obtained best performances with these respective orders. The mean accuracies[2] for the FM targets are shown in Table 1.

For RR010 the re-scored predictions have on average slightly worse accuracies than the original ones while for CORNET the re-scoring improves the accuracy for $L/5$ and $L/10$. Anyway, if we look at the detailed accuracies, we can note that, as a general trend, the re-scored predictions improve good predictions and make bad predictions worst. This trend is much more visible if we repeat the test on the whole set of FM+TBM targets (data not shown). In some cases, the improvement we obtain by re-scoring is notable, as in the case of the difficult T0353 target (which is well predicted by both RR010 and CORNET). This behavior seems to be consistent with what discussed above: high-scored pairs imply a higher re-scoring for all pairs in their neighborhood so, when there is

---

[2] We used as sequence length the number of contacts which fall into the domain boundaries as specified at *http://predictioncenter.gc.ucdavis.edu/casp7/meeting_docs/shorts.html*. Moreover, since some targets have not L/x native contacts for seq. sep. $\geq y$ we always compute the accuracy for the $min\{L/x, \#(\text{native contacts for seq. sep.} \geq y)\}$ best scored pairs.

Table 1. Mean accuracies (seq. sep $\geq$ 24) for RR010 and CORNET predictions on CASP7 FM targets and for the re-scored predictions with fragment method of order 3 (RR010+F3) and order 4 (CORNET+F4)

Prediction	L	L/2	L/5	L/10
RR010	13.42	16.12	22.90	26.42
RR010+F3	10.93	15.07	21.77	25.03
CORNET	11.59	12.48	14.12	17.59
CORNET+F4	10.25	12.36	17.85	20.05

a consistent number of bad high-scored pairs, the re-scoring obtained passing through fragment contact maps is more likely to have poor performances.

4   Conclusion

In this paper we considered fragment contact maps. A fragment contact map of order w of some protein P is a binary symmetric matrix which contains complete information about all contacts between all P's fragments of length exactly w. Fragment contact maps are coarse-grained approximations of residue contact maps. The order of the approximation depends on the fragment length parameter (order). From fragment contact maps it is possible to recover contact predictions with a high level of accuracy also in presence of a high percentage of (random) error. We tested experimentally the error tolerance of contact maps with respect to the error tolerance of fragment contact maps (for orders from 3 to 10). We tested percentages of random error from 35% to 45% (with step 2). In all tested cases, under the same amount of error, the performances of fragment contact maps are better than those of contact maps. The error tolerance of fragment contact maps largely depends on the order parameter. With orders 6, 7 it is possible to obtain accuracy performances which are comparable with the state-of-art in contact prediction also for 43% of random error (with this amount of error the shape of the original map is almost completely lost by visual inspection). As a simple concrete application of our method we showed how to implement a fragment contact predictor from a residue contact predictor (we remark that this is just a simple experiment and that fragment contact maps should be predicted with more clever techniques). From a such predicted fragment contact map is then possible to obtain back a new contact prediction. We used this technique to re-score the predictions of the best performing group, RR010, at last CASP7 experiments and the predictions of the CORNET predictor on the same test set. As a general trend, our simple method has the effect to improve already good predictions and to further decrease bad predictions: this can sometimes improve the mean accuracy (as in the case of CORNET).

References

1. Grana, O., Baker, D., MacCallum, R.M., Meiler, J., Punta, M., Rost, B., Tress, M.L., Valencia, A.: CASP6 assessment of contact prediction. Proteins 61(suppl. 7), 214–224 (2005)
2. Izarzugaza, J.M., Grana, O., Tress, M.L., Valencia, A., Clarke, N.D.: Assessment of intramolecular contact predictions for CASP7. Proteins 69(suppl. 8), 152–158 (2007)
3. Fariselli, P., Olmea, O., Valencia, A., Casadio, R.: Prediction of contact maps with neural networks and correlated mutations. Protein Eng. 14, 835–843 (2001)
4. Shackelford, G., Karplus, K.: Contact prediction using mutual information and neural nets. Proteins 69(suppl. 8), 159–164 (2007)
5. Lesk, A.: Introduction to Bioinformatics. Oxford University Press, Oxford (2006)
6. Pollastri, G., Vullo, A., Frasconi, P., Baldi, P.: Modular DAG-RNN Architectures for Assembling Coarse. Protein Structures J. Comp. Biol. 13(3), 631–650 (2006)
7. Vassura, M., Margara, L., Di Lena, P., Medri, F., Fariselli, P., Casadio, R.: Reconstruction of 3D Structures From Protein Contact Maps. IEEE/ACM Transactions on Computational Biology and Bioinformatics 5(3) (July-September 2008)
8. Vassura, M., Margara, L., Di Lena, P., Medri, F., Fariselli, P., Casadio, R.: FT-COMAR: fault tolerant three-dimensional structure reconstruction from protein contact maps. Bioinformatics (2008), doi:10.1093/bioinformatics/btn115
9. Vassura, M., Di Lena, P., Margara, L., Medri, F., Fariselli, P., Casadio, R.: Fault Tolerance for Large Scale Protein 3D Reconstruction from Contact Maps. In: Giancarlo, R., Hannenhalli, S. (eds.) WABI 2007. LNCS (LNBI), vol. 4645, pp. 25–37. Springer, Heidelberg (2007)
10. Vendruscolo, M., Domany, E., Park, K.: Towards an energy function for the contact map representation of proteins. Proteins 40, 237–248 (2000)
11. Vendruscolo, M., Kussell, E., Domany, E.: Recovery of protein structure from contact maps. Folding and Design 2, 295–306 (1997)
12. Vullo, A., Frasconi, P.: A Bi-Recursive Neural Network Architecture for the Prediction of Protein Coarse Contact Maps. In: Proceedings of the 1st IEEE Computer Society Bioinformatics Conference, Stanford Univ. (August 2002)

Analysis of Kernel Based Protein Classification Strategies Using Pairwise Sequence Alignment Measures

Dino Franklin[1], Somdutta Dhir[2], and Sándor Pongor[2]

[1] Federal University of Goiás, Campus de Catalão, 75705-220 Catalão GO, Brazil
[2] International Centre for Genetic Engineering and Biotechnology, Padriciano 99, 34012 Trieste, Italy

Abstract. We evaluated methods of protein classification that use kernels built from BLAST output parameters. Protein sequences were represented as vectors of parameters (e.g. similarity scores) determined with respect to a reference set, and used in Support Vector Machines (SVM) as well as in simple nearest neighbor (1NN) classification. We found, using ROC analysis, that aggregate representations that use aggregate similarities with respect to a few object classes, were as accurate as the full vectorial representations, and that a jury of 6 1NN-based aggregate classifiers performed as well as the best SVM classifiers, while they required much less computational time.

1 Introduction

Automated classification of protein sequences are important issues in computational biology as human annotation is costly and time consuming. Automated protein function assignment methods are usually based on sequence similarity. A classical method of assessing the similarity of two proteins is sequence comparison via Dynamic Programming, but due to the large size of current databases, fast heuristic algorithms, such as BLAST [1] became the method of choice.

Liao and Noble [2] developed an efficient kernel method for detecting remote protein homology based on pairwise sequence comparison [2]. Every protein was represented as a vector of sequence similarity scores calculated with respect to a reference set of sequences (we term this method *full approach*). One can then train an SVM with these vectors, and use it to classify unknown proteins, represented as vectors built from similarity scores calculated with respect to the same reference dataset. Since this approach maps the proteins to high dimensional vectors, it has a high discriminative power at the expense of a relatively high computational cost.

In the so-called *aggregate approach* developed by the ICGEB Group for the SBASE library of protein domain sequences [3], the reference set is first aggregated into groups and a protein is then represented as a vector of similarities with respect to a set of groups. Because of the aggregation, the vectors are much smaller than the vectors used by the *full approach*). The success of this approach

F. Masulli, R. Tagliaferri, and G.M. Verkhivker (Eds.): CIBB 2008, LNBI 5488, pp. 222–231, 2009.

depends on how the sequences are grouped, and how the similarities with respect to the groups are defined; however, it potentially allows a substantial reduction in the computation time. This strategy was employed to a combination of various BLAST output parameters (*raw-score, hsp, etc.*) that were used to train SVM classifiers used in the SBASE search engine [3].

In this paper we compare the full vector representation with various levels of aggregate representations. We show that the performance of the full approach can be reached using aggregate strategies, and further improvements are possible employing a jury of Nearest Neighbor (1NN) classifiers that use different BLAST output parameters.

2   Methods

2.1   BLAST Output

BLAST (acronym for Basic Local Alignment Search Tool) is the most widely used alignment algorithm for protein sequence comparison [1]. The fundamental unit of BLAST, the *hsp* (High-scoring Segment Pair), is a pair of sequence segments (taken from the two aligned sequences) whose local alignment score exceeds a cutoff score. *msp* (Maximal-scoring Segment Pair) is the *hsp* with the highest score in a pairwise comparison of two sequences, given a set of *hsps* and a scoring system. For each *hsp* BLAST also outputs a raw score (*rs*), a bit-score (*bs*) which is independent of the scoring-matrix; and an *e-value* which estimates the number of *hsps* having a *score* S (or higher) to have occurred by chance. In this paper, *msp* and *hsp* are interchangeable because we only consider the best *hsp*.

2.2   Datasets

LN is a protein sequence dataset especially designed for testing classifiers [2]. This dataset of 4,352 protein sequences was obtained from SCOP v.1.53 after removing similar sequences using an e-value threshold of 10^{-25} . The 54 classification tasks were obtained by selecting families with at least 5 members for positive test set and superfamilies with at least 10 members outside the family for positive training set.

SCOP95 is a protein dataset obtained from SCOP database v.1.69 by considering proteins with sequence identity lower than 95%. The 11,944 entries are distributed in 246 classification tasks.

SCOP40mini is a subset of SCOP95, but the sequences are 40% similar at most. Using a strategy similar to [2], 1,357 protein sequences are divided in 55 tasks.

CATH95 is a protein dataset obtained from proteins in CATH v.3.0.0 with sequence identity below 95%. The tasks were derived selecting similarity groups (S) with at least 5 members, and groups with at least 10 members outside S but within the same homology group (H) for the positive set. CATH95 comprehends 165 classification tasks and 11,373 protein sequences.

SCOP40mini and LN are non redundant datasets. The benchmark datasets used in this work are available in:

1) `http://www.cs.columbia.edu/compbio/svm-pairwise` [2];
2) `http://net.icgeb.org/benchmark/` [7]

2.3   Classification Algorithms and Performance Assessment

In this study, we have performed classification simulations using Nearest Neighbor (1NN) and Support Vector Machines (SVM) [4]. 1NN assigns to an object the class of the closest training example in the feature space. SVM is a classical algorithm that computes an optimal hyperplane which separates the positive from the negative training sets. SVM classification is based on the query position regarding this hyperplane. In this work we used the SVM algorithm as implemented in Ref.[9].

Classification performance was characterized using the Receiver Operating Characteristic (ROC curve) using the ROCR package [6]. ROC curve is a plot of true positive rate vs. false positive rate through every prediction threshold value. The Area Under ROC Curve (AUC) is 0.5 for random prediction, and it is 1 when all positive examples have higher predicted values than the negative ones. The Average Area under the ROC curve is an average of AUC values calculated from a number of classification tasks defined on a larger database [2,5,6].

3   Results

3.1   Comparing Various Aggregation Strategies

Suppose a set of proteins is divided into classes, folds, superfamilies, and families as in the diagram presented in Figure 1. A protein can then be represented either as a vector of similarity scores, calculated with respect to all proteins in the set, or as a smaller vector, composed of similarity scores calculated with respect to protein families. Using vectors calculated with respect to families, superfamilies, folds and classes, we obtain smaller and smaller vectors. These vectors can then be used in classifiers using a classical machine learning arrangement as described in [2,7]. In Figure 1, the black squares represent the remote family to be detected.

Figure 2 shows the classifier performance using full pairwise representation for the SCOP40mini dataset. The results show that the classifier performance is usually better if one uses *rs* as kernel, followed by *e-value* and *hsp*. The tests have also shown that classification performance is task and dataset dependent.

The Unary Aggregation is the simplest possible representation which is based exclusively on the positive training set. It does not consider the negative examples when modeling the protein. In a previous paper , we have shown that using the Binary Aggregation, based on separately aggregated positive and negative training sets, considerably improves the performance of a classifier [5]. The most conspicuous result in Figure 4 is the excellent performance of the binary strategy based on the maxima of the positive and negative training sets. For most of the tests, the binary strategy, which is the Ref.[3] approach, has yielded the best AAUCs, sometimes even better than the full pairwise approach.

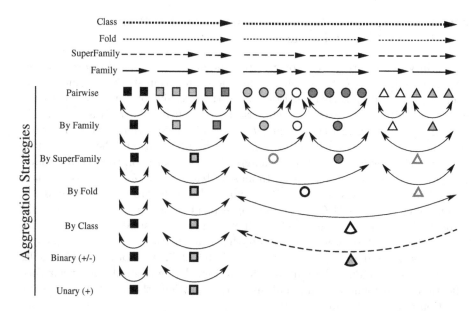

Fig. 1. Diagram of possible aggregations of the protein sequence similarities representation based on SCOP for superfamily classification. On the top, the arrows show the classes, folds, superfamilies, and families division. The proteins similarities are represented by geometric figures related to their folds, and shaded according to their families. The aggregations (maximum or average) are represented by arrowed arcs. On the left side, the aggregation criteria are shown. See text for further details.

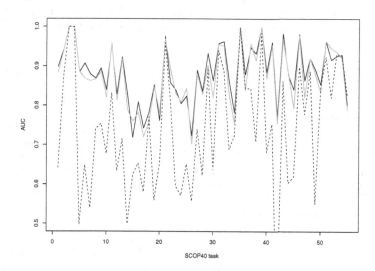

Fig. 2. AUC for every task of the SCOP40mini dataset for three BLAST output parameters: rs (solid line), hsp (gray line) and e-$value$ (dashed line). The tests were done using full pairwise representation [1] and an SVM with linear kernel. The parameters were optimized by simple grid search.

3.2 Correlation and Influence on Classification Performance of BLAST Output Parameters

The utility of various BLAST output parameters (*raw-score, bit-score, e-value, hsp-length* and *bit-score*) as well as different aggregation strategies (maximum, average) for building kernel functions was estimated via correlation analysis and classification tests on several benchmark datasets, using the 1NN algorithm.

The results in Table 1 show that *bit-score* and *e-value* are the BLAST output parameters that apparently contain most of the information. The classification performance is generally better when aggregates are based on the maximal similarity instead of average similarity. In most of tests, *raw-score* and *bit-score* have yielded the best AUCs. This is in fact expected as they are linearly related.

The tendency of AAUCs is in good agreement with the correlation analysis. Table 2 shows the average correlation between BLAST similarity measures for the LN dataset. It is interesting to note that some features are quite correlated, e.g. for bs_{max} and rs_{max}, the correlation is 0.99217. A priori, AAUCs results suggest that the higher the mutual information, the more similar are their classification performances. This analysis can be particularly relevant for feature selection in a multiple input approach.

Table 3 shows the classification performances of 1NN on different kernel spaces obtained from combined BLAST output parameters, e.g. $rs/len(query)$, raw-score divided by the length of the query. In some cases, a feature that has a high AUC in maximum aggregation, has a low AUC using average aggregation and vice-versa. This is because the positive and negative training sets are mapped to 2 scalars.

Finally, Figure 3 shows the performance (AAUCs) of SVM classifiers with a linear kernel using several single and combined BLAST output parameters as kernels and different representation strategies for the SCOP40mini dataset.

Table 1. Average Area Under Curve (AAUC) for 1NN classification based on Single BLAST output parameters

Aggregation Strategy: Maximum

Dataset	hsp	raw-score (rs)	BLAST output bit-score (bs)	e-value (ev)
LN (54)*	0.7840	0.7992	0.7992	0.7917
SCOP95 (250)	0.6971	0.6986	0.6986	0.6974
CATH95 (165)	0.9786	0.9789	0.9789	0.9787

Aggregation Strategy: Average

Dataset	hsp	raw-score (rs)	BLAST output bit-score (bs)	e-value (ev)
LN (54)	0.7640	0.7903	0.7902	0.7325
SCOP95 (250)	0.6936	0.6967	0.6967	0.6933
CATH95 (165)	0.9770	0.9788	0.9788	0.9755

* *Values in parenthesis mean quantity of tasks in the dataset.*

Table 2. Average of Correlation between BLAST output parameters for the LN dataset

	hsp_{max}	rs_{max}	bs_{max}	ev_{max}	hsp_{avg}	rs_{avg}	bs_{avg}	$Avg(log(ev))$
hsp_{max}	1							
rs_{max}	0.7253	1						
bs_{max}	0.702	0.9945	1					
ev_{max}	-0.6105	-0.8868	-0.8952	1				
hsp_{avg}	0.8758	0.7864	0.7808	-0.7029	1			
rs_{avg}	0.6214	0.9371	0.9632	-0.8981	0.7532	1		
bs_{avg}	0.6037	0.9235	0.9554	-0.8876	0.7370	0.9975	1	
$Avg(log(ev))$	-0.5421	-0.8696	-0.9146	0.8461	-0.6800	-0.9703	-0.9849	1

Table 3. Average Area Under Curve (AAUC) using 1NN and different kernel functions from combined BLAST output parameters for maximum or average aggregates

Combined BLAST output	LN (54)	SCOP95 (250)	CATH95 (165)
NSD *	0.7774	0.6975	0.9761
$Max(hsp/len(query))$	0.7086	0.6934	0.9784
$Max(hsp/len(subject))$	0.7828	0.6976	0.9788
$Avg(hsp/len(query))$	0.6900	0.6892	0.9770
$Avg(hsp/len(subject))$	0.7802	0.6946	0.9773
$Max(rs/len(query))$	0.5909	0.6872	0.9768
$Max(rs/len(subject))$	0.7744	0.6976	0.9786
$Max(rs/rs(query))$	0.5935	0.6873	0.9769
$Max(rs/rs(subject))$	0.7738	0.6976	0.9786
$Avg(rs/len(query))$	0.8135	0.6850	0.9736
$Avg(rs/len(subject))$	0.7942	0.6949	0.9782
$Avg(rs/rs(query))$	0.8187	0.6850	0.9737

* *NSD means number of neighbors. Max and Avg mean maximal and average aggregation. len() is for the sequence length of subject or query.*

Based on the results presented in Table 3 and Figure 3, we selected the best performing features for building combined classifiers. Among these, (NSD) defined as the number of significant similarities (similarity score above a certain threshold) is particularly interesting as is is not well correlated with the other parameters $Max(hsp/len(subject))$, $Avg(rs/len(query))$, etc.

Based on these results, we have trained SVM classifier that inputs 6 different BLAST output parameters as features. Although an optimal feature set is dataset dependent, the results in Figure 5 show that using a set of features can improve the classification performance as compared with an individual feature, but not significantly. Especially, the performance improvement is only a bit better than using rs_{max} alone.

3.3 Complexity Analysis

The total time for running a protein classification using a machine learning (ML) algorithm consists of training time and test time. The training time has three

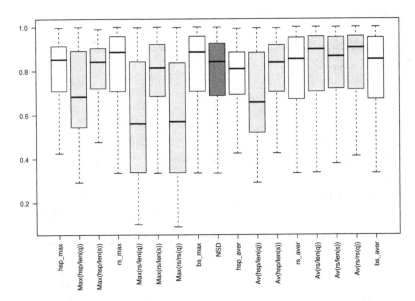

Fig. 3. Boxplot of AAUCs (1NN) for some single and combined BLAST output parameters for the LN dataset. The boxes for the combined output parameters are shown in light gray.

components: a) the alignment i.e. the BLAST running time; b) an aggregation time necessary to format the data to be used by the ML, and c) the learning time taken by the ML algorithm.

BLAST itself has three basic steps: a pre-alignment step, where all high scoring complementary k-words of the query sequence are selected and searched in the database; and the actual alignment phase, when the matched regions are expanded. Supposing that N is the length of the query, M is the length of the database, and k is the word length and, since there are 20^k possible words, the complexity of BLAST in terms of big $\mathcal{O}$ notation is given by $\mathcal{O}(N) + \mathcal{O}(N.20^k) + \mathcal{O}(M) + \mathcal{O}(M.N)$, i.e., $\mathcal{O}(M.N)$. Although, it has the same complexity as DP algorithms, BLAST is much more efficient ($\sim$ 500 times) since it discards non significant local alignments.

Concerning SVMs, standard training algorithms have time complexity of $\mathcal{O}(n^3)$, and space complexity of $\mathcal{O}(n^2)$, where n is the size of input data [8] $- n$ is the amount of sequences in the training set. Although there are improved algorithms that can reduce the time complexity to $\mathcal{O}(n^{2.3})$, and to $\mathcal{O}(n)$ with parallelization, our tests were performed in R language using SMO-type decomposition method [9]. The time complexity for the SVM training phase is $n_{SV}.Dn.n$, and $n_{SV}.Dn$ for test phase, where n_{SV} is the number of support vectors and Dn is the dimension of the input data. In a similar way, the space complexity for the training phase is n^2, and $n_{SV}.Dn$ for the test phase.

The total learning computational complexity of the full pairwise approach is $\mathcal{O}(n^3)$, since it comprehends obtaining the kernel using BLAST, $\mathcal{O}(n^2.M^2)$, and

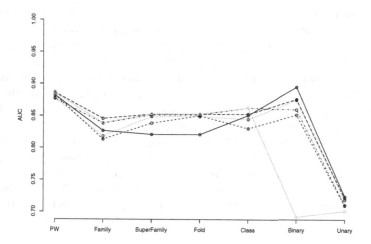

Fig. 4. Performance (AAUCs) of an SVM classifier with a linear kernel for different inputs – rs_{max} (solid), NSD (dashed), $Max(hsp/len(subject))$ (dotted), $Max(hsp/len(query))$ (dot-dashed), $Max(rs/len(subject))$ (long dash), and $Avg(rs/len(subject))$ (gray) – and representation strategies for the SCOP40mini dataset

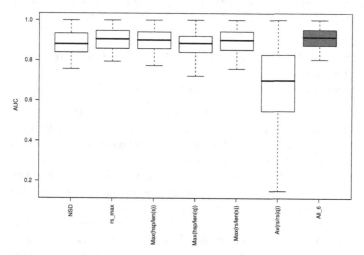

Fig. 5. Boxplot of AUCs for classifiers based on a linear SVM for the SCOP40mini dataset using 6 different similarity measures and a combination of them

the SVM training, $n^2.n_{SV}$. For the groupwise approaches the BLAST complexity remains the same, but an aggregation step with complexity $\mathcal{O}(n^2)$ is added to reduce the full pairwise matrix. Furthermore, the SVM training complexity step diminishes to $n_g.n.n_{SV}$ because of the reduction of the kernel matrix. In conclusion, the computational complexities for groupwise approaches are still polynomial, but one degree lower, $\mathcal{O}(n^2)$. It is noteworthy that using 1NN requires no training at all.

The first step of test phase consists of aligning a query to every sequence in training set using BLAST. This step has time complexity of $n.M^2$. BLAST running time are constant, independent of the algorithm or representation used. The complexity of the aggregation step is $\mathcal{O}(n)$. The SVM test complexity is given by $n^2.n_{SV}$ when using full pairwise strategy, and it is $n.n_{SV}$ for groupwise approaches. Hence, the computational complexity for test phase can be given by $\mathcal{O}(n^2)$ for the full approach, and $\mathcal{O}(n)$ for the aggregate ones. For test phase, the computational complexity using 1NN is equivalent to SVM.

The SVM time complexity used here was based on a linear kernel. When using a radial kernel, for example, the complexity increases proportionally to n_{SV}.

Briefly, the more efficient the aggregate representation, the lower is the learning complexity. Usually, running time for the test phase is smaller than that for the training phase. As expected, the most resource intensive strategy is the full approach.

3.4   A Protein Classifier Committee Machine

Based on the insights from analysis and tests performed here, we have designed a simple heuristic predictor that is based on a few BLAST output parameters. The predictor works like a voting system of 1NN predictors predicting protein function based on sequence similarity. The algorithm description is the following:

Given a query, one runs BLAST against the training set in order to obtain the categories of:

a) the protein (in the training set) which has higher raw-score (rs);
b) the protein which has higher hsp;
c) the proteins which have the five highest bit-score (bs).

The predictor uses the categories, i.e. functions of (a) and (b) as votes. A third vote comes from an aggregate feature that is given by the majority function from the five proteins in (c). The third vote is a variation of the NSD approach, adapted in such a way that categories with few instances are not underweight.

The basic idea is that the majority vote suggests the query function. If no majority is obtained, the Committee Machine follows the NSD vote, because it has yielded the best results during tests in SBASE [3].

The computational efficiency of the predictor is excellent. First, it does not require a training time and, secondly, it only needs running BLAST and ordering its output during tests. For example, if SVMs were used instead, it would require further training/tests for a set of SVMs equivalent to the amount of categories in the dataset.

In spite of its simplicity this 1NN-based Committee Machine has proved to be as accurate as the SVM classifier, providing the best performance on the SCOP40mini dataset. It consists of a set of linear SVMs that uses 6 different BLAST output parameters as inputs. Since the Committee Machine can not be assessed by ROC Curve, the classifiers performances are compared in terms of True Positive Rate (TPR) and True Negative Rate (TNR), in Table 4.

Table 4. Average Classification in terms of True Positive and Negative Rate for a set of 54 SVMs using a linear kernel and 6 BLAST output parameters as input vs. the simple Committee Machine for the SCOP40mini dataset

Classifier	TPR	TNR
54 SVM(All6)	0.5532	0.9446
1NN-Committee Machine	0.5067	0.9777

4   Conclusions

We have assessed the performance and computational complexity of protein classifiers based on kernel spaces obtained from pairwise sequence alignment. The binary representation strategy has presented one of the best classification performances, and also one of the minimum computational time cost.

We compared different BLAST output parameters in a feature selection analysis. The analysis showed that the feature performances are dataset dependent and there is no unanimously best similarity measures applicable to all datasets. However, some combined features have shown very good performance and could be taken into account when using a reduced approach.

Finally, we have shown that it is possible to design a simple and efficient protein classifier using a combination of separate classifiers into a voting system.

References

1. Altschul, S., Madden, T., Schaffer, A., Zhang, J., Zhang, Z., Miller, W., Lipman, D.: Gapped BLAST and PSI-BLAST: A New Generation of Protein Database Search Programs. Nucleic Acid Research 25, 3389–3402 (1997)
2. Liao, L., Noble, W.S.: Combining Pairwise Sequence Similarity and support Vector machines for Detecting Remote Protein Evolutionary and Structural Relationship. Journal of Computational Biology 10, 857–868 (2003)
3. Murvai, J., Vlahovicek, K., Barta, E., Pongor, S.: The SBASE Protein Domain Library, Release 8. 0: A collection of Annotated Protein Sequence Segments. Nucleic Acid Research 29, 58–60 (2001)
4. Vapnik, V.N.: Statistical Learning Theory. John Wiley & Sons, New York (1998)
5. Kaján, L., Kertesz-Farkas, A., Franklin, D., Ivanova, N., Kocsor, A., Pongor, S.: Application of a Simple Log-likelihood Ratio Approximant to Protein Sequence Classification. Bioinformatics 22, 2865–2869 (2006)
6. Sing, T., Sander, O., Beerenwinkel, N., Lengauer, T.: ROCR: Visualizing Classifier Performance in R. Bioinformatics 21(20), 3940–3941 (2005)
7. Sonego, P., Pacurar, M., Dhir, S., Kertézs-Farkas, A., Kocsor, A., Gaspari, Z., Leunissen, J., Pongor, S.: A Protein Classification Benchmark Collection for Machine Learning. Nucleic Acid Research 236, D232–D236 (2006)
8. Tsang, I., Kwork, J., Cheung, P.-M.: Core Vector Machines: FAST SVM Training on Very Large Data Sets. Journal of Machine Learning Research 6, 363–392 (2005)
9. Fan, R.-E., Chen, P.-H., Lin, C.-J.: Working Set Selection Using the Second Order Information for Training SVM. Journal of Machine Learning Research 6, 1889–1918 (2005)

Topology Preserving Neural Networks for Peptide Design in Drug Discovery

Jörg D. Wichard[1], Sebastian Bandholtz[2], Carsten Grötzinger[2], and Ronald Kühne[1]

[1] FMP Berlin, Robert-Rössle-Str. 10, 13125 Berlin, Germany
wichard@fmp-berlin.de
http://www.fmp-berlin.de
[2] Charité Universitätsmedizin Berlin,
Department of Hepatology and Gastroenterology,
Augustenburger Platz 1, D-13353 Berlin, Germany
http://www.charite.de

Abstract. We describe a construction method and a training procedure for a topology preserving neural network (TPNN) in order to model the sequence-activity relation of peptides. The building blocks of a TPNN are single cells (neurons) which correspond one-to-one to the amino acids of the peptide. The cells have adaptive internal weights and the local interactions between cells govern the dynamics of the system and mimic the topology of the peptide chain. The TPNN can be trained by gradient descent techniques, which rely on the efficient calculation of the gradient by back-propagation. We show an example how TPNNs could be used for peptide design and optimization in drug discovery.

1 Introduction

Building artificial neural networks with problem specific topology has a long tradition. In particular for pattern recognition the topology of the retina served as a template for the famous Perceptron (Rosenblatt [1], Minsky and Papert [2]) and later Fukushima's Cognitron [3] and Neogognitron [4]. Further examples of neural networks with particular topology are the self-organizing map of Kohonen [5] and the cellular neural network (CNN) of Chua et al. [6].

The idea of translating the topology of a chemical compound into a molecular graph network (MGN) was developed recently [7,8,9]. In a MGN a compound can be described as graph where each atom is a node and each chemical bond is an edge. In this way a chemical structure is translated into cellular network topology: Each atom becomes a cell, each bond a local interaction between the cells and the weights of the cells depend on element and bond types. The MGN works in this case as a discrete time cellular neural network [10] and the training is based on stochastic gradient descent as described in [11].

In the current work we propose a simpler strategy that is dealing with the particular structure of peptides. Instead of building an MGN based on the atoms of the peptide, we use neurons that represent the amino acids of the peptide as building blocks. This has two reasons: First of all, it reduces the complexity of the problem. The MGN approach works fine for small molecules but in the case of peptides there are on average 16 Atoms

F. Masulli, R. Tagliaferri, and G.M. Verkhivker (Eds.): CIBB 2008, LNBI 5488, pp. 232–241, 2009.

per amino acid which leads to large MGNs even for small peptides. As a consequence learning rates are quite poor and the adaption of weights is slow and time consuming. The second reason is the intrinsic structure of the problem. Our method should optimize the properties of peptides with respect to several pharmacological requirements. In this terms it is straightforward to build models for amino acids.

The TPNN is a tool for peptide design in drug discovery. Therefor it is necessary, that it captures the fundamental items of the underlying sequence-activity relation of the peptides under investigation. It is difficult to run any statistical method that could learn a correlation from a few datapoints in such a high dimensional space as the space of possible peptides: There are for example 10^{20} different possible decapeptides from the 20 natural amino acids. Many attempts have been made to utilize artificial neural networks in order to generate predictive models of quantitative sequence-activity relationships between a set of molecular descriptors and the related activity of the peptides. An overview could be found in Weekes and Fogel [12] or in Terfloth and Gasteiger [13].

Earlier works from Schneider and Wrede [14] proposed the use of neural networks and computer-based evolutionary search for peptide design. Their work was criticized by Darius and Rojas [15] who questioned the statistical significance of their results in general. But Schneider et al. [16] demonstrated later that their method was able to generate novel peptides with substantial biological activity in real world experiments. Another group reported the effective application of a tailored genetic algorithm for the de novo construction of peptidic thrombin inhibitors [17].

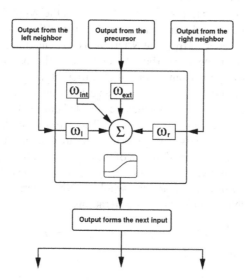

Fig. 1. This figure shows a sketch plot of a single cell which forms the building block of the topology preserving neural network. The internal state of the cell is defined by w_{int}. The inputs from the left and the right elements of the chain are given by w_l and w_r. The input from the precursor cell goes with w_{ext}. An individual cell has the weight vector $\boldsymbol{w} = (w_{int}, w_l, w_r, w_{ext})$. The sum of all inputs passes the activation function and forms the output for the next iteration.

We propose a new strategy that constructs a network architecture which mimics and preserves the topology of the peptide chain. The amino acids of the peptide chain are represented by cells with adapable weights (see figure 1). Training the network means adapting the weights of the cells that correspond to the amino acids of the peptide. The adapted cells are the building blocks of new *virtual* peptides and the resulting TPNN-model defines the fitness function in a genetic algorithm that is used to search for optimal peptides with respect to the desired pharmacological properties.

This paper is structured as follows: In the next Section we define the single cells which form the building blocks of the network and the translation process from an amino acid chain into a TPNN. Section 3 describes the training of the TPNN and in section 4 we give an example how TPNNs are used for peptide design and optimization in drug discovery.

2   Peptides and Topology Preserving Neural Networks

Peptides are short polymers of amino acids that are linked with an amide bond. Short in this sense means that the peptide chains are short enough to be made synthetically from the constituent amino acids. Among the variety of amino acids are the 20 proteinogenic amino acids the most important ones. They build the proteins of almost all known life forms. Nevertheless is our approach not restricted to he proteinogenic amino acids and can be extended to any set of *non-standard amino acids* which can help to improve the metabolic stability of the peptides.

A peptide has a representation as a string S wherein the symbols denote the amino acids. By convention a three letter code is used to name the most common ones. In general the sequence is reported from the N-terminal end containing free amino group to the C-terminal end containing free carboxyl group. We further assume that the peptides are composed of amino acids from a pool of M different individuals and we call this pool the *alphabet*. In a TPNN is each amino acid from the alphabet represented as a single cell with four individual weights that are adjusted during the network training. In figure 1 we show a schematic plot of such a cell. The internal weight of the cell is ω_{int}, the inputs from the left and the right neighboring cells are connected with the weights ω_l and ω_r. The input from the former iteration of the cell goes with ω_{ext}. The four weights are combined in the weight vector $\omega = (\omega_{int}, \omega_l, \omega_r, \omega_{ext})$. The cells are connected to form a chain as shown in figure 2 and the whole TPNN is iterated through time. The state of the $i - th$ TPNN cell y_i^t evolves for iterations $t = 0, \ldots, T - 1$ according to

$$x_i^{t+1} = \omega_{int} + y_{i-1}^t \omega_l + y_{i+1}^t \omega_r + y_i^t \omega_{ext} \tag{1}$$

$$y_i^{t+1} = \sigma(x_i^{t+1}), \tag{2}$$

wherein the activation function $\sigma(x)$ is a hyperbolic tangent with an additional linear term, that does not fully saturate for input values x far from the origin

$$\sigma(x) = \tanh(x) + \lambda x \quad \text{with } \lambda \ll 1. \tag{3}$$

This feature ensures a non-vanishing derivative of the training error which is important for training algorithms that use gradient information to adjust the weights [18]. The

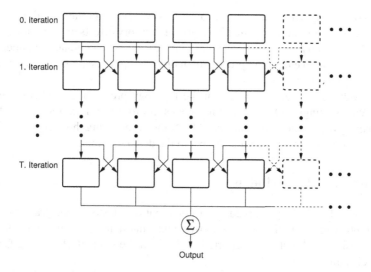

Fig. 2. The figure shows the linkage between the individual cells that mimic the chain topology of the peptide and the propagation of the states in a TPNN over time. After the last iteration, the mean over all cell outputs gives the final result (i.g. the activity of the peptide under investigation).

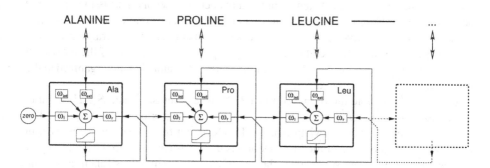

Fig. 3. This example shows how to translate the amino acid sequence $S = (Ala, Pro, Leu, \ldots)$ into a TPNN: The peptide starts with the amino acids *ALANINE-PROLINE-LEUCINE-…* and the cells that correspond to these amino acids are connected in the same order, thus preserving the topology of the amino acid chain of the peptide. The open ends of the TPNN get zeros as input.

number of iterations T is set to be the average length of the sequences under investigation. Translating a peptide into a TPNN is straightforward: The cells that correspond to the amino acids are connected in the same order as in the peptide as shown in figure 3.

The modelling task of the TPNN is to find a mapping between the peptide and an associated pharmacological property, in general the activity. This is called structure-activity relation or in case of peptides also sequence-activity relation. The biological activity of chemical compounds is usually measured in assays to establish the level of inhibition of particular signal transduction or a metabolic pathway. Chemicals can also be biologically active in terms of toxicity. The most important candidates in drug discovery are those compounds that have good inhibitory effects on specific targets and that have low toxicity. Our approach helps to identify these candidates and opens a way to implement an active learning procedure in order to optimize peptides with respect to the relevant properties.

The novelty of our approach lies in the topology preservation network structure, wherein we use single cells as building blocks in order to mimic the chain of amino acids forming the peptide. Training the TPNN means adapting the weights of the cells that represent the amino acids with respect to the desired output.

3 Training TPNN Models

The training of a TPNN is based on a combination of stochastic gradient descend and back propagation with several improvements that make the training of the shared weights feasible and that were reported in detail in the context of training CNNs for pattern recognition [19].

In stochastic gradient descent, the true gradient is approximated by the gradient of the loss function which is evaluated on a single training sample. The network weights are then adjusted by an amount proportional to this approximate gradient. A training sample consists of two parts: The first part is the peptide sequence S that is a composition of the M possible amino acids taken from the alphabet. The second part is the measured activity that could be a continuous value or a class label, for example the classes *active*, *weak-active* or *non-active*. Let's assume that we have a collection of N training samples $\{S_n, a_n\}_{n=1,...,N}$ and the weights $\omega^i = (\omega^i_{int}, \omega^i_l, \omega^i_r, \omega^i_{ext})_{i=1,...,M}$ of the individual cells that correspond to the M different amino acids in the alphabet are organized in the weight vector $\Omega = (\omega^1, \ldots, \omega^M)$.

Let $f(S_i, \Omega)$ denote the output of the TPNN for a given sequence S_i. This output value has to be compared to the training label a_i by means of a *loss function*. The loss function measures the deviation of the TPNN output from the desired value a_i. In optimization usually a quadratic loss function is used, basically due to the simplicity of the resulting derivatives. We propose the use of an ϵ–insensitive loss function λ_ϵ. The advantages of this strategy as well as the functional form of λ_ϵ are described in section 3.1. The training error $E(\Omega, \epsilon)$ is simply the loss averaged over the entire training set

$$E(\Omega, \epsilon) := \sum_{i=1}^{N} \lambda_\epsilon(a_i - f(S_i, \Omega)).$$

(4)

3.1 The ϵ–Insensitive Loss Function

The training error is calculated with the ϵ–*insensitive squared loss function*

$$\lambda_\epsilon(\xi) := \begin{cases} 0 & : \xi \leq \epsilon \\ (\xi - \epsilon)^2 & : \xi > \epsilon. \end{cases}$$

(5)

It calculates the loss contribution of the training samples. The output of the TPNN has zero loss and gradient if it lies inside the ϵ–margin of the desired output. This forces

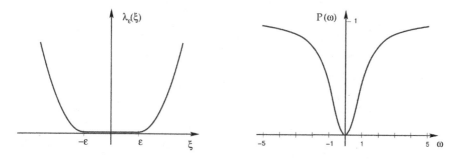

Fig. 4. On the left side: The ϵ-insensitive quadratic loss function. The right side shows the the penalty function for a single network weight $P(\omega) = \frac{\omega^2}{1+\omega^2}$.

the training algorithm to focus on the training samples that are not properly explained by the current model rather than adjusting the network weights by gradient steps of already correctly learned samples. The use of ϵ–insensitive loss functions in the context of learning problems is described in greater detail by Vapnik [20].

3.2 Weight Decay

Weight decay is a regularization method that penalizes large weights in the network. The *weight decay penalty term* causes the insignificant weights to converge to zero. This results in a model with a minimum number of free parameters, according to the principle of Occam's razor also known as the *law of parsimony*. It tells us to prefer the simplest of all equally good models. The penalty term is given by

$$P(\Omega) = \gamma \sum_{i=1}^{N} \frac{\omega_i^2}{1 + \omega_i^2}, \tag{6}$$

where Ω denotes the weight vector of the TPNN and the regularization parameter $\gamma = 0.001$ is small. The penalty term is added to the training error and contributes to the gradient. In figure 4 we show an one dimensional example.

3.3 Stochastic Gradient Descent and Backpropagation

Training a learning machine is put to effect by minimizing the training error as defined in Equ. 4 with respect to the network weights Ω. The method of training a TPNN presented in this paper is based on a *stochastic gradient descend* or *on-line learning*, which is a common method for training Artificial neural networks (see LeCun et al. for an overview [18]). The gradient of the entire training error from equ. 4 and the penalty term from equ. 6 is a sum of terms of the form

$$\frac{\partial}{\partial \Omega} \left[\lambda_\epsilon(a_i - f(\mathbf{S}_i, \Omega)) + P(\Omega) \right]. \tag{7}$$

Thus, a classical gradient descent or batch learning algorithm first computes all these terms and then performs a gradient step into the direction given by their sum. The

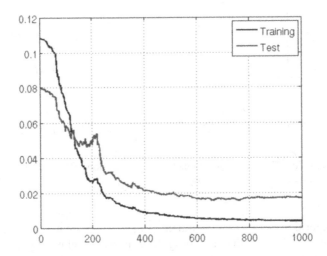

Fig. 5. Example for the convergence of training and test errors in a typical TPNN training of 1000 epochs

stochastic gradient descent performs a series of very small consecutive steps, determining each step direction from the gradient of an individual summand only. After each step, the new parameter set Ω is re–inserted into the loss before the next gradient is computed. In other words, we devise an update rule for the parameters of the form

$$\Omega_i = \Omega_{i-1} - \delta\Omega_i, \tag{8}$$

with $i = 1 \ldots N$, wherein N is the number of training samples. The update $\delta\Omega_i$ depends on the i-th training sample only and is given by

$$\delta\Omega_i = \mu\frac{\partial}{\partial\Omega}\left[\lambda_\epsilon(a_i - f(\boldsymbol{S}_i, \Omega_{i-1})) + P(\Omega_{i-1})\right]. \tag{9}$$

We calculate the update $\delta\Omega_i$ with the standard error back-propagation technique as it is used for the common feed-forward multilayer perceptron [18]. The parameter μ is controlling the stepsize of the gradient descend. The initial step size is already small (around $\mu = 0.01$) and it is decreased after each training epoch with a constant factor. This is necessary to achieve a slow convergence of the weights. Note that in each training step only a few selected values of the entire weight vector Ω are adjusted, namely the ones that correspond to amino acids that appear in the sequence of the training sample.

A sweep through the whole data set is called one *epoch*. Up to 1000 epochs are necessary to train the TPNN in a typical experimental setup (20 amino acids in the alphabet, peptides of length 5 to 10, 30-50 training samples from measurements as a starting set). In figure 5 we show the training and test errors for a simple example. The task was to learn the molecular weight of peptides of length 8 from only 20 examples in the training set. The slow convergence of the error is a consequence of the relative small stepsize in the gradient descent, but it is necessary in order to get overall convergence of the weights.

3.4 Building Ensembles

A common way to improve the performance of neural networks in regression or classification tasks is ensemble building [21]. It is well known, that neural network ensembles perform better in terms of generalisation than single models would do [22,23]. An ensemble of TPNNs consists of several single TPNN models that are trained on randomly chosen subsets of the training data and the training starts with random weight initializations. This ensures the diversity of the resulting models which is the key issue in ensemble building. To compute the output of the ensemble for one input sequence, the output variables of all TPNNs belonging to the ensemble are averaged.

4 TPNN Models and Peptide Design

The TPNN was developed to model the sequence activity relation of peptides. If we assume that a TPNN learns some important properties of peptides with respect to the pharmacological assay under investigation then we could use this TPNN for the optimization of peptides in terms of biological activity. The general way is to build a proper model that is further used as a fitness measure in an optimization algorithm. In the case of peptide design, any standard evolutionary algorithm [24,25,26,27] could be used to explore the high dimensional space of possible peptides.

Similar attempts have been reported by Schneider and Wrede [14,28] and Schneider et al. [16] termed *simulated molecular evolution*. In their approach an artificial neural network is used to model the locally encoded amino acid sequence features of the peptides. In contrast to our approach, the neural networks are used in a conventional way as nonlinear function approximation, while we explicitly use the chain topology of the peptide to construct the TPNN model.

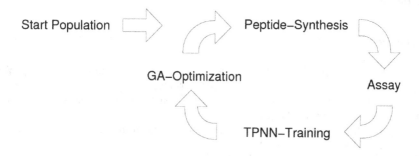

Fig. 6. The TPNN-based peptide optimization requires several cycles. The start population is generated based on experts knowledge concerning the target, i.g. natural ligands, known ligands or compounds that bind to similar targets and variations thereof. The start population is tested in the biological assay and the results are used to train the TPNN. The trained TPNN is used as fitness function in a GA in order the generate new suggestions for peptide sythesis. The results of the biological assay are included in the next TPNN training cycle and so on.

A typical cycle for peptide design works as follows (see figure 6). A start population of peptides is selected, based on prior knowledge of the target under investigation. If there are no natural ligands that bind to the target, one could look for ligands that bind to similar targets and build a small peptide library based on that. The sequence strings of the peptides together with the measurements from the biological assay deliver the data for TPNN training. The adopted cells are the building blocks for the assembling of new (virtual) peptides and the resulting TPNN is the fitness function in a genetic algorithm (GA) that is generating new suggestions for peptide synthesis based on the learned structure-activity relation. These suggestions are synthesized and tested in the biological assay and the results are included in the next TPNN training cycle. This process is repeated several times and improves the peptides from one cycle to the next.

5   Conclusion and Outlook

We developed a new topology preserving network architecture - the TPNN - that mimics the chain structure of peptides. Training a TPNN means to adopt the weights of the cells that correspond to the amino acids of the peptide and thus learning the underlying structure-activity relation of the peptide. The trained cells work as building blocks for the assembling of new (virtual) peptides and the resulting TPNN works as fitness function in GA based search of the sequence space.

Thus the concept of TPNN is the core of a novel peptide optimization process for drug discovery. This is work in progress. We developed and tested our method with some toy problems and in-house data from earlier peptide screening campaigns. The experiments for a inhibition of a specific receptor started recently and in the first cycle we have promising results.

Acknowledgments

The authors would like to thank the members of the Molecular Modelling Group at FMP Berlin.

References

1. Rosenblatt, F.: The perceptron: A probabilistic model for information storage and organization in the brain. Psychological Review 65, 386–408 (1958)
2. Minsky, M., Papert, S.: Perceptrons: An Introduction to Computational Geometry. MIT Press, Cambridge (1969/1988) (expanded edn.)
3. Fukushima, K.: Cognitron: A self-organizing multilayered neural network. Biological Cybernetics 20(3-4), 121–136 (1975)
4. Fukushima, K.: Neocognitron: A self-organizing neural network model for a mechanism of pattern recognition unaffected by shift in position. Biological Cybernetics 36(4), 193–202 (1980)
5. Kohonen, T.: Self-organized formation of topologically correct feature maps. Biological Cybernetics 43(1), 59–69 (1982)
6. Chua, L., Yang, L.: Cellular neural networks: Theory. IEEE Trans. on Circuits and Systems 35, 1257–1272 (1988)

7. Merkwirth, C., Lengauer, T.: Automatic generation of complementary descriptors with molecular graph networks. Journal of Chemical Information and Modeling 45(5), 1159–1168 (2005)
8. Ogorzałek, M., Merkwirth, C., Wichard, J.: Pattern recognition using finite-iteration cellular systems. In: Proceedings of the 9th International Workshop on Cellular Neural Networks and Their Applications, pp. 57–60 (2005)
9. Merkwirth, C., Ogorzałek, M.: Applying CNN to cheminformatics. In: Proceedings of the ISCAS, pp. 2918–2921 (2007)
10. Harrer, H., Nossek, J.: Discrete-time cellular neural networks. Int. J. Circuit Theory and Applications 20, 453–467 (1992)
11. Merkwirth, C., Bröcker, J., Ogorzałek, M., Wichard, J.: Finite Iteration DT-CNN - New Design and Operation Principles. In: Proceedings of the ISCAS, Vancouver, Canada, vol. 5, pp. 504–507 (2004)
12. Weekes, D., Fogel, G.: Evolutionary optimization, backpropagation, and data preparation issues in QSAR modeling of HIV inhibition by HEPT derivatives. Biosystems 72(1-2), 149–158 (2003)
13. Terfloth, L., Gasteiger, J.: Neural networks and genetic algorithms in drug design. Drug Discovery Today 6(suppl. 2), 102–108 (2001)
14. Schneider, G., Wrede, P.: The rational design of amino acid sequences by artificial neural networks and simulated molecular evolution: De novo design of an idealized leader peptidase cleavage site. Biophysical Journal 66, 335–344 (1994)
15. Darius, F., Rojas, R.: Simulated molecular evolution or computer generated artifacts? Biophysical Journal 67, 2120–2122 (1994)
16. Schneider, G., Schrodl, W., Wallukat, G., Müller, J., Nissen, E., Ronspeck, W., Wrede, P., Kunze, R.: Peptide design by artificial neural networks and computer-based evolutionary search. Proceedings of the National Academy of Sciences 95(21), 12179–12184 (1998)
17. Kamphausen, S., Höltge, N., Wirsching, F., Morys-Wortmann, C., Riester, D., Goetz, R., Thürk, M., Schwienhorst, A.: Genetic algorithm for the design of molecules with desired properties. Journal of Computer-Aided Molecular Design 16(8-9), 551–567 (2002)
18. LeCun, Y., Bottou, L., Orr, G., Müller, K.: Efficient BackProp. In: Orr, G., Müller, K. (eds.) NIPS-WS 1996. LNCS, vol. 1524, pp. 9–50. Springer, Heidelberg (1998)
19. Wichard, J., Ogorzałek, M., Merkwirth, C., Bröcker, J.: Finite iteration DT-CNN with stationary templates. In: Proceedings of the 8th IEEE International Workshop on Cellular Neural Networks and their Applications, Budapest, Hungary, pp. 459–464 (2004)
20. Vapnik, V.: The Nature of Statistical Learning Theory. Springer, New York (1999)
21. Hansen, L., Salamon, P.: Neural network ensembles. IEEE Trans. on Pattern Analysis and Machine Intelligence 12(10), 993–1001 (1990)
22. Geman, S., Bienenstock, E., Doursat, R.: Neural networks and the bias/variance dilemma. Neural Computation 4, 1–58 (1992)
23. Perrone, M.P., Cooper, L.N.: When networks disagree: Ensemble methods for hybrid neural networks. In: Mammone, R.J. (ed.) Neural Networks for Speech and Image Processing, pp. 126–142. Chapman-Hall, Boca Raton (1993)
24. Fogel, L.J., Owens, A.J., Walsh, M.J.: Artificial Intelligence through Simulated Evolution. John Wiley, New York (1966)
25. Rechenberg, I.: Evolutionsstrategie: Optimierung technischer Systeme nach Prinzipien der biologischen Evolution. Frommann Verlag, Stuttgart (1973) (in German)
26. Holland, J.H.: Adaptation in natural and artificial systems. MIT Press, Cambridge (1975)
27. Koza, J.R.: Genetic programming: On the programming of computers by means of natural selection. MIT Press, Cambridge (1992)
28. Schneider, G., Wrede, P.: Artificial neural networks for computer-based molecular design. Progress in Biophysics and Molecular Biology 70(3), 175–222 (1998)

A Machine Learning Approach to
Mass Spectra Classification
with Unsupervised Feature Selection

Michele Ceccarelli[1,2,3], Antonio d'Acierno[4,*], and Angelo Facchiano[4]

[1] Department of Biological and Environmental Sciences, University of Sannio
Via Port'Arsa 11, 82100, Benevento, Italy
ceccarelli@unisannio.it
[2] Research Center on Software Technologies, University of Sannio
Via Traiano 1, 82100, Benevento, Italy
[3] Bioinformatics Core, Biogem, Ariano Irpino, Italy
[4] Institute of Food Sciences, Italian National Research Council
Via Roma 52 A/C, Avellino, Italy
{antonio.dacierno,angelo.facchiano}@isa.cnr.it

Abstract. Mass spectrometry spectra are recognized as a screening tool
for detecting discriminatory protein patterns. Mass spectra, however, are
high dimensional data and a large number of local maxima (a.k.a. *peaks*)
have to be analyzed; to tackle this problem we have developed a three-
step strategy. After data pre-processing we perform an unsupervised feature selection phase aimed at detecting salient parts of the spectra which
could be useful for the subsequent classification phase. The main contribution of the paper is the development of this feature selection and
extraction procedure grounded on the theory of multi-scale spaces. Then
we use support vector machines for classification. Results obtained by
the analysis of a data set of tumor/healthy samples allowed us to correctly classify more than 95% of samples. ROC analysis has been also
performed.

1 Introduction

SELDI-TOF (Surface-Enhanced Laser Desorption and Ionization Time-Of-Flight)
technology is considered a modified form of MALDI-TOF (Matrix-Assisted Laser
Desorption and Ionization Time-Of-Flight). According to these techniques, proteins are co-crystallized with UV-absorbing compounds, then a UV laser beam
is used to vaporize the crystals, and ionized proteins are then accelerated in an
electric field. The analysis is then completed by the TOF analyzer. Differences
in the two technologies, which reside mainly in the sample preparation, make
SELDI-TOF more reliable for biomarkers discovery and other proteomic studies
in biomedicine.

The proteomic characterization by means of TOF (both SELDI and MALDI)
technologies of samples from individuals is considered to carry information about

---

* Corresponding author.

F. Masulli, R. Tagliaferri, and G.M. Verkhivker (Eds.): CIBB 2008, LNBI 5488, pp. 242–252, 2009.
© Springer-Verlag Berlin Heidelberg 2009

the healthy or pathological state of the individual. In fact, samples as serum, plasma, and other kinds of extracts contain proteins for which the covalent structure may be modified in specific pathologies, which may induce modifications as glycation or methylation, which imply the addition of a small molecule to the protein, or may alterate and prevent expected modifications. In any of these cases, the proteomes of samples by an healthy individual and an affected individual should be discernible, being their mass profile altered. Therefore, among the thousands of proteins and peptides present in a serum sample, which represent its proteome, few key signals may be significant markers of the pathological state, and their search within the proteome represents a still open field of research.

Data produced by mass spectrometry (the spectra) are represented by a (typically) very large set of measures representing the quantity of biomolecules having specific mass-to-charge (m/z) ratio values. Given the high dimensionality of spectra, given their different length and since they are often affected by errors and noise, preprocessing techniques are mandatory before any data analysis.

After preprocessing (to correct noise and reduce dimensionality), several statistical and artificial intelligence based technologies could be used for mining these data. In [1], for example, genetic algorithms and self organizing maps were used to distinguish between healthy women and those affected by ovarian cancer. Support Vector Machines (combined with Particle Swarm Optimization) have been used [6] to distinguish cancer patients from non-cancer controls. Principal Component Analysis has been used for dimensionality reduction followed by linear discriminant analysis on SELDI spectra of human blood serum [8]. Several statistical methods have been compared in [9]. Nearest centroid classification has been used in several applications; in [9], for example, it is used for protein mass spectrometry while ant colony optimization has been used [7] for peak selection from MALDI-TOF spectra. Independent component analysis has been recently used [10] for the extraction of protein signals profiles.

In this paper we first describe a method we have implemented to extract features describing the spectra. Principal Component Analysis has been then used to further reducing features dimensionality and finally a Support Vector Machine (SVM) has been applied for classification obtaining very interesting results. The experimental data analyzed were derived from a study on women affected and unaffected by ovarian cancer. The serum samples were analyzed, as described in detail in the article by Liotta et al. ([2]), by mass spectrometry techniques, in particular by SELDI-TOF.

The paper is organized has follows. In section 2 we describe data preprocessing, features extraction/reduction and classification; then, in section 3, we describe experimental results while final discussion, some conclusions and future work are the concerns of section 4.

2 Data Preparation and Classification

2.1 Data Preprocessing

Before the feature selection phase, there is a preprocessing step aimed at homogenization and correction of the spectra data.

The spectral data produced by a single laser shot in a mass spectrometer consists of a vector of counts. Each count represents the number of ions hitting the detector during a small, fixed interval of time. A complete spectrum is acquired within tens of milliseconds, so a typical spectrum is a vector containing between 10000 and 100000 entries. In practice, most mass spectrometers produce spectra by averaging the counts over many individual laser shots. Thus, the raw data produced by running a sample through a mass spectrometer can best be thought of as a time series vector containing tens of thousands of real numbers. Unless an entry in the vector is known to represent an actual count of the number of ions, it is usually just called an intensity and is assumed to be measured in continuous arbitrary units. Peaks in a plot of the intensity as a function of time represent the proteins or peptides that are present in the sample. A typical data set arising in a clinical application of mass spectrometry contains tens or hundreds of spectra; each spectrum contains many thousands of intensity measurements representing an unknown number of protein peaks. Any attempt to make sense of this volume of data requires extensive low-level processing in order to identify the locations of peaks and to quantify their sizes accurately. Inadequate or incorrect pre-processing methods, however, can result in data sets that exhibit substantial biases and make it difficult to reach meaningful biological results.

In our experiments we applied the following preprocessing steps:

- *resampling:* Gaussian kernel reconstruction of the signal in order to have a set of d-dimensional vectors with equally spaced mass/charge values;
- *baseline correction:* removes systematic artifacts, usually attributed to clusters of ionized matrix molecules hitting the detector during early portions of the experiment, or to detector overload;
- *normalization:* corrects for differences in the total amount of protein desorbed and ionized from the sample plate.

All the above steps were implemented and applied by using the MATLAB programming environment and the Bioinformatics Toolbox.

2.2   Feature Selection

The feature selection and description is crucial for mass spectrometry since subsequent analysis are performed only on the selected features. Several methods have been proposed which often rely on biased data sets and can reach biological conclusion difficult to be interpreted [11,12]. Peak detection is the traditional method for extracting features and several techniques to identify peaks among the background noise have been proposed (see for example [13]). Recently model based approaches have been revaluated for the phase of feature selection of mass spectra data [14] claiming that this approach can give a better representation of the MS signals by incorporating information about peaks shapes ans isotropic distributions. The models based methods typically perform a huge number of regressions to fit signal models to spectra. Here we adopt a hybrid method which is fast just as the peak selection methods, and at the same time tries to model the average spectrum at various scales. The basic principle adopted for the selection of features relies on the scale space theory of signal analysis [15,16]. The main idea of a scale-space representation is to generate a one-parameter family of derived signals in which the fine-scale information is successively suppressed. The main property of a scale-space analysis is the *causality* principle, which states that each feature at a given scale must have a cause at a previous scale. This principle preserves peaks or other feature to be artificially introduced through scales and forces the analysis to be from finer scale to coarser scales.

We assume that the peaks can be (in some way) profitably used to describe the spectrum itself; when, however, two peaks are too close they should be considered as a single maximum. Therefore the multiscale analysis can help in observing the same signal at coarser scales for feature detection and signal matching purposed [15] as the scale increases the signal becomes coarser. Here we adopt a linear scale space which is implemented through a Gaussian kernel [16]. In particular, here the scale is just the maximum width, σ, of the Gaussian kernel used to filter the signal. As can be seen in figure 1, some peaks collapses in a single local maximum. Clearly, as σ increases, we have smother versions of our spectra. Our feature selection phase is based on the mean of the signals at the maximum chosen scale. In particular, the local maxima of the mean of the smoothed signals are considered as the locations of the considered peaks to be used as features, see figure 2. Finally, each spectrum will be described by the mean value assumed by the original spectrum in window centered in each of the selected local maxima.

As the value of sigma increases the number of extracted features decreases, for example, at a scale of 0.1 we have 156 local maxima of the mean smoothed spectrum, whereas at scale 1.0 we have just 23 components. Therefore, as a last feature extraction step we perform a principal component analysis for dimensionality reduction (see figure 3).

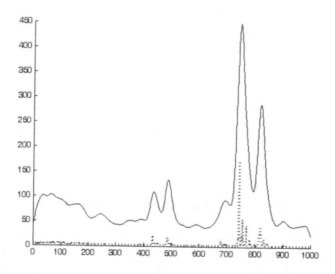

Fig. 1. A spectrum (dotted) end its regularized version

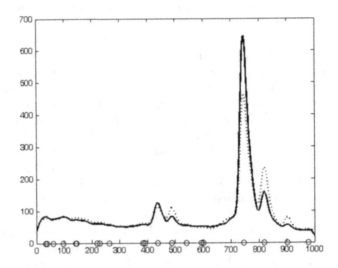

Fig. 2. The curve obtained summing regularized spectra of cancer data (dotted) and the one one obtained summing the healthy data. Circles on the m/z axis represent local maxima points.

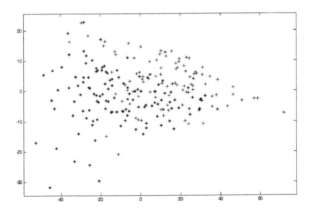

Fig. 3. The data we have to classify: the first component against the second one

2.3 Classification

A *learning machine* is any of such functions' estimation algorithm. The quality of the machine is evaluated in terms of the mean classification error as the number of training samples goes to infinity. The machine acts as a classifier and what we want from such a classifier is that it generalizes well, that is that it has a good balance between the *capacity* and the *accuracy* of the classifications performed, the *capacity* being the possibility of learning any given set of labels and the *accuracy* being the right classification's percentage on the training data. Rather than a classifier that has the best training accuracy, we want a classifier that do not over-fits training data and which generalize well to unseen new data. There's an extensive theoretical work on probability bounds on the actual risk of misclassification of a pattern recognition learning machine and bounds on generalization performances are the subject of many institutional text books (i.e. [3]).

Support Vector Machine (SVM) is a technique proposed [17] for Pattern Recognition and Data Mining classification tasks. While at present there exists no general theory that guarantees good generalization performances of SVM, but only probability bounds on its performance accuracy, there is a growing interest in this technique due to ample literature demonstration of good performances in various heterogeneous fields [18].

The main advantage of SVM over, for example, Feed Forward Neural Networks, is that it has no local minima problems and that it has less parameters to choose. So, although there is still much study to do, i.e. in how to choose the kernel and how to extend the method to the multi-label case [20], we found reasonable to test the use of this method with our data. Results reported below show very good generalization performances with our proteomic data, giving another empirical argument to its potential powerful generalization properties.

Linear SVM. In the case of linearly separable patterns on two-classes vectors it is straightforward to show the basic ideas of SVM: given a set of points in $\Re^k$ and a two-classes labels vector, SVM aims to find a linear surface that splits them in two groups according to the indicated labels, in the best possible way. Intuitively, if data are linearly separable (that is if it exists *at least* one hyperplane that splits them in two group), the problem becomes how to define and how to find *the best* possible hyperplane to do it. The SVM answer is that the best possible hyperplane is the one that maximizes the *margin*, that is the one that has maximal distance from *both* sets of points (Figure 4).

To be more rigorous, the problem of finding the hyperplane that has maximal *margin* among two sets of point differently labeled can be formulated as a constrained quadratic optimization problem

$$min(\frac{1}{2}\|w\|^2) \tag{1}$$

subject to:

$$y_i(w^T x_i + b) \geq 1 \tag{2}$$

where $y_i \in \{-1, 1\}$ are classes labels, w is the normal to the hyperplane and $2/\|w\|$ is the margin, that is the distance between both sets of points.

Its dual formulation with Lagrange Multipliers is the following:

$$max(L_D = \sum_i \alpha_i - \frac{1}{2} \sum_{i,j} \alpha_i \alpha_j y_i y_j x_i x_j) \tag{3}$$

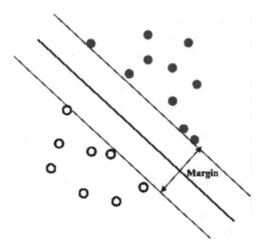

Fig. 4. Maximal margin principle on a set of two classes linearly separable samples

subject to:

$$\sum_i \alpha_i y_i = 0; \quad \alpha_i \geq 0 \tag{4}$$

where α_i are Lagrange Multipliers. This problem leads to the solution:

$$w = \sum_i \alpha_i y_i x_i \tag{5}$$

$$\sum_i \alpha_i y_i = 0; \quad \alpha_i \geq 0 \tag{6}$$

We can note looking at the solution that the α_i are nonzero only for data lying on the marginal hyperplanes [3]. This fact has an important advantage: even removing a random subset of points not lying on the marginal hyperplanes the classification's result remains unchanged, so the technique is robust against perturbation of non-marginal points. The name of "support vectors" given to these points is due to this important property.

It can even happen that no hyperplane separates the given points. In this cases it is nonetheless possible to apply the SVM method assigning a penalty to each point in the wrong class, the weight of the penalty being a user-defined parameter. The optimization problem becomes to maximize:

$$L_D = \sum_i \alpha_i - \frac{1}{2} \sum_{i,j} \alpha_i \alpha_j y_i y_j x_i x_j \tag{7}$$

subject to:

$$\sum_i \alpha_i y_i = 0; \quad 0 \leq \alpha_i \leq C \tag{8}$$

where C is the "cost" of misclassified points. The only difference from the optimal hyperplane case is that the α_i now have an upper bound of C.

Nonlinear SVM. To generalize further, we could consider surfaces that are not linear and work on a different model or we could project all data non linearly in another space (possibly with infinite dimension) where they are linearly separable and perform the classification linearly in this new space, maintaining the method almost unchanged. In practice if we look at the optimization's problem formulation, we see that data appear only in the form of dot products $x_i \cdot x_j$ and hence data transformed through a function $\Phi : \Re^k \mapsto \Gamma$ (where Γ is a space of dimension $h \geq k$) appear also in the form of dot products $\Phi(x_i) \cdot \Phi(x_j)$. If we consider a dot product function $K(x_i, x_j) = \Phi(x_i) \cdot \Phi(x_j)$ it is possible, in line of principle, to substitute K in formulae and to compute the solution without even knowing the form of function Φ. Such a dot product function is called *Kernel*. Whatever function that satisfies dot product's constraints can be used as *Kernel* function, and there is an active field o research in the choice of the most suitable kernel for a given problem [4].

Another aspect of SVM is that the above discussion about *binary* classifiers can be easily extended to multiclass problems. The two most common approaches to the extension to the multi-label case are the 'one-against-one' and the 'one-against-all' [5,20]. For multiclass-classification with k levels, $k > 2$, we have used the 'one-against-one' approach, in which $k(k-1)/2$ binary classifiers are trained and the appropriate class is found by a voting scheme.

3   Experimental Results

In this paper we tested support vector machines [21] using radial basis functions as kernel functions. In the following we will first describe the tuning of the system and then we will analyze in details the performance of our classificator.

3.1   System Tuning

The whole performance of our approach of course depends on (at least) four parameters: the variance σ of the Gaussian function, the size w of the window, the number n of principal components to be considered and the variance v of the RBF kernel functions; thus we decided to perform several tests having in mind to tune the whole system. Namely, we tested the classification performance having σ varying in the interval $[0.1:1]$, the window size varying in the interval $[1:21]$, the number of principal components used for classification varying in the interval $[2:10]$ and v varying in the interval $[0.5:5]$. For each parameters' quadruplet k−fold (with $k = 10$) cross validation has been used to test the generalization performance; here, as it is well known, the data set is divided in k subsets and k trials are performed using one of the subsets as test sets and the remaining $k - 1$ subsets as training set. Each test has been repeated 10 times, so deriving that each quadruplet has been tested 100 times: in this phase, the mean correct classification value has been clearly used as quality measure.

The best mean correct classification rate (97%) has been obtained having $\sigma = 0.1$, a size of the window equal to 3, using 8 components and having $v = 3$.

3.2   ROC Analysis

As it is well known, a receiver operating characteristics (ROC) graph is a technique for visualizing and selecting classifiers based on their performance [19]. As it is well known, a ROC graph is a two-dimensional graph where on the X axis is plotted the *false positive rate (FPR)* and the *true positive rate (TPR)* is plotted on the Y axis, where:

$$TPR = \frac{Positives\ correctly\ classified}{Total\ positives} \tag{9}$$

and

$$FPR = \frac{Negatives\ correctly\ classified}{Total\ negatives} \tag{10}$$

Many classifiers are designed to produce just a class decision (say true or false) and so a single confusion matrix thus deriving that these classifiers are represented by a single point in the ROC space. Some classifiers, like SVM classifiers, produce, on the other hand, a *score* that is then interpreted as the degree to which a given instance is a member of a class. Scoring classifiers are typically used with a threshold to produce a binary classifier; for each threshold we have confusion matrix and so a point in the ROC space, thus deriving a ROC curve.

According to [19], and considering the *best* system (i.e. the tuned one), we start from the 10 test sets $T_1, T_2, \ldots, T_{10}$ using 10-folds cross validation. We perfomed this set 10 times so obtaining 10 ROC graphs merged using *vertical slicing* [19]. Here each ROC graph (R_i) is treated as a function $TPR = R_i(FPR)$; the vertically averaged ROC curve is defined as the function:

$$\widehat{R}(FPR) = mean(R_i(FPR)) \tag{11}$$

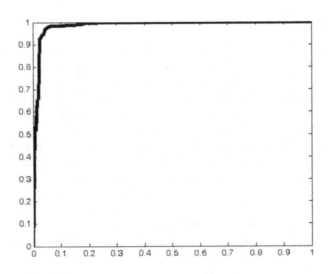

Fig. 5. The vertically averaged ROC curve

Figure 5 shows the obtained graph. The area under the curve (AUC), proved to be equal the probability that the classifier ranks a randomly chosen positive instance higher than an randomly chosen negative instance, for our system is 0.986.

4 Discussion

In this paper we presented a three-steps strategy for classifying SELDI spectra. After pre-processing we described features' extraction and classification by means of SVMs obtaining very interesting results.

The features' extraction we propose is worth to be emphasized; even if in the paper it has been described without a strong theoretical formalism, the process we are using is heavily based on multi-scale analysis and the results we have obtained demonstrate the goodness of such an approach.

Several open problems need to be addressed. First of all, our method has to be tested with different data sets. Another problem we have to face is to compare our approach with other proposed solutions; this is not a trivial problem since in many cases data sets are not available and/or the software is not easy to be obtained, to be recoded or even compiled or simply used. Last, the brute force method we have used to tune the system could be easily improved using exact, heuristic or AI based techniques.

Acknowledgments

This work has been partially supported by the CNR project Bioinformatics and by Programma Italia-USA "Farmacogenomica Oncologica" Prog. No. 527/A/3A/5.

References

1. Petricoin, E.F., et al.: Use of proteomic patterns in serum to identify ovarian cancer. The Lancet 359, 572–577 (2002)
2. Liotta, L.A., et al.: High-resolution serum proteomic features for ovarian cancer detection. Endocrine-Related Cancer 11, 163–178 (2004)
3. Vapnik, V.: The Nature Of Statistical Learning Theory. Springer, New York (1995)
4. Cristianini, N., Taylor, J.S.: Kernel Methods for Pattern Analysis. Cambridge University Press, Cambridge (2004)
5. Ulrich, H.G.K.: Advances in kernel methods: support vector learning. MIT Press Cambridge, Cambridge (1999)
6. Ressom, H.W., et al.: Particle swarm optimization for analysis of mass spectral serum profiles. In: GECCO 2005: Proceedings of the 2005 conference on Genetic and evolutionary computation (2005)
7. Ressom, H.W., et al.: Peak selection from MALDI-TOF mass spectra using ant colony optimization. Bioinformatics 23, 619–626 (2007)

8. Lilien, R., Farid, H., Donald, B.: Probabilistic Disease Classification of Expression-Dependent Proteomic Data from Mass Spectrometry of Humsn Serum. Journal of Computational Biology (January 2003)
9. Wu, B., et al.: Comparison of statistical methods for classification of ovarian cancer using mass spectrometry data. Bioinformatics 19, 1636–1643 (2003)
10. Mantini, D., et al.: Independent component analysis for the extraction of reliable protein signal profiles from MALDI-TOF mass spectra. Bioinformatics 24, 63–70 (2008)
11. Baggerly, K., et al.: Reproducibility of SELDI-TOF protein patterns in serum: comparing datases from different experiments. Bioinformatics 20, 777–785 (2007)
12. Sorace, J.M., Zhan, M.: A data review and reassessment of ovarian cancer serum proteomics profiling. BMC Bioinformatics 4, 24–32 (2003)
13. Tibshirani, R., et al.: Sample classification from protein mass spectrometry, by peack probability contrasts. Bioinformatics 20, 3034–3044 (2004)
14. Karin Noy, K., Fasulo, D.: Improved model based, platform independent feature extraction for mass spectrometry. Bioinformatics 23, 2528–2535 (2007)
15. Witkin, A., Terzopoulos, D., Kass, M.: Signal matching through scale space. International Journal of Computer Vision, 133 (1987)
16. Lindeberg, T.: Scale-Space Theory in Computer Vision. Kluwer Academic Publisher, Dordrecht (1994)
17. Boser, B., Guyon, I., Vapnik, V.: A training algorithm for optimal margin classifiers. In: Proceedings of the Fifth Annual workshop on Computational Learning Theory (1992)
18. Schoelkopf, B., et al.: Comparing Support Vector Machines with Gaussian Kernels to Radial Basis Function Classifiers. IEEE Transactions on Signal Processing 45, 2758–2765 (1997)
19. Fawcett, T.: An introduction to ROC analysis. Pattern Recogn. Lett. 27, 861–874 (2006)
20. Hsu, C.V., Lin, C.J.: A comparison of methods for multi-class support vector machines. IEEE Transactions on Neural Networks 13, 415–425 (2002)
21. Burges, C.J.C.: A Tutorial on Support Vector Machines for Pattern Recognition. Data Mining and Knowledge Discovery 121, 121–167 (1998)

Liver i-Biopsy™ and the Corresponding Intelligent Fibrosis Scoring Systems: i-Metavir F and i-Ishak F

Alexandru George Floares[1,2]

[1] SAIA – Solutions of Artificial Intelligence Applications,
Str. Vlahuta, Bloc Lama C, Ap. 45, Cluj-Napoca, 400310, Romania
[2] IOCN – Oncological Institute Cluj-Napoca, Department of Artificial Intelligence,
Str. Republicii Nr. 34-36, Cluj-Napoca, 400015, Romania

Abstract. An important goal of modern medicine is to replace invasive, painful procedures with non-invasive techniques for diagnosis. We investigated the possibility of a knowledge discovery in data approach, based on computational intelligence tools, to integrate information from various data sources - imaging data, clinical and laboratory data, to predict with acceptable accuracy the results of the biopsy. The resulted intelligent systems, tested on 700 patients with chronic hepatitis C, based on C5.0 decision trees and boosting, predict with 100% accuracy the fibrosis stage results of the liver biopsy, according to two largely accepted fibrosis scoring systems, Metavir and Ishak, with and without liver stiffness (FibroScan®). We also introduced the concepts of intelligent virtual biopsy or i-Biopsy™ and that of i-scores. To our best knowledge i-Biopsy™ outperformed all similar systems published in the literature and offer a realistic opportunity to replace liver biopsy in many important medical contexts.

1 Introduction

An important goal of modern medicine is to replace invasive procedures with non-invasive techniques for diagnosis. Non-invasive techniques do not penetrate mechanically, nor break the skin or a body cavity, i.e., they don't require an incision into the body or the removal of biological tissue. An important invasive procedure is biopsy and the main non-invasive techniques are diagnostic images and diagnostic signals. Usually, the information content of the non-invasive imaging techniques is lower than the that of the invasive techniques. As a consequence, the accuracy of the diagnostic based on these non-invasive techniques alone is lower.

The main question addressed in our studies is: can we extract and integrate information from various (non-invasive) data sources, e.g. imaging data, clinical and laboratory data, to reach an acceptable 95%-100% diagnostic accuracy? For any usual or traditional medical approach, the answer to this important question is *NO*. This is because our brain is very skilled in dealing with images, but the

F. Masulli, R. Tagliaferri, and G.M. Verkhivker (Eds.): CIBB 2008, LNBI 5488, pp. 253–264, 2009.

information content of the medical images does not seem to be enough for a diagnostic, and is less skilled in dealing simultaneously with many variables, and this seems to be the solution to the problem.

A knowledge discovery in data or data mining approach, based on computational intelligence tools, could be the foundation for a positive answer to the above important medical question. The extraction and integration of information from various data sources is indeed possible, and the diagnostic accuracy of the resulted intelligent systems could even reach 100%.

To illustrate this thesis, we present an overview of some of our recent investigations, and also some new results, on building intelligent systems capable to predict the results of liver biopsy [1], [2]. We used several non-invasive approaches - routine laboratory tests and basic ultrasonographic features - with and without liver stiffness measurement by transient elastography (FibroScan®), to build intelligent systems for staging liver fibrosis in chronic hepatitis C. The fact that we reached similar results for prostate biopsy (manuscript in preparation) corroborate our believe that this approach can become a standard one.

2   Biomedical Background

The hepatitis C virus is one of the most important causes of chronic liver disease. It accounts for about 15% of acute viral hepatitis, 60% to 70% of chronic hepatitis, and up to 50% of cirrhosis, end-stage liver disease, and liver cancer. An estimated 150-200 million people worldwide are infected with hepatitis C.

Liver fibrosis can accompany almost any chronic liver disease arising as a result of wound repair. It is the net result of the balance between fibrinogenesis - production of extracellular matrix - and fibrolysis - degradation of extracellular matrix (see [3] for a recent review of liver fibrosis). Progressive fibrosis of the hepatic parenchyma leads to cirrhosis, nodule formation, altered hepatic function and risk of liver-related morbidity and mortality. Evaluating the degree of fibrosis is valuable for the following medical reasons [4]:

1. The actual stage of fibrosis will indicate the likelihood of response to treatment; advanced stages generally have an inferior response rate.
2. If fibrosis progression is slow, treatment with antiviral therapy may be less urgent.
3. The approximate time to the development of cirrhosis, the end-stage liver disease, can be estimated.

Liver biopsy is the gold standard for grading the severity of disease and staging the degree of fibrosis and permanent architectural damage. Liver biopsy is invasive and usually painful; complications severe enough to require hospitalization can occur in approximately 4% of patients [5]. In a review of over 68,000 patients recovering from liver biopsy, 96% experienced adverse symptoms during the first 24 hours of recovery. Hemorrhage was the most common symptom, but infections also occurred. Side effects of the biopsies included pain, tenderness, internal bleeding, pneumothorax, and rarely, death [6].

Table 1. Liver biopsy: Metavir and Ishak fibrosis scoring systems

Stage	Metavir System	Ishak System
0	No fibrosis	No fibrosis
1	Periportal fibrosis expansion	Fibrous expansion of some portal areas,with or without short fibrous septae
2	Portal-Portal septae > 1 septum	Fibrous expansion of most portal areas,with or without short fibrous septae
3	Portal-Central septae	Fibrous expansion of most portal areas with occasional Portal-Portal bridging
4	Cirrhosis	Fibrous expansion of portal areas with marked bridging (Portal-Portal or Portal-Central)
5		Marked bridging (Portal-Portal or Portal-Central) with occasional nodules (incomplete cirrhosis)
6		Cirrhosis

Two main scoring systems are used for staging liver fibrosis degree: Metavir F or Ishak F [7] (see Table 1). For routine reporting and clinical follow-up, the simpler Metavir scoring system could be used. For research purposes the more detailed Ishak system could more suitable. Some limitations of these scoring systems should be emphasized. Some times hepatic fibrosis is not be homogenous throughout the liver and the liver specimen obtained by biopsy may not accurately reflect the overall average degree of fibrosis. The reliability of the assessment of fibrosis stage increases with the size of the liver sample. In most studies, a minimum length of 10 mm is required.

Transient elastography (FibroScan®) is an ultrasound imaging technique used to quantify hepatic fibrosis in a totally non-invasive and painless manner. It performs well in identifying severe fibrosis or cirrhosis, but is less accurate in identifying lower degrees of fibrosis.

3 Intelligent Systems for Liver Fibrosis Stage Prediction – A Knowledge Discovery in Integrated Medical Data Approach

3.1 Data Integration and Data Preprocessing

One of the key aspect of this studies consists in integrating various medical data: clinical, imaging and lab data. Our experiments showed that isolated data sources do not contain enough information for building accurate intelligent systems. The main problems we found, in mining the medical data bases used in these studies, were related to the small number of patients relative to the number of features, and to the extend of missing data. However, comparing to similar medical studies we investigated large datasets, with hundreds of patients.

The order of the pre-processing steps is important. Due to the above mentioned problems, one should avoid as much as possible to eliminate patients (records) form the analysis during data pre-processing, and try to eliminate uninformative features first. Most of the commercial data mining software packages will simply eliminate the cases with missing values. If feature selection is performed first, even without using sophisticated methods for missing data imputation, the number of eliminated cases is smaller. For a recent exhaustive collection of feature selection methods see [8]. Feature selections was performed in three steps:

1. Cleaning. Unimportant and problematic features and patients were removed.
2. Ranking. The remaining features were sorted and ranks were assigned based on importance.
3. Selecting. The subset of features to use in subsequent models was identified.

In data cleaning, the following variables were removed:

1. Variables that have more than 70% missing values.
2. Categorical variables that have a single category counting for more than 90% cases.
3. Continuous variables that have very small standard deviation (almost constants).
4. Continuous variables that have a coefficient of variation CV < 0.1 (CV = standard deviation/mean).
5. Categorical variables that have a number of categories greater than 95% of the cases.

For ranking the features, an important step of feature selection, also important for understanding the biomedical problem, we used a simple but effective method which considers one feature at a time, to see how well each feature alone predicts the target variable. For each feature, the value of its importance is calculated as $(1 - p)$, where p is the p value of the corresponding statistical test of association between the candidate feature and the target variable. The target variable was categorical with more than two categories for all investigated problems, and the features were mixed, continuous and categorical.

For categorical variables, the p value was based on Pearson's Chi-square test of independence between X, the feature under consideration with I categories, and Y target variable with J categories. The Chi-square test involves the difference between the observed and expected frequencies. Under the null hypothesis of independence, the expected frequencies are estimated by $\widehat{N} = N_i \cdot N_j / N$. Under the null hypothesis, Pearson's chi-square converges asymptotically to a chi-squared distribution χ_d^2 with degree of freedom $d = (I-1)(J-1)$, and the p value is equal with the probability that $\chi_d^2 > X^2$, where $X^2 = \sum_{i=1}^{I} \sum_{j=1}^{J} (N_{ij} - \widehat{N}_{ij})^2 / \widehat{N}_{ij}$. The categorical variables were sorted first by p value in the ascending order, and if ties occurred they were sorted by chi-squared in descending order. If ties still occurred, they were sorted by degree of freedom d in ascending order.

For the continuous variables, p values based on the F statistic are used. For each continuous variable a one-way ANOVA F test is performed to see if all the different classes of Y have the same mean as X. The p value based on F statistic is calculated as the probability that $F(J-1, N-J) > F$, where $F(J-1, N-J)$ is a random variable that follows and F distribution with degrees of freedom $J-1$ and $N-J$, and

$$F = \frac{\sum_{j=1}^{J} N_j \left(\overline{x}_j - \overline{\overline{x}}\right)^2 / (J-1)}{\sum_{j=1}^{J} (N_j - 1)\, s_j^2 / (N-J)}$$

If the denominator for a feature was zero, the p value of that feature was set to zero. The features were ranked first by sorting them by p value in ascending order, and if ties occurred, they were sorted by F in descending order. If ties still occurred, they were sorted by N in descending order.

Based on the features' importance $(1 - p)$, with p calculated as explained above, we ranked and grouped features in three categories:

1. important features, with $(1 - p)$ between 0.95 and 1,
2. moderately important features, with $(1 - p)$ between 0.90 and 0.95, and
3. unimportant features, with $(1 - p)$ less than 0.90.

Some of the categorical features and also the target categorical variable have imbalanced distributions, and this can cause some modeling algorithms to perform poorly. We tested the influence on the prediction accuracy of several methods for dealing with imbalanced data. For a recent review and comparison between various methods see [9]. Because the number of patients is small relative to the number of features, a very common situation in biomedical data bases, we only used over-sampling methods and not undersampling methods. We also found that simple techniques such as random oversampling perform better than the "intelligent" [9] sampling techniques.

3.2 Intelligent Systems as Ensemble of Classifiers

For modeling, we first tested the fibrosis prediction accuracy of various methods:

1. Neural Networks
2. C5.0 decision trees
3. Classification and Regression Trees
4. Support Vector Machines,
5. Bayesian Networks.

The gastroenterologists and also the ultrasonographists preferred white-box algorithms, e.g., decision trees. As a consequence, both accuracy and intelligibility of the resulted intelligent systems recommended the use of C5.0 decision trees; we used 10-fold cross-validation as a robust method to determine the average accuracy. C5.0 algorithm performed his own feature selection from the features' set remaining in the analysis after preprocessing.

To improve the predictive power of classifiers learning systems we tried both Breiman's bagging [10] and Freund and Schapire's boosting [11] ensemble methods. Both form a set of classifiers that are combined by voting, bagging by generating replicated bootstrap samples of the data, and boosting by adjusting the weights of training cases. While both approaches improve predictive accuracy, boosting showed sometimes greater benefit. Unfortunately, boosting doesn't always help, and when the training cases are noisy, boosting can actually reduce classification accuracy. Naturally, it took longer to produce boosted classifiers, but the results often justified the additional computation. Boosting should always be tried when peak predictive accuracy is required, especially when unboosted classifiers are already quite accurate.

Boosting combines many low-accuracy classifiers (weak learners) to create a high-accuracy classifier (strong learner). We used a boosting version called *AdaBoost*, with reweighting; AdaBoost comes from ADAptive BOOSTing [11].

Suppose we are given the training set data $(X_1, F_1), \ldots, (X_n, F_n)$, where n is the number of patients, the input $X_i \in \Re^p$ represents the p selected features in the preprocessing steps (image, laboratory data, etc.), and the categorical output F_i is the fibrosis stage according to one of the two scoring systems Metavir F and Ishak F (see Table 1), and assumes values in a finite set $\{F0, F1, \ldots, Fk\}$, were $k = 5$ for Metavir F (from Metavir F0 to Metavir F4) and $k = 7$ for Ishak F (from Ishak F0 to Ishak F6). The goal is to find a classification rule $F(\mathbf{X})$ from the training data, so that given a new patient's input vector $\mathbf{X}$, we can assign it a fibrosis degree F from $\{F0, F1, \ldots, Fk\}$ according to the corresponding scoring systems. Moreover, we want to find the best possible classification rule achieving the lowest misclassification error rate. We assumed that the patients' training data are independently and identically distributed samples from an unknown distribution. Starting with the unweighted training sample, the AdaBoost builds a classifier which can be a neural network, decision tree, etc., that produces class labels - fibrosis degree. If a training data point (patient) is misclassified, the weight of that training patient is increased (boosted). A second classifier is built using the new weights, which are now different. Again, misclassified training patients have their weights boosted and the procedure is repeated. Usually, one may build hundred of classifiers this way. A score is assigned to each classifier, and the final classifier is defined as the linear combination of the classifiers from each stage.

With the above notations, and notating with I an indicator function, a compact description of the AdaBoost algorithm used is the following:

1. Initialize the patient weights $\omega_i = 1/n, i = 1, 2, \ldots, n$.
2. For $m = 1$ to M:

    (a) Fit a classifier $F^{(m)}(\mathbf{x})$ to training patients using weights ω_i.
    (b) Compute

$$err^m = \sum_{i=1}^{n} \omega_i I(F_i \neq F^{(m)}(\mathbf{X}_i)) / \sum_{i=1}^{n} \omega_i.$$

(c) Compute

$$\alpha^{(m)} = \log \frac{1 - err^{(m)}}{err^{(m)}}.$$

(d) Set

$$\omega_i \leftarrow \omega_i \cdot \exp(\alpha^{(m)} \cdot I(F_i \neq F^{(m)}(\mathbf{X}_i)), i = 1, 2, \ldots, n.$$

(e) Re-normalize ω_i.

3. Output

$$F(\mathbf{X}) = \arg\max \sum_{m=1}^{M} \alpha^{(m)} \cdot I(F^{(m)}(\mathbf{X}) = k).$$

While for two-class classification problems AdaBoost could be very successful in producing accurate classifiers, the multi-class classification is more involved, and some technical criteria must be satisfied and experiments need to be done.

The final intelligent systems are the result of a more detailed data mining predictive modeling strategy which is patented now, consisting mainly in:

1. Extracting and integration information from various medical data sources, after a laborious preprocessing:

   (a) cleaning features and patients,
   (b) various treating of missing data,
   (c) ranking features,
   (d) selecting features,
   (e) balancing data.

2. Testing various classifiers or predictive modeling algorithms.
3. Testing various methods of combining classifiers.

3.3 Intelligent Virtual Biopsy and Intelligent Scoring Systems

In our opinion, replacing painful, invasive, risky and/or costly procedures with intelligent systems, taking as inputs integrated data from non-invasive, routine, or cheap medical procedures, techniques and tests, and producing as output 95-100% similar results with the replaced techniques, will be an important medical trend in the near future. For this reason, we intend to outline some general ideas and introduce some terms and concept to characterize this new exciting enterprize. These will be illustrated with the results of the our investigations.

The central new concept is *Intelligent Virtual Biopsy* (IVB), which designates an intelligent system capable to predict, with an acceptable accuracy (e.g., 95-100%), the results given by a pathologist, examining the tissue samples from real biopsies, expressed as scores of a largely accepted scoring system. As a shorter alternative term we suggested intelligent biopsy or *i-Biopsy*^{TM} were the term intelligent indicates that the system is based on artificial or computational intelligence. To predict the pathologist's scores, the intelligent systems take as inputs and integrate various non-invasive biomedical data.

Also, to distinguish between the scores of the real biopsy, and their counterparts predicted by the i-Biopsy^{TM}, we proposed the general term of *i-scores* belonging to *i-scoring systems*. In the gastroenterological context of these investigations, we have the following correspondences:

1. Liver intelligent virtual biopsy (IVB), or liver *i-Biopsy*^{TM} is the intelligent system corresponding to the real liver biopsy.
2. The *i-Metavir F* and *i-Ishak F* correspond to the two liver fibrosis scoring systems Metavir F and Ishak F, respectively.
3. The *i-scores* are the values predicted by the intelligent systems for the fibrosis scores.

From a biomedical point of view, the most important general characteristics of the i-Biopsy^{TM}and the i-scores are exemplified for liver biopsy and the Metavir F and Ishak F scores:

1. I-Metavir F and i-Ishak F scores have exactly the same biomedical meaning as Metavir-F and Ishak-F, scoring the same pathological features (see Table 1).
2. I-Metavir F and i-Ishak F scores are obtained in a non-invasive, painless, and riskless manner, as opposed to Metavir-F and Ishak-F,
3. Unlike real biopsy, i-Biopsy^{TM}can be used to evaluate fibrosis evolution because it can be repeated as many time as needed.
4. In the early stages of liver diseases, when the symptoms are not harmful but treatment is most effective, the doctors hesitate to indicate the biopsy and the patients to accept it, but i-Biopsy^{TM}can be performed.
5. Sometimes, the result of the i-Biopsy^{TM}could be superior to that of real biopsy because, in building the intelligent systems, the results of the technically incorrect biopsies were eliminated from the data set.

4   Results

In the first study, the distribution of the 255 chronic hepatitis C patients was the following:

1. 5.1% (13/255) patients had Metavir F0 fibrosis score.
2. 33.73% (86/255) patients had Metavir F1 fibrosis score.
3. 42.75% (109/255) patients had Metavir F2 fibrosis score.
4. 18.42% (47/255) patients had Metavir F3 fibrosis score.
5. Metavir F4 patients were excluded being less than 5%.

At the end of the preprocessing stage, besides liver stiffness (FibroScan®), the relevant features for predicting liver fibrosis, according to Metavir scoring system, were: age, aspartate aminotransferase, gamma-glutamyl-transpeptidase, cholesterol, triglycerides, thickness of the gallbladder wall, spleen area and perimeter, left lobe and caudate lobe diameter, liver homogeneity, posterior attenuation of the ultrasound, liver capsule regularity, spleen longitudinal diameter, the maximum subcutaneous fat, perirenal fat.

Combining all these features, the intelligent system was able to predict each fibrosis stage with 100% accuracy. In the mean time, more than 500 patients with chronic hepatitis C were investigated and the accuracy of the intelligent system was 100% for all Metavir scores, including Metavir F4, which was excluded in the first study. We intend to try to reduce the number of features to at most ten without sacrificing the accuracy, because some of our investigations showed that this is possible [1].

To build the intelligent system for Ishak fibrosis score prediction, we investigated a lot of 276 chronic hepatitis C patients, having the following distribution:

1. 3.26% (9/276) had Ishak F0 fibrosis score.
2. 18.48% (51/276) had Ishak F1 fibrosis score.
3. 17.39% (48/276) had Ishak F2 fibrosis score.
4. 43.48% (120/276) had Ishak F3 fibrosis score.
5. 13.04% (36/276) had Ishak F4 fibrosis score.
6. 4.35% (12/276) had Ishak F5 fibrosis score.
7. Patients with Ishak F6 fibrosis scores were excluded from the study being too few.

The relevant features for predicting liver fibrosis were: caudate lobe diameter, left lobe diameter, liver capsule regularity, liver homogeneity, thickness of the abdominal aortic wall, steatosis (ultrasonographic), cholesterol, sideremia, and liver stiffness. The boosted C5.0 decision trees were able to predict each Ishak fibrosis stage with 100% accuracy.

We also wanted to investigate if it is possible to build an intelligent system, capable to predict fibrosis scores according to Metavir F scoring system, without using apparently a key source of information - the liver stiffness measured with FibroScan®. Such an intelligent system could be useful to those gastroenterology clinics having ultrasound equipment but not the expensive FibroScan®.

After feature selection, the relevant features for this situation were: cholesterol, caudate lobe diameter, thickness of the abdominal aortic wall, aspartate aminotransferase, preperitoneal fat thickness, splenic vein diameter, time averaged maximum velocity in hepatic artery, time averaged mean velocity in hepatic artery, flow acceleration in hepatic artery, hepatic artery peak systolic velocity. Combining these 10 attributes, the boosted C5.0 decision trees were able to predict each fibrosis stage, according to Metavir scoring system, with 100% accuracy, even without liver elastography.

All these models have 100% accuracy; at the moment of writing this paper, the intelligent systems were tested on 528 patients with chronic hepatitis C.

5   Discussions

The reasons for the relative disproportion between the number of patients and the number of features is that, at the beginning of these investigations, our multi-disciplinary team tried to define a large number of potentially important features. We intended to use a *data-driven* approach avoiding as much as possible restrictive a priori assumptions. Usually, this opens the door for potential surprises, e.g., previously unknown and unexpected relationships between fibrosis and various other biomedical features. There were some unexpected findings (results not shown) but they need further investigations.

While the data related problems are not so serious as in mining genomics or proteomics data, the fact that the difficulties are not so evident could be a trap. This apparent simplicity was responsible for some initially poor results, but a careful pre-processing increased the accuracy of the predictions with 20% to 25%.

A short comment about the meaning of 100% diagnostic accuracy seems to be necessary, because it confused many physicians who say that 100% accuracy is not possible in medicine. The meanings will be made clear more easy by means of examples. We have proposed intelligent systems predicting the fibrosis scores resulted from liver biopsy with 100% accuracy. Usually, an invasive liver biopsy is performed and a pathologist analyzes the tissue samples and formulates the diagnostic, expressed as a fibrosis score. The pathologist may have access to other patient's data, but usually these are not necessary for the pathological diagnostic. Moreover, in some studies it is required that the pathologist knows nothing about the patient. His or her diagnostic can be correct or wrong for many reasons, which we do not intend to analyze here. On the contrary, for the intelligent system some of the clinical, imaging and lab data of the patient are essential, because they were somehow incorporated in the system. They were used like features to train the system, and they are required for a new, unseen patient, because the i-Biopsy^{TM}is in fact a relationship between these inputs and the fibrosis scores.

Intelligent systems do not deal directly with diagnostic *correctness*, but with diagnostic prediction accuracy. In other words, the intelligent system will predict, in a non-invasive and painless way, and without the risks of the biopsy, a diagnostic which is 100% identic with the pathologist diagnostic, if the biopsy is performed. While the accuracy and the correctness of the diagnostic are related in a subtle way, they are different concepts. An intelligent system will use the information content of the non-invasive investigations to predict the pathologist diagnostic, without the biopsy. The correctness of the diagnostic is a different matter, despite the fact that a good accuracy is almost sure related with a correct diagnoses, but we will not discuss this subject.

The accuracy of the diagnosis, as well as other performance measures like the area under the receiver operating characteristic (AUROC), for a binary classifier system [12], are useful for intelligent systems comparison. From the point of view of accuracy, one of the most important medical criterions, to our best knowledge the proposed liver intelligent virtual biopsy or i-Biopsy^{TM}system outperformed the most popular and accurate system, FibroTest [13] commercialized by Biopredictive company. The liver i-Biopsy^{TM}presented in this paper is based on a five classes classifier, more difficult to build than binary classifiers; we also build binary classifiers as decision trees with 100% accuracy and mathematical models (work in progress, results not shown). Despite the fact that AUROC is only for binary classifiers, loosely speaking a 100% accuracy n classes classifier is equivalent with n binary classifiers with AUROC $= 1$ (maximal). In [13], a total of 30 studies were included which pooled 6,378 subjects with both FibroTest and biopsy (3,501 chronic hepatitis C). The mean standardized AUROC was 0.85 (0.82-0.87).

Moreover, in some circumstances the result of the liver i-Biopsy^{TM}could be superior to that of real biopsy. When building the intelligent system, the results of the potentially erroneous biopsies, which are not fulfilling some technical requirements, were eliminated from the data set. Thus, the i-Biopsy^{TM}predicted results correspond only to the results of the correctly performed biopsies, while some of the real biopsy results are wrong, because they were not correctly performed. Due to the invasive and unpleasant nature of the biopsy, is very improbable that a patient will accept a technically incorrect biopsy to be repeated. Unlike real biopsy, i-Biopsy^{TM}can be used to evaluate fibrosis evolution, which is of interest in various biomedical and pharmaceutical studies, because, being non-invasive, painless and without any risk, can be repeated as many time as needed. Also, in the early stages of liver diseases, often the symptoms are not really harmful for the patient, but the treatment is more effective then in more advanced fibrosis stages. The physician will hesitate to indicate an invasive, painful and risky liver biopsy, and the patients is not so worried about his or her disease to accept the biopsy. However, i-Biopsy^{TM}can be performed and an early start of the treatment could be much more effective.

Moreover, we have obtained high accuracy results for other liver diseases, like chronic hepatitis B and steatohepatitis, for other biopsy findings, like necroinflammatory activity and steatosis (results not shown), and also for prostate biopsy in prostate cancer. These corroborate our believe that this approach can become a standard one.

References

1. Floares, A.G., Lupsor, M., Stefanescu, H., Sparchez, Z., Badea, R.: Intelligent virtual biopsy can predict fibrosis stage in chronic hepatitis C, combining ultrasonographic and laboratory parameters, with 100% accuracy. In: Proceedings of The XXth Congress of European Federation of Societies for Ultrasound in Medicine and Biology (2008)
2. Floares, A.G., Lupsor, M., Stefanescu, H., Sparchez, Z., Serban, A., Suteu, T., Badea, R.: Toward intelligent virtual biopsy: Using artificial intelligence to predict fibrosis stage in chronic hepatitis c patients without biopsy. J. Hepatol. 48(2) (2008)
3. Friedman, S.: Liver fibrosis. J. Hepatol. 38(suppl.1), 38–53 (2003)
4. Rossi, E., Adams, L.A., Bulsara, M., Jerey, G.P.: Assessing liver fibrosis with serum marker models. Clin. Biochem. Rev. 28(1), 310 (2007)
5. Lindor, A.: The role of ultrasonography and automatic-needle biopsy in outpatient percutaneous liver biopsy. Hepatology 23, 1079–1083 (1996)
6. Tobkes, A., Nord, H.J.: Liver biopsy: Review of methodology and complications. Digestive Disorders 13, 267–274 (1995)
7. Bedosa, P., Poynard, T.: An algorithm for the grading of activity in chronic hepatitis c. the metavir cooperative study group. Hepatology 2(24), 289–293 (1996)
8. Guyon, I., Gunn, S., Nikravesh, M., Zadeh, L.: Feature Extraction: Foundations and Applications. In: Studies in Fuzziness and Soft Computing. Springer, Heidelberg (August 2006)

9. Hulse, J.V., Khoshgoftaar, T.M., Napolitano, A.: Experimental perspectives on learning from imbalanced data. In: Proceedings of the 24 th International Conference on Machine Learning, Corvallis, OR (2007)
10. Breiman, L.: Bagging predictors. Machine Learning 24(2), 123–140 (1996)
11. Freund, Y., Schapire, R.E.: A decisiontheoretic generalization of on-line learning and an application to boosting. Journal of Computer and System Sciences 55(1), 119–139 (1997)
12. Fawcett, T.: Roc graphs: Notes and practical considerations for researchers. Technical report, HP Laboratories, Palo Alto, USA (2004)
13. Poynard, T., Morra, R., Halfon, P., Castera, L., Ratziu, V., Imbert-Bismut, F., Naveau, S., Thabut, D., Lebrec, D., Zoulim, F., Bourliere, M., Cacoub, P., Messous, D., Muntenau, M., de Ledinghen, V.: Meta-analyses of Fibrotest diagnostic value in chronic liver disease. BMC Gastroenterology 7(40) (2007)

An Extension of the TIGR M4 Suite to Preprocess and Visualize Affymetrix Binary Files

Mario Cannataro[1], Maria Teresa Di Martino[2], Pietro Hiram Guzzi[1],
Pierosandro Tagliaferri[2], Pierfrancesco Tassone[2], Giuseppe Tradigo[1],
and Pierangelo Veltri[1]

[1] Bioinformatics Laboratory, Department of Experimental Medicine and Clinic,
University Magna Græcia of Catanzaro, 88100 Catanzaro, Italy
[2] Medical Oncology Unit and Referral Center for Genetic Counselling and Innovative
Treatments, T. Campanella Cancer Center, 88100 Catanzaro, Italy

Abstract. DNA microarrays are used to efficiently measure levels of
expression of genes by enabling the scan of the whole genome in a single
experiment, through the use of a single chip. In Human specie, a microarray analysis allows the measurement of up to 30000 different genes
expressions for each sample. Data extracted from chips are preprocessed
and annotated using vendor provided tools and then mined. Many algorithms and tools have been introduced to extract biological information
from microarray data, nevertheless, they often are not able to automatically import raw data generated by recent arrays, such as the Affymetrix
ones. The paper presents a software tool for the automatic summarization and annotation of Affymetrix binary data. It is provided as an extension of TIGR M4 (TM4), a popular software suite for microarray
data analysis, and enables the operator to directly load, summarize and
annotate binary microarray data avoiding manual preprocessing. Preprocessed data is organized in annotated matrices suitable for TM4 analysis
and visualization.

1 Introduction

A main goal of genomics is to infer knowledge about genes and relationships
among them and, more in general, about disease development. DNA microarrays technology enables large scale genomics through the investigation of the
expression of genes. For instance, recent developments of Affymetrix[1] chips, i.e.
the Human Gene 1.0 ST chip, interrogates the whole mRNA transcript. Each
chip uses a set of probes to bind different genes that have been tagged with
a fluorescent marker. Chips are then introduced into scanners that light them
using a laser source producing a set of images.

A typical workflow for analysing microarray data comprises three main phases
[13,9], as depicted in Figure 1: *(i)* preprocessing, *(ii)* statistical-data mining analysis, and *(iii)* biological interpretation. Preprocessing aims to transform images

---

[1] http://www.affymetrix.com

F. Masulli, R. Tagliaferri, and G.M. Verkhivker (Eds.): CIBB 2008, LNBI 5488, pp. 265–274, 2009.

Fig. 1. Workflow of Microarray Data Analysis

coming from instrument into suitable numerical data taking account of noise. Statistical and data mining analysis aim to extract biological meaningful genes on the basis of their expression values. All selected genes are finally analyzed on the basis of biological considerations in order to evidence the involved molecular mechanisms. Each step of this workflow usually employs different visualization algorithms and tools to present data in a friendly way.

A lot of tools are able to process microarray data, one of them is the *TM4* tool [14], a comprehensive suite consisting of four main applications. A Microarray Data Manager (*MADAM*) which stores and retrieves data from a database. Spotfinder, an image quantification tool able to read color array images, remove noise and extract relevant features. A Microarray Data Analysis System (*MIDAS*) that reads gene data and performs several analyses. In particular it implements normalization, gene filtering, gene grouping and data mining. A main drawback of this tool is that it is not able to execute the first preprocessing phase, i.e. the summarization, and it needs a preliminary transformation phase. When using Affymetrix array, this phase can be performed by using the Affymetrix Power Tools (*APT*) [1]. Such tools are a set of command line programs provided by Affymetrix that are able to transform binary *CEL* files into numerical data, implementing main summarization algorithms.

This paper presents μ-CS (Microarray CEL Files Summarizer), an extension of *TM4* able to realize the data summarization and annotation in an automatic way. The μ-CS architecture comprises three main modules: (*i*) a wrapper of the Affymetrix Power Tools, (*ii*) a library manager which is able to find the needed libraries to realize summarization, and (*iii*) an annotation manager that is able to find updated annotations, i.e. the information about genes to realize annotation. This way the user can directly manage these data using the TM4 Suite without wasting time and avoiding possible mistakes while manually summarizing and annotating data.

The paper is structured as follows: Section 2 introduces the main steps performed during microarray analysis. Section 3 presents μ-CS, the main contribution of this work. A case study showing the application of the μ-CS tool to preprocess some microarray data that is further analyzed through TM4 is presented in Section 4. Section 5 describes some related works and Section 6 concludes the paper.

2   Microarray Data Analysis Workflow

The preprocessing of microarray data can be structured as a pipeline of sequential steps as depicted in Figure 2. As data feeds into next steps, it becomes more

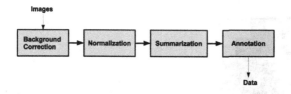

Fig. 2. Microarray Preprocessing Workflow

and more refined. The current state of the art employs a four-step data preprocessing phase:(i) Background Correction, (ii) Normalization, (iii) Summarization, and (iv) Annotation. This strategy starts with a background correction phase [12,16] which removes the background noise. This phase aims to remove non-specific signals, e.g. signals caused by non-specific DNA that is uncorrectly bound to a different probe. For these aims a lot of algorithms have been introduced. They employ different estimators of background that reveal the noise and subtract it from signal.

Normalization consists in reducing the bias among chips and within different regions of the same chip [6,10], aiming at removing non-biological variability within a dataset. Common causes of such variability are: variable loading of DNA onto arrays, mixing of DNA across different areas of the array, variability in the efficacy of the labeling reactions among different arrays. Algorithms for normalization of microarrays can be divided into two major classes: (i) Within-array normalization algorithms, that seek to remove variability within a single array, and (ii) Between-array normalization algorithms, which seek to remove variability among a set of arrays.

Summarization combines multiple preprocessed probe intensities to a single expression value. All arrays employ more than one probe for each genes as introduced before. Summarization takes into account of all the probes for the same genes and average them by enhancing the signal-to-noise ratio. For Affymetrix arrays summarization is usually done as a single step with normalization using a set of Affymetrix libraries which store the topological information about probes. There exist different summarization algorithms such as: the Robust Multi-array Average (RMA) [11,7] and the Probe Logarithmic Intensity Error ($PLIER$) [2]. All of these algorithms are based on several assumptions on the data distribution and they require a set of specific libraries in order to correctly access binary data.

Considering, for instance, Affymetrix arrays, the RMA uses a parametric model to describe the relationships between the different measurements and the probeset intensities. PLIER is based on a probe affinity parameter, which represents the strength of a signal produced at a specific concentration for a given probe. The probe affinities are calculated using data across arrays. The error model employed by PLIER assumes that the error is proportional to the observed intensity, rather than to the background-subtracted intensity. The Affymetrix Power Tools (APT) are a set of tools implementing low level algorithms for working with Affymetrix GeneChip arrays. They are able to read a set of *CEL* files and produce a data matrix as depicted in Figure 3. In order to perform the

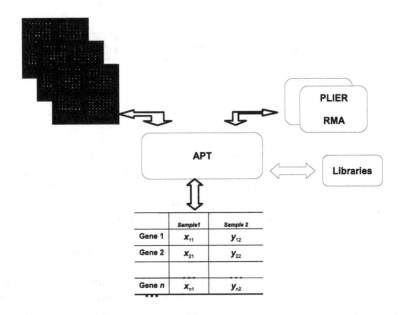

Fig. 3. Summarization of CEL Files

normalization and the summarization, APT need: (i) a set of binary *CEL* files, (ii) a model algorithm, e.g. PLIER, and (iii) the correct libraries that enable the correct interpretation of the images. Finally, a process known as *annotation*, associates to each probe its known annotations such as gene Symbol or Gene Ontology [8] by matching probes to public databases or knowledge bases. Often annotation files are provided by the chip manifacturer and contain different levels of annotation, e.g. database identifier, description of molecular function, associated protein domains. The annotation of Affymetrix data can be conducted by using the APT tools by providing an annotation file that depends on the kind of the used chip and on the specie.

3 μ-CS

The Microarray CEL Files Summarizer (μ-CS) is an extension of the TM4 suite allowing users to directly manage Affymetrix binary data. The realized tool receives a set of binary files as input, summarizes them, i.e. converts them into a matrix, and integrates the generated data with annotations. The resulting matrix is finally managed by the MIDAS module of TM4. The architecture of the proposed tool, as depicted in Figure 4(a), comprises the following modules: (i) the **APT Wrapper**, (ii) the **Libraries Manager**, and (iii) the **Annotations Manager**. The APT Wrapper is able to locally invoke the Affymetrix Power Tools executable without user intervention. Both Libraries Manager and Annotations Manager store all the needed data on two databases: **LibraryDB** and

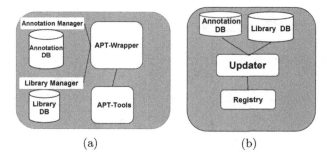

Fig. 4. Architecture of μ-CS

AnnotationDB. The APT Wrapper receives the job requests from the main application, manages them and then invokes the original instance of the APT. An instance of APT is invoked whenever summarization requests are received. After the jobs completion, the wrapper reads the output files, organizes them in a data structure, and transparently sends it to the main application, i.e. TM4. The LibraryDB stores all the needed libraries to parse binary files while the AnnotationDB stores the annotation files. Vendors periodically update these data so the need to realize an automatic update manager arises. An ad-hoc module, depicted in Figure 4(b), periodically verifies the availability of updates by connecting to the Affymetrix repositories. If it finds available updates for the LibraryDB or for the AnnotationDB, it downloads them updating the specific database. An ad-hoc registry keeps trace of versions for both databases. In such a way operators can transparently use the latest annotation libraries. They can also install their own libraries and manually update the registry.

4   Case Study: Preprocessing and Visualizing Gene 1.0 ST Data

This Section shows the functionalities of μ-CS through a case study: the analysis of a Human Gene 1.0 dataset freely available for download on the Affymetrix web site. This dataset contains various mixture levels of two tissues: brain and heart from Human samples. We selected 10 arrays from these to perform our study.

Initially the operator has to launch the TM4 suite as usual. The extension presented in the paper is accessible within the classic menu of the loader application. At this point the user has to choose the file type that he/she wants to load, in this case *CEL* files, as depicted in Figure 5. As explained in the previous sections, the preprocessing tools need two main libraries, one for correctly accessing *CEL* files named CDF libraries, and another one for the annotation of data results. Thus the user has to manually select the appropriate libraries for the chosen array. Those libraries are provided by the Affymetrix repository accessible through the Affymetrix website and normally they have to be searched and downloaded manually by the user. Moreover, if new versions are released the user should be aware of that and should download and use them.

Fig. 5. Selecting the source CEL files

Fig. 6. Selection of Libraries

At this point, μ-CS will visualize a selection window containing the main options that the user can select. In particular, as depicted in Figure 6, the user can choose the summarization method. Moreover he/she can select the appropriate libraries needed to parse *CEL* files and to annotate the genome. The tool automatically presents the installed libraries that are up to date with the vendor. Each preprocessing step has different options, among those the type of algorithm used and its own parameters. At this step the researcher has to select the preferred algorithm and its parameters. For instance, considering the normalization algorithm, it can choose, for these arrays, to employ the RMA or the PLIER algorithm and related parameters. μ-CS then reads the binary files and invokes the APT executable by using the user's specified parameters.

	Sample 1	Annotation
Gene 1	x_{11}	gi:010011, go:molecular function
Gene 2	x_{21}	gi:012002, go:molecular function
...	...	...
Gene n	x_{n1}	

Fig. 7. Example of Annotated File

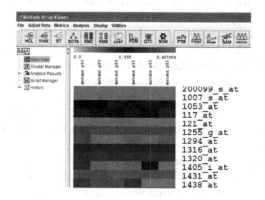

Fig. 8. Visualizing gene expression data in TM4

When the summarization and annotation are completed, the module sends results to the TM4 suite for subsequent analysis. Figure 7 shows a fragment of a summarized and annotated *CEL* file. Each row contains an identifier of the gene as provided by the manufacturer, its expression level and a set of specific information, (e.g. the identifier in public databases, the Gene Ontology annotations) for that gene.

Finally, the user can visualize the data by selecting the preferred way, e.g. heat map as depicted in Figure 8, and analyze it. The current version of μ-CS is implemented as a plug-in of TM4, and it is currently tested in a joint collaboration between the University of Catanzaro Bioinformatics Laboratory and Tommaso Campanella Cancer Center [5,4].

5   Related Work

Although the existence of several commercial and noncommercial tools, there is a growing number of freely available, open source software for analyzing and visualizing microarray data. By using such tools both scientists and operators can validate the implemented algorithms, modify them or extend the whole system to add new functionalities if needed. We here report some of the most used systems.

The already introduced *TM4* is a comprehensive suite of tools. The **Microarray Data Manager** (*MADAM*), which is able to store data into a database compliant to the Minimal Information About a Microarray Experiment (*MIAME*) [3] format. It is designed to load and retrieve microarray data from a MySQL database, but with minimum effort an adaptor to a different database can easily be written. **Spotfinder**, an image quantification tool that reads images coming from 2 color arrays, removes noise and extracts relevant features, made available for further analyses. The **Microarray Data Analysis System** (*MIDAS*) that reads gene data in a tabular form and performs several analysis. In particular it implements normalization, both local and global, filtering of genes based on their expression, genes grouping in clusters and data mining. The **TIGR Multiexperiment Viewer** (*MeV*) is able to cluster data and visualize discovered clusters. The *MeV* system can also handle several input file formats such as flat ASCII text files (*txt*) and Genepix (*gpr*) files, merging together different experiments. These experiments can be analyzed by using different algorithms. A main drawback of this system is the lack of native management for the Affymetrix data file format. The TM4 provides the Express Converter software that is not able to process Affymetrix data while it is useful to convert images coming from two color experiments.

Bioconductor[2] is a programming environment for bioinformatics developed on top of the R programming language[3], an open source version of the S-Plus commercial software. It is structured with different packages and offers many open source applications for microarray data management and analysis. It has normalization functions as well as filtering, visualization and data analysis. One of them, called *OneChannelGui* [15], is able to parse Human Gene 1.0 ST data wrapping the Affymetrix Power Tools. OneChannelGUI is an extension of the *affylmGUI* package providing a graphical interface for Bioconductor libraries. Embedded libraries provide quality control, noise removal, feature selection and statistical analysis features for single channel microarrays. OneChannelGUI can also manage Affymetrix *CEL* data but, compared with *MeV*, it lacks in visualization capabilities.

6   Conclusion

The study of gene expression data is nowadays an important research strategy in biology and medicine. Microarray technology enables the investigation of such reality and uses chips that are able to scan the whole genome, e.g. the Affymetrix Human Gene 1.0 ST array. The visual analysis of these data helps the researcher to conduct biological relevant analysis. There exist different tools able to manage these data such as the TIGR MeV. A main drawback of this tool is that it cannot directly manage Affymetrix binary data. In this paper we proposed μ-CS, an extension of this tool that natively summarizes and annotates Affymetrix binary data by wrapping existing summarization and annotation tools, the Affymetrix

---

[2] http://www.bioconductor.org
[3] http://www.r-project.org

Power Tools, and related libraries. Moreover, μ-CS is able to automatically update the needed libraries by locating and downloading them from the vendor repository. In this way, users of the MeV tool can directly manage Affymetrix binary data without worrying about locating and invoking the proper preprocessing tools and chip-specific libraries. Different preprocessing tools and libraries can also be installed by the user. Future work will regard the extension of the system to support the management and summarization of different arrays and its implementation as a service.

Acknowledgments

Authors are grateful to Andrea Greco for his work in developing μ-CS.

References

1. Affymetrix Power Tools (APT),
   `http://www.affymetrix.com/partners/programs/programs/developer/tools/powertools.affx`
2. Affymetrix. Guide to Probe Logarithmic Intensity Error (plier) estimation, `http://www.affymetrix.com/support/technical/technotes/plier_technote.pdf`
3. Brazma, A.: Minimum Information About a Microarray Experiment (MIAME)-toward standards for microarray data. Nat. Genet. 29(4), 365–371 (2001)
4. Cannataro, M., Di Martino, M.T., Guzzi, P.H., Tassone, P., Tagliaferri, P., Tradigo, G., Veltri, P.: A tool for managing Affymetrix binary files through the TIGR TM4 suite. In: International Meeting of the Microarray and Gene Expression Data Society, Riva del Garda, Italy, September 1-4 (2008)
5. Di Martino, M.T., Guzzi, P.H., Ventura, M., Pietragalla, A., Neri, P., Bulotta, A., Calimeri, T., Barbieri, V., Caraglia, M., Veltri, P., Cannataro, M., Tassone, P., Tagliaferri, P.: Whole gene expression profiling shows a differential transcriptional response to cisplatinum in BRCA-1 defective versus brca1-reconstituted breast cancer cells. In: Annual Meeting of Associazione Italiana di Oncologia Medica (AIOM), Verona (2008)
6. Fujita, A., Sato, J.R., Rodrigues, L.O., Ferreira, C.E., Sogayar, M.C.: Evaluating different methods of microarray data normalization. BMC Bioinformatics 7, 469 (2006)
7. Harbron, C., Chang, K.-M.M., South, M.C.C.: Refplus: an R package extending the RMA algorithm. Bioinformatics (July 2007)
8. Harris, M.A., Clark, J., Ireland, A., Lomax, J., Ashburner, M., Foulger, R., Eilbeck, K., Lewis, S., Marshall, B., Mungall, C., Richter, J., Rubin, G.M., Blake, J.A., Bult, C., Dolan, M., Drabkin, H., Eppig, J.T., Hill, D.P., Ni, L., Ringwald, M., Balakrishnan, R., Cherry, J.M., Christie, K.R., Costanzo, M.C., Dwight, S.S., Engel, S., Fisk, D.G., Hirschman, J.E., Hong, E.L., Nash, R.S., Sethuraman, A., Theesfeld, C.L., Botstein, D., Dolinski, K., Feierbach, B., Berardini, T., Mundodi, S., Rhee, S.Y., Apweiler, R., Barrell, D., Camon, E., Dimmer, E., Lee, V., Chisholm, R., Gaudet, P., Kibbe, W., Kishore, R., Schwarz, E.M., Sternberg, P., Gwinn, M., Hannick, L., Wortman, J., Berriman, M., Wood, V., Tonellato, P., Jaiswal, P., Seigfried, T., White, R.: The Gene Ontology (GO) database and informatics resource. Nucleic Acids Res. 32(Database issue), 258–261 (2004)

9. Hibbs, M.A., Dirksen, N.C., Li, K., Troyanskaya, O.G.: Visualization methods for statistical analysis of microarray clusters. BMC Bioinformatics 6 (2005)
10. Irizarry, R.A., Bolstad, B.M., Collin, F., Cope, L.M., Hobbs, B., Speed, T.P.: Summaries of affymetrix genechip probe level data. Nucleic Acids Res. 31(4) (February 2003)
11. Irizarry, R.A., Hobbs, B., Collin, F., Beazer-Barclay, Y.D., Antonellis, K.J., Scherf, U., Speed, T.P.: Exploration, normalization, and summaries of high density oligonucleotide array probe level data. Biostat. 4(2), 249–264 (2003)
12. Rocke, D., Durbin, B.: A model for measurement error for gene expression arrays. J. Comput. Biol. 8(6), 557–569 (2001)
13. Rubinstein, B.I.P., McAuliffe, J., Cawley, S., Palaniswami, M., Ramamohanarao, K., Speed, T.P.: Machine learning in low-level microarray analysis. SIGKDD Explor. Newsl. 5(2), 130–139 (2003)
14. Saeed, A.I., Sharov, V., White, J., Li, J., Liang, W., Bhagabati, N., Braisted, J., Klapa, M., Currier, T., Thiagarajan, M., Sturn, A., Snuffin, M., Rezantsev, A., Popov, D., Ryltsov, A., Kostukovich, E., Borisovsky, I., Liu, Z., Vinsavich, A., Trush, V., Quackenbush, J.: TM4: a free, open-source system for microarray data management and analysis. Biotechniques 34(2), 374–378 (2003)
15. Sanges, R., Cordero, F., Calogero, R.A.: OneChannelGUI: a graphical interface to Bioconductor tools, designed for life scientists who are not familiar with R language. Bioinformatics 469 (September 2007)
16. Tu, Y., Stolovitzky, G., Klein, U.: Quantitative noise analysis for gene expression microarray experiments. Proceedings of the National Academy of Sciences 99(22), 14031–14036 (2002)

A Supervised Learning Technique and Its Applications to Computational Biology

Mario R. Guarracino[1], Altannar Chinchuluun[2], and Panos M. Pardalos[2,*]

[1] High Performance Computing and Networking Institute, National Research
Council, Italy
mario.guarracino@icar.cnr.it
[2] Center for Applied Optimization, Department of Industrial and Systems
Engineering, University of Florida, Gainesville, FL, 32611-6595 USA
{altannar,pardalos}@ufl.edu

Abstract. The problem of classifying data in spaces with thousands of
dimensions have recently been addressed in literature for its importance
in computational biology. An example of such applications is the analysis
of genomic and proteomic data. Among the most promising techniques
that classify such data in lower dimensional subspace, Top Scoring Pairs
has the advantage of finding a two-dimensional subspace with a simple decision rule. In the present paper we show how this technique can
take advantage from the utilization of incremental generalized eigenvalue
classifier to obtain higher classification accuracy with a small training set.

Keywords: Classification, Top Scoring Pair, generalized eigenvalue classification, gene expression data.

1 Introduction

Classification is one of the major data mining tasks which have many application in different fields including management science, finance, economics and
biomedicine. Classification techniques are widely used for data analysis and prediction in computational biology [7,15,17,18]. Among those, binary classification
methods based on statistical learning theory [25] are among the most effective
techniques. They split a set of points in two classes, with respect to a training set
of points whose membership is known for each class. These techniques have been
applied to gene expression microarray data for class prediction such as detecting
disease, identifying tumors or predicting treatment response. There are several
well known binary classification techniques available, including neural networks
[4], decision trees [14] and support vector machines [6]. Each of those has some
limits. Neural networks and decision trees provide very accurate models for the
training data, which do not easily generalize. Support vector machines (SVM),
on the other hand, provide more generalizable models, but the computational
complexity can be overwhelming for problems with large training datasets. Recently a new technique for the determination of decision rules suitable for gene

* Research of the third author is partially supported by NSF and Air Force grants.

F. Masulli, R. Tagliaferri, and G.M. Verkhivker (Eds.): CIBB 2008, LNBI 5488, pp. 275–283, 2009.

expression data analysis has been proposed [8]. This binary classification technique determines a separation of the two classes based on the over expression of one gene with respect to another. The gene pair is chosen to maximize the classification accuracy on the training set. This technique has been successfully used for the differentiation of gastrointestinal stromal tumors and leiomyosarcomas [21]. From a geometrical point of view, the obtained separation represents the bisectrix of the first and third quadrant of the space spanned by the two genes. It is clear that in such subspace it is possible to increase the accuracy using a more general separating line, which does not need to pass through the origin. To this extend, we decided to use the Regularized Generalized Eigenvalue Classifier (ReGEC) [10] to improve the classification accuracy within the subspace. Results on public domain microarray datasets show that the combination of both classifier outperforms each technique.

The present paper is organized as follows. In Section 2, we discuss some classification techniques including Top Scoring Pair and Generalized Eigenvalues classifications, and propose a classification technique based on these techniques. The results of the numerical experiments of the method on some existing biomedical data are reported in Section 3 and future work is discussed in Section 4.

2   The Algorithm

Let us consider a gene expression profile of Q genes $\{1, 2, ..., Q\}$. Suppose that the training set C consists of N profiles or arrays, $C = \{x_1, x_2, ..., x_N\}$, where the column vector $x_k = \{x_{1,k}, ..., x_{Q,k}\}$ represents the Q expression values for the k-th sample. Each sample x_n has a true class label A ($class_1$) or B ($class_2$). If we assume that expressions of a sample and its class label as random variables denoted by $\mathbf{X}$ and $\mathbf{Y}$, then the elements of the labeled training set represent independent and identically distributed samples from the underlying probability distribution of $(\mathbf{X}, \mathbf{Y})$.

2.1   Top Scoring Pair

Geman et al. [8] introduced a simple decision rule, called Top Scoring Pair (TSP), a binary classifier for gene expression data. The method focuses on detecting "marker gene pairs" (i, j) for which there is a significant difference in the probability of the event $x_{.,i} < x_{.,j}$ from class 1 to class 2. In other words, the probabilities $p_{i,j}(class_t) = P(x_{.,i} < x_{.,j} | y = class_t)$, $t = 1, 2$ are estimated by the relative frequencies of occurrences of $x_{.,i} < x_{.,j}$ within profiles and over samples. Let $\Delta_{ij} = |p_{ij}(1) - p_{ij}(2)|$ denote the "score" of gene pair (i, j). Then the method seeks a pair with the largest score denoted Δ_{max}. If multiple gene pairs achieve the same top score Δ_{max}, then a secondary score based on the rank differences in each sample in each class is used. For each top scoring pair (i, j), the "average rank difference"

$$\sigma_{i,j}(class_m) = \frac{\sum\limits_{n \in Class_t} (x_{i,n} - x_{j,n})}{|Class_t|}, \quad t = 1, 2$$

is computed. Here, $Class_t = \{n|y_n = class_t, \ n \in \{1, \ldots, N\}\}$ and $|Class_t|$ is its capacity.

The 'rank score' of gene pair (i, j) is then defined to be $\Sigma_{ij} = |\sigma_{ij}(class_1) - \sigma_{ij}(class_2)|$. Finally, the pair with the largest rank score from those pairs with the top score Δ_{max} is chosen.

Once the top-scoring pair (i, j) is selected, if $p_{ij}(class_1) > p_{ij}(class_2)$, the TSP classifier h_{TSP} for a new profile x_{new} is defined as follows:

$$y_{new} = h_{TSP}(x_{new}) = \begin{cases} class_1, & if \ x_{i,new} < x_{j,new}, \\ class_2, & Otherwise. \end{cases} \tag{1}$$

Of course, the decision rule is reversed if $p_{ij}(c_1) \leq p_{ij}(c_2)$.

Price et al. [21] used TSP classifier to differentiate gastrointestinal stromal tumors and leiomyosarcomas and obtained an accuracy of 99.3%. on the microarray samples and an estimated accuracy of 97.8% on future cases. In general, TSP method provides a very simple classifier that is independent of data normalization and avoids overfitting and most importantly an easy prediction phase.

Later, Tan et al. [24] proposed an extension of the TSP classifier called k-TSP classifier. The k-TSP classifier considers k disjoint Top Scoring Pairs (k-TSP). Learning and Prediction of k-TSP are similar to that of TSP classifier. It sorts gene pairs from the largest to the smallest according to their scores Δ_{ij}. If tie occurs, it breaks them using the secondary score Σ_{ij}. The classifier uses the first k top scoring disjoint gene pairs from the list. In the prediction phase, the k-TSP classifier employs an unweighted majority voting procedure to obtain the final prediction. Price et al. also reported that, for the data used in their paper [21], the TSP classifier outperformed k-TSP classifier.

2.2 Classification Based on Generalized Eigenvalues

Generalized eigenvalue classification methods [10,13] have been introduced as an alternative to SVM algorithms. They have the advantage of classifying data that are not linearly separable and a very simple formulation, that leads to a straight forward implementation. Nevertheless, their classification accuracy is comparable with those obtained by SVM, at a lower execution time. In the next, we briefly introduce these methods.

Mangasarian et al. [13] proposed to classify two sets of points A and B using two hyperplanes, each closest to one set of points, and furthest from the other. Let $x'w - \gamma = 0$ be a hyperplane in $\mathbb{R}^m$. In order to satisfy the previous condition, for points in A represented as rows, the hyperplanes can be obtained by solving the following optimization problem:

$$\min_{w, \gamma \neq 0} \frac{\|Aw - e\gamma\|^2}{\|Bw - e\gamma\|^2}, \tag{2}$$

where e is the unit vector of appropriate length.

The hyperplane for the B can be obtained by minimizing the inverse of the objective function in (3). Now, let

$$G = [A \quad -e]^T[A \quad -e], \quad H = [B \quad -e]^T[B \quad -e], \quad z = [w' \ \gamma]', \tag{3}$$

then equation (2), becomes:

$$\min_{z \in \mathbb{R}^m} \frac{z'Gz}{z'Hz}. \tag{4}$$

The expression in (4) is the Raleigh quotient of the generalized eigenvalue problem $Gx = \lambda Hx$. The stationary points are obtained at and only at the eigenvectors of (4), where the value of the objective function is given by the eigenvalues. When H is positive definite, the Raleigh quotient is bounded and it ranges over the interval determined by minimum and maximum eigenvalues [19]. H is positive definite under the assumption that the columns of $[B \quad -e]$ are linearly independent. The inverse of the objective function in (4) has the same eigenvectors and reciprocal eigenvalues. Let $z_{min} = [w_1' \quad \gamma_1]'$ and $z_{max} = [w_2' \quad \gamma_2]'$ be the eigenvectors related to the eigenvalues of smallest and largest modulo, respectively. Then $x'w_1 - \gamma_1 = 0$ is the closest hyperplane to the set of points in A and the furthest from those in B and $x'w_2 - \gamma_2 = 0$ is the closest hyperplane to the set of points in B and the furthest from those in A.

In order to regularize the problem, we can solve:

$$\min_{w,\gamma \neq 0} \frac{\|Aw - e\gamma\|^2 + \delta\|\tilde{B}w - e\gamma\|^2}{\|Bw - e\gamma\|^2 + \delta\|\tilde{A}w - e\gamma\|^2}, \tag{5}$$

where the $\tilde{A}$ and $\tilde{B}$ represent the diagonals of A and B, respectively.

Choosing the eigenvectors related to the new minimum and maximum eigenvalue, we obtain solutions that are close to the ones of the original problem, resulting in proximity planes P_1 and P_2.

A point x is classified using the distance

$$dist(x, P_i) = \frac{|x'w_i - \gamma_i|}{\|w_i\|}. \tag{6}$$

and the class of a point x is determined as

$$class(x) = argmin_{i=1,2}\{dist(x, P_i)\}. \tag{7}$$

We denote the generalized eigenvalue classification by ReGEC. This method has several advantages. Indeed, the computational kernels for eigenvalue problems can be found in many problem solving environments, such as Matlab, R and Weka, and are available for many programming languages in form of software libraries, for both multicore and multicomputer architectures. This means ReGEC implementation is very simple and consists only of few lines of code. Furthermore, the planes evaluated by ReGEC do not need to be parallel, as in the SVM formulation, which produces separations of the space that can be useful also for nonlinearly separable datasets. Finally, nonlinear transformations, generally called kernel functions, can still be used to transform the problem in higher dimensional data spaces and obtain higher separability.

Incremental Selection of Subsets. Incremental subset selection permits to construct a small set of points that retains the information of the entire training set and provides comparable accuracy results. A kernel built from a smaller subset is computationally more efficient in predicting new elements, compared to the one that uses the entire training set. Furthermore, a smaller set of points reduces the probability of over-fitting the problem. Finally, as new points are available, the cost to retrain the algorithm decreases if the influence of those new points on classification is only evaluated with respect to that subset, rather than the whole training set.

The algorithm takes an initial set of points $C \supset C_0 = A_0 \cup B_0$ and the entire training set C as input, where A_0 and B_0 are sets of points in C_0 that belong to the two classes A and B. We refer to C_0 as the *incremental subset*. Let $\Gamma_1 = C \backslash C_0$ be the initial set of points that can be included in the incremental subset. ReGEC classifies all of the points in the training set C using the kernel from C_0. Let P_{A_0} and P_{B_0} be the hyperplanes found by ReGEC, R_0 be the classification accuracy and M_0 be the points that are misclassified. Then, among the points in $\Gamma_1 \cap M_0$ the point that is farthest from its respective hyperplane is selected, i.e.

$$x_1 = x_i : \max_{x \in \{\Gamma_1 \cap M_0\}} \left\{ dist(x, P_{class(x)}) \right\}, \tag{8}$$

where $class(x)$ returns A or B depending on the class of x. This point is the candidate point to be included in the incremental subset. This choice is based on the idea that a missclassified point very far from its plane is either needed in the classification subset because its information can improve accuracy, or it is an outlier. We update the incremental set as $C_1 = C_0 \cup \{x_1\}$. Then, we classify the entire training set C using the points in C_1 to build the kernel. Let the classification accuracy be R_1. If $R_1 > R_0$ then we keep the new subset; otherwise we reject the new point, that is $C_1 = C_0$. In both cases $\Gamma_2 = \Gamma_1 \backslash \{x_1\}$. The algorithm repeats until $|\Gamma_k \cap M_{k-1}| = 0$ at some k^{th} iteration. The s initial points are the training points closest the centroids determined by a simple k-*means* algorithm applied to each class. In [5], it has been proven that the k-mean based selection criteria gives the best performance in term of stability and accuracy, with respect to random selection of initial points. We denote this algorithm by IReGEC.

2.3 TSP-IReGEC Algorithm

We recall that the computational complexity of the eigenvalue problem increases with the number of points and features in the training set. Therefore, an alternative approach is needed to construct a small subset of points and features of the training set.

We note that TSP finds two genes and classifies points through bisection of the first quadrant in the two dimensional space as shown in the Figure 1. However, as shown in Figure 1 (a), the bisection of the first quadrant may not be able to classify all the points. Fortunately, there are other lines that can perfectly classify the points (see Figure 1 (b)). Our idea is to find one of those lines by using generalized eigenvalue classification method.

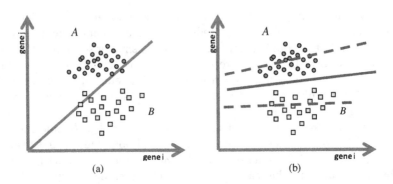

Fig. 1. The training set in two dimensional space corresponding to top scoring genes by (a) TSP classification (b) IReGEC classification

The proposed feature selection algorithm is based on TSP and the incremental generalized eigenvalue classification IReGEC. TSP method filters a pair of features providing the higher Δ_{ij} score, as described in Section 2.1. After selecting a the top scoring pair, the IReGEC is applied in order to define the incremental subset. Solving IReGEC with few features is very fast and therefore overall computational is improved. The algorithm can be summarized in Algorithm 1.

Algorithm 1. TSP-IReGEC(C, C_0)

1: *Select with TSP the pair of features*
2: $\Gamma_1 = C \setminus C_0$
3: $\{R_0, M_0\} = Classify(C, C_0)$
4: $k = 1$
5: **while** $|\Gamma_k \cap M_{k-1}| > 0$ **do**
6:     $x_k = x : \max_{x \in \{M_{k-1} \cap \Gamma_k\}} \{dist(x, P_{class(x)})\}$
7:     $\{R_k, M_k\} = Classify(C, \{C_{k-1} \cup \{x_k\}\})$
8:     **if** $R_k >= R_{k-1}$ **then**
9:         $C_k = C_{k-1} \cup \{x_k\}$
10:     **end if**
11:     $\Gamma_k = \Gamma_{k-1} \setminus \{x_k\}$
12:     $k = k + 1$
13: **end while**

We recall that, TSP provides clinically relevant two genes that has the potential to aid in the development of efficient cancer diagnosis and in the selection of appropriate therapies. On the other hand, IReGEC finds a small set of points that retains the information of the entire training set and provides comparable accuracy results to other classification methods. Therefore, we can logically claim that TSP-IReGEC provides the most clinically important two genes along with a set of important samples that represent the training set.

3   Accuracy Results

TPS-IReGEC has been implemented with Matlab 7.3.0. Results are calculated using an Intel Xeon CPU 3.20GHz, 6GB RAM running Red Hat Enterprise Linux WS release 3. Matlab function *eig* for the solution of the generalized eigenvalue problem is used as computational kernel of ReGEC.

TSP-IReGEC is tested on publicly available benchmark data sets. Data has been obtained from k-TSP Program Download Page [24]. Dataset characteristics and references are reported in Table 1. Results regarding its performance in terms of classification accuracy are presented. In Table 2 accuracy results are reported for the datasets of Table 1 for null classification and various methods. They represent the accuracy for Leave One Out Cross Validation (LOOCV). Results for TSP are taken from [24] and used for comparison. Results for IReGEC and TSP-IReGEC have been evaluated automatically choosing the value of the regularization parameter during the LOOCV among $\{1.e-2, 1.e-3, 1.e-4, 1.e-5\}$. We have used the same pruning algorithm to speed up LOOCV as described in the accompanying material of [24].

For all datasets, the number of initial points is fixed to $k = 2$, except for DLBCL ($k = 3$). We note that TSP-IReGEC is more accurate than TSP in three cases out of four. On average, TSP is less accurate than TSP-IReGEC. Further experimental results with more classification methods and data sets need to be done.

Table 1. Dataset characteristics

Dataset	Platform	No. of genes (P)	No. of samples (N)	Reference
Leukemia	Affy	7129	25 (AML) 47 (ALL)	(Golub et al., 1999)
CNS	Affy	7129	25 (C) 9 (D)	(Pomeroy et al., 2002)
DLBCL	Affy	7129	58 (D) 19 (F)	(Shipp et al., 2002)
Prostate2	Affy	12 625	38 (T) 50 (N)	(Stuart et al., 2004)

Table 2. Accuracy results

Method / Dataset'	Leukemia	CNS	DLBCL	Prostate2	Avarage
NULL	0.6527	0.7352	0.7532	0.5681	0.6728
IReGEC	1.0000	0.7941	0.8831	0.6591	0.8341
TSP-IReGEC	0.9861	0.8529	0.9691	0.7046	0.8782
TSP	0.9380	0.7790	0.9810	0.6760	0.8435

4   Conclusions and Future Work

In this paper, we proposed to use Top Scoring Pair algorithm in conjunction with IReGEC. The resulting technique is on average, and in the majority of the cases, more accurate than both TSP and IReGEC. We will investigate on how to generalize the TSP to use a more general line, than the bisector of first and third quadrant, to classify points in the subspace spanned by each pair, using only the points selected by the incremental procedure. Finally we will research if it is possible to lower the computational complexity of the pair search.

References

1. Alon, A., Barkai, N., Notterman, D.A., Gish, K., Ybarra, S., Mack, D., Levine, A.J.: Broad Patterns of Gene Expression Revealed by Clustering Analysis of Tumor and Normal Colon Tissues Probed by Oligonucleotide Arrays. Proc. Natl. Acad. Sci. USA 96, 6745–6750 (1999)
2. Bennet, K., Campbell, C.: Support Vector Machines: Hype or Hallelujah? SIGKDD Explorations 2(2), 1–13 (2000)
3. Bennett, K., Mangasarian, O.: Robust Linear Programming Discrimination of Two Linearly Inseparable Sets. Optimization Methods and Software 1, 23–34 (1992)
4. Bishop, C.M.: Neural Networks for Pattern Classification. Oxford Press (1995)
5. Cifarelli, C., Guarracino, M.R., Seref, O., Cuciniello, S., Pardalos, P.M.: Incremental Classification with Generalized Eigenvalues. Journal of Classification 24(2), 205–219 (2007)
6. Cortes, C., Vapnik, V.: Support Vector Machines. Machine Learning 20, 273–279 (1995)
7. Floudas, C., Pardalos, P.M. (eds.): Optimization in Computational Chemistry and Molecular Biology. Kluwer Academic Publishers, Dordrecht (2000)
8. Geman, D., d'Avignon, C., Naiman, D.Q., Winslow, R.L.: Classifying Gene Expression Profiles from Pairwise mRNA Comparisons. Statistical Applications in Genetics and Molecular Biology 3, Article 19 (2004)
9. Golub, T.R., Slonim, D.K., Tamayo, P., Huard, C., Gaasenbeek, M., Mesirov, J.P., Coller, H., Loh, M., Downing, J.R., Caligiuri, M.A., Bloomfield, C.D., Lander, E.S.: Molecular Classification of Cancer: Class Discovery and Class Prediction by Gene Expression Monitoring. Science 286, 531–537 (1999)
10. Guarracino, M.R., Cifarelli, C., Seref, O., Pardalos, P.M.: A Classification Algorithm based on Generalized Eigenvalue Problems. Optimization Methods and Software 22(1), 73–81 (2007)
11. Hedenfalk, I., Duggan, D., Chen, Y., Radmacher, M., Simon, R., Meltzer, P., Gusterson, B., Esteller, M., Raffeld, M., Yakhini, Z., Ben-Dor, A., Dougherty, E., Kononen, J., Bubendorf, L., Fehrle, W., Pttalunga, S., Gruvberger, S., Loman, N., Johannsson, O., Olsson, H., Wilfond, B., Sauter, G., Kallioniemi, O.P., Borg, A., Trent, J.: Gene-Expression Profiles in Hereditary Breast Cancer. The New England Journal of Medicine 344, 539–548 (2001)
12. Iizuka, N., Oka, M., Yamada Okabe, H., Nishida, M., Maeda, Y., Mori, N., Takao, T., Tamesa, T., Tangoku, A., Tabuchi, H., Hamada, K., Nakayama, H., Ishitsuka, H., Miyamoto, T., Hirabayashi, A., Uchimura, S., Hamamoto, Y.: Oligonucleotide Microarray for Prediction of Early Intrahepatic Recurrence of Hepatocellular Carcinoma after Curative Resection. The Lancet 361, 923–929 (2003)
13. Mangasarian, O.L., Wild, E.W.: Multisurface Proximal Support Vector Classification via Generalized Eigenvalues. Technical Report 04-03, Data Mining Institute (September 2004)
14. Mitchell, T.M.: Machine Learning. McGraw Hill, New York (1997)
15. Mondaini, R.P., Pardalos, P.M. (eds.): Mathematical Modelling of Biosystems. Springer, Heidelberg (2008)
16. Nutt, C.L., Mani, D.R., Betensky, R.A., Tamayo, P., Cairncross, J.G., Ladd, C., Pohl, U., Hartmann, C., McLaughlin, M.E., Batchelor, T.T., Black, P.M., von Deimling, A., Pomeroy, S.L., Golub, T.R., Louis, D.N.: Gene Expression-based Classification of Malignant Gliomas Correlates better with Survival than Histological Classification. Cancer Res. 63(7), 1602–1607 (2003)

17. Pardalos, P.M., Principe, J. (eds.): Biocomputing. Kluwer Academic Publishers, Dordrecht (2002)
18. Pardalos, P.M., Boginski, V., Vazakopoulos, A. (eds.): Data Mining in Biomedicine. Springer, Heidelberg (2007)
19. Parlett, B.N.: The Symmetric Eigenvalue Problem. SIAM, Philadelphia (1998)
20. Pochet, N.L.M.M., Janssens, F.A.L., De Smet, F., Marchal, K., Suykens, J.A.K., De Moor, B.L.R.: Macbeth: A Microarray Classification Benchmarking Tool. Bioinformatics 21(14), 3185–3186 (2005)
21. Price, N.D., Trent, J., El-Naggar, A.K., Cogdell, D., Taylor, E., Hunt, K.K., Pollock, R.E., Hood, L., Shmulevich, I., Zhang, W.: Highly Accurate Two-Gene Classifier for Differentiating Gastrointestinal Stromal Tumors and Leiomyosarcomas. Proceedings of the National Academy of Sciences of the United States of America 104(9), 3414–3419 (2007)
22. Schölkopf, B., Smola, A.-J.: Learning With Kernels: Support Vector Machines, Regularization, Optimization, and Beyond. MIT Press, Cambridge (2002)
23. Singh, D., Febbo, P.G., Ross, K., Jackson, D.G., Manola, J., Ladd, C., Tamayo, P., Renshaw, A.A., D'Amico, A.V., Richie, J.P., Lander, E.S., Loda, M., Kantoff, P.W., Golub, T.R., Sellers, W.R.: Gene Expression Correlates of Clinical Prostate Cancer Behavior. Cancer Cell 1(2), 203–209 (2002)
24. Tan, A.C., Naiman, D.Q., Xu, L., Winslow, R.L., Geman, D.: Simple Decision Rules for Classifying Human Cancers from Gene Expression Profiles. Bioinformatics 21, 3896–3904 (2005)
25. Vapnik, V.: The Nature of Statistical Learning Theory. Springer, Heidelberg (1999)
26. van't Veer, L.J., Dai, H., Van De Vijver, M.J., He, T.D., Hart, A.A.M., Mao, M., Peterse, H.L., Van Der Kooy, K., Marton, M.J., Witteveen, A.T., Schreiber, G.J., Kerkhoven, R.M., Roberts, C., Linsley, P.S., Bernards, R., Friend, S.H.: Gene Expression Profiling Predicts Clinical Outcome of Breast Cancer. Nature 415, 536 (2002)

A Visualization ToolKit Based Application for Representing Macromolecular Surfaces

Paolo Cozzi, Ivan Merelli, and Luciano Milanesi

Institute for Biomedical Technologies – National Research Council,
Via F.lli Cervi 93, 20090 Segrate (Milano), Italy
{paolo.cozzi,ivan.merelli,luciano.milanesi}@itb.cnr.it

Abstract. Macromolecular structural analysis has been used for a long time, in particular for proteins, with the aim of correlating three dimensional structures with possible functions and interactions. An innovative approach is to tackle these aspects by working on macromolecular surfaces instead of using amino acid or atomic configurations directly, because protein interaction is mainly driven by the macromolecular external characteristics. However, it is difficult to work with surfaces because applications for molecular visualization lack appropriate functions to manipulate these data structures properly. In this paper we describe a visualization pipeline to represent molecular surfaces in three-dimensional space. The pipeline relies on the Visualization Toolkit libraries, and provides a high level functions and easy user interface. This software has been developed in the context of a surface analysis project, but its classes and methods can be used and integrated in a wide range of applications.

Keywords: Molecular Surfaces, Molecular visualization, Visualization Toolkit.

1 Introduction

One of the main targets of proteomics is to understand protein functionalities by establishing homologies with well-annotated proteins. Considering that similar proteins can have the same function deriving from the same ancestor, a series of methods relying on finding homology by sequence similarity has been developed. However, when the similarity between the sequences is very low, the analysis of protein structures can be more significant [1], and in particular a method based on the identification of molecular surfaces similarity can provide non trivial results. For example, cofactors, substrates and regulatory elements tend to bind in clefts on the surface. For such cases, the analysis of clefts by comparison with proteins of known function could provide fair indicators of what the protein might do.

Another difficult task of proteomics is to understand protein interactions, since molecular mechanisms involve systemic effects produced by interactions between proteins with different functions. In this context the study of molecular

F. Masulli, R. Tagliaferri, and G.M. Verkhivker (Eds.): CIBB 2008, LNBI 5488, pp. 284–292, 2009.

surfaces can be interesting because molecular surfaces are involved in molecular recognition, and their description can shed light on how proteins interact and how molecular processes take place. Camacho argues that from the physical point of view macromolecular interactions occur in two different stages, in particular when proteins are involved [2]: there is a first stage of molecular recognition, where molecules diffuse near each other until the interface patches come sufficiently close, then the binding stage begins, when high affinity atomic interactions are formed by modification of the side-chain and backbone conformations of the molecules. For this reason we think that molecular surfaces are important to better understand interaction between proteins by finding complementarities, and for functional analysis by finding similarity between surfaces of different proteins.

In a previous work [3], we proposed a method of computer vision oriented to robotics to compare molecular surfaces by describing them using a set of images of local description. This representation enables the correlation of the surface simply by analyzing the similarity between the images from the two macromolecules. More precisely, the core of the matching procedure consist in establishing point-to-point correspondences between meshes by correlating images which describe the local topology of surfaces. The surfaces can then be aligned by minimizing the distances between points from which the corresponding images were defined. In this context, a method able to visualize results of alignments is very useful both in testing the application and to verify the surfaces similarity. To accomplish this task, a reliable surface representation is needed: vertices and polygons have to be properly visualized in the three-dimensional space and the observer should look at the scene from a significant point of view to understand how surfaces are similar.

2 Background

In our approach, the search for surface similarity can be done by representing two given surfaces using triangular meshes and trying to superimpose them relying on the correspondences found by comparison between surfaces. This comparison can not be performed trivially by means of the Euclidean distance assessment between meshes, since surfaces could be similar but not identical. Furthermore, considering two mesh instances of the same object, points could be positioned in different places, and this elucidates how the matching procedure can be difficult for non identical surfaces. Moreover, it is necessary to make the algorithm independent from the reference system to allow fast comparisons for any mutual orientation.

A suitable solution for this task has been identified in a method of computer vision oriented to robotics applications, which describes shapes using object-centered systems of reference defined by coordinates of each vertex and its related normal. Through this local system of coordinates it is possible to obtain a set of images for each macromolecule, which represents the surface exhaustively. By using these images, the algorithm is able to establish point-to-point correspondences that are filtered and clustered to achieve complete surface matches.

Therefore, an application for representing surfaces and their correlation in a graphic way can be very helpful. There is a lot of software available for molecular visualization, but none that can easily handle surfaces data or correlations between vertices. Clearly, rather than develop a new software for molecular visualization a better approach is to represent surface data in a well-established framework, which should provide enough flexibility to highlight the corresponding vertices on surfaces.

Thus, we thought that ray tracing techniques can represent molecular surfaces in a very informative way. Ray tracing techniques simulate the interaction of light with objects by following the path of each light ray from the viewer's eyes into the three dimensional space, and then generates an image like a photo in a real world. When a ray intersects an object, it has to be determined if that point is being lit by a light source, typically by tracing a ray from the point of intersection towards the light source. On the other hand, ray tracing is computationally very intensive and it is difficult to define a good point of view focused on surface's similarity. Moreover, it is difficult for the observer to have a good idea of the three dimensional object represented by observing a static image.

3   The Visualization Pipeline

These problems can be solved by allowing the observer to interact with the scene, changing the point of view or rotating the object represented. To tackle this issue, we decided to develop an application on the top of the Visualization ToolKit (VTK), an open source system for 3D computer graphics and scientific data visualization [4]. The Visualization Toolkit is a collection of C++ class libraries with different interface layers, which allow users to exploit the fast function libraries implemented in C++ with the flexibility and the extensibility of a scripting language like python. Simply stated, the visualization implies that data are properly transformed and represented. Transformation is the process of converting data from its original form into graphics primitives and computer images. Representation includes both the internal data structures used to store the data and the graphics primitives used to display data. The process of converting the raw data in structures and their transformation in graphics primitives is usually referred as the visualization pipeline (Fig. 1).

Surface data such as vertices and polygons are stored in VTK's special objects, while local correspondences between similar surfaces can be recorded to make textures on the surfaces. Afterwards objects have to be transformed in graphics primitives (the vtkPolyDataMapper) and finally an actor (vtkActor) can be generated. Several vtkActors can be passed to the render window, where lights and shadow are calculated for each object in the scene. In the last step of the visualization pipeline, the render window object is provided to the user, enabling for him the possibility to interact with the scene and to modify the point of view of the world represented [5].

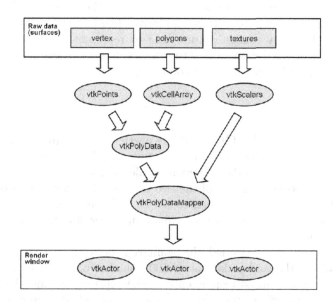

Fig. 1. In visualization Pipeline raw data such vertex, polygons and textures are converted in VTK data structures to obtain a vtkActor, the key element in VTK visualization

4   Implementation

On the top of VTK, we implemented a system which creates classes and specific methods to read surface data and to manage object in a transparent way. All process needed for data visualization are done automatically by a python interface layer, which implements a few classes and methods for converting raw data in a vtkActor and changing attributes such as colours or opacity. This solution simplifies the access to the VTK complex structures, providing a more flexible and userfriendly environment for dealing with surface data. This simplified collection of methods can be called directly by the python terminal or by a user developed script.

The system implemented provides a specific class to deal with each particular type of data involved in the surface analysis. Surface data are recorded in the OFF file format, a specific text file which defines vertices and connections which compose the polygonal mesh. Correlations between surfaces are stored in text files as vertices coordinates and values of corresponding correlations. The appropriate class for input data is automatically instantiated according to the provided file. Then all the process for reading and processing the file, instantiating the VTK object, and filling them with data are done automatically. A more generic class inherits methods from surface's and correlation's classes, and provides some additional methods to interact with the scene represented, like changing the colour and the opacity of the object or highlighting and labelling the similar vertices directly on surfaces to improve the visibility of the scene and to better understand the surface similarity. The last class allows the user to instantiate the VTK render window and to represent the surface data as VTK actors, providing the possibility to interact with the scene represented by enlarging a detail or by changing the point of view.

5   Results

Methods implemented in this framework are particularly suitable for working on data generated by a correlation processes between surfaces. In particular, when two surfaces are matched, a data correlation file is produced which reports the correspondences between vertices of the two surfaces. This file can be processed by the system and the matching vertices can be highlighted on the surface. This type of representation facilitates the evaluation of results while searching for surface similarities.

For example, we have studied the similarity between the surface of the lysozyme (PDB:1UIB) and a dataset of 1009 patches obtained by calculating the molecular surfaces in correspondence of the prosite domains identified by running interproscan [6] on the non redundant chain set of protein structures from the Protein Data Bank [7] that didn't include the query protein. The five highest ranking solutions found are reported in Table 1. The first two solutions seem to have a good alignment, but inspecting results with the system implemented a different conclusion can be drawn. In detail, the first solution found is the prosite domain PS00128 obtained from a homologous protein (PDB:3LZT), and that domain identifies lactalbumine proteins and lysozime enzymes. Once similarity between surfaces has been computed, the user can evaluate the results of the alignment by parsing the data and building a vtkActor calling only a few methods of the system implemented, and all the operations needed for the visualization are done automatically.

In Fig. 2 a detail of two matched surfaces from the first result are represented: the correlated vertices are highlighted in red and indexed on the image. The result of the superimposition of two surfaces is reported in Fig. 3. In this figure

Table 1. Top ranking solutions for the similarity analysis of the protein 1UIB (lysozime), sorted by the percentage of patch vertices closer to 1 Å from the protein vertices (Overlap column). For each solution the prosite identifier for the domain, the prosite description, the RMSD distance between aligned vertices, the mean value of correlation between vertex's images and the number of vertices found as similar is reported.

prosite Name	Description	RMSD	Mean	Vertices	Overlap
PS00128	LACTALBUMINE_LYSOZYME	0.526	0.946	100	71.669
PS00206	TRANSFERRIN_2	1.350	0.917	10	55.483
PS00128	LACTALBUMINE_LYSOZYME	0.540	0.941	7	48.384
PS00790	RECEPTOR_TYR_KIN_V_1	1.000	0.942	7	46.059
PS00678	WD_REPEATS	1.668	0.951	10	44.021

Fig. 2. In this figure the correlations between vertex on surfaces of the protein 1UIB (on the left) and on a patch obtained by extracting the prosite domain PS00128 on the protein 3LZT (on the right) are presented. It is interesting to note that 1UIB also presents the same functional domain and that it was correctly detected.

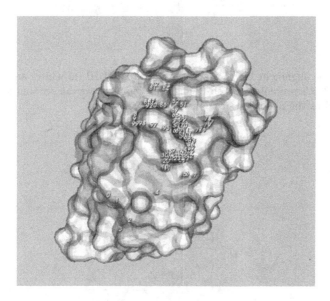

Fig. 3. Here the surface alignment between surface of protein 1UIB (in blue) and the patch obtained by the prosite domain PS00128(in white) is presented. Changing the opacity makes the surface transparent and the image becomes more informative.

we can see how the most similar surface found is well aligned to the query protein, and this proves that the solution is good. In Fig. 4 we can see that the second solution is not as good as the first solution found.

Another example of the application of the Visualization Toolkit for visualizing molecular surfaces concerns an analysis of complementarities between two proteins. In this case, surfaces for protein thermitase (2TEC_E) and its inhibitor englin C (2TEC_I) have been calculated and compared searching for complementarities.

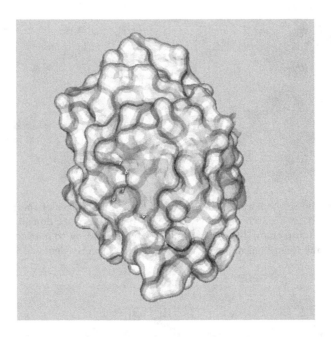

Fig. 4. Here the alignment between surface of protein 1UIB (in white) and the prosite domain PS00206 is presented. By observing this alignment we can see that this solution is not as good as the first solution found (PS00128).

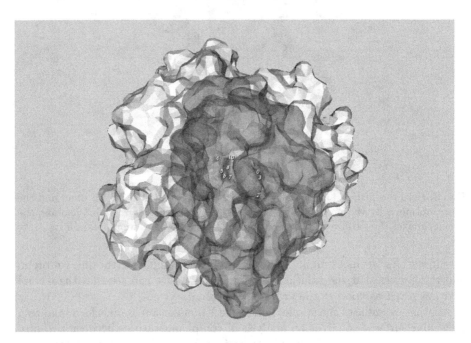

Fig. 5. In this figure are presented the results of the alignment for the protein 2TEC_E and its inhibitor 2TEC_I, with the vertices found as similar in evidence

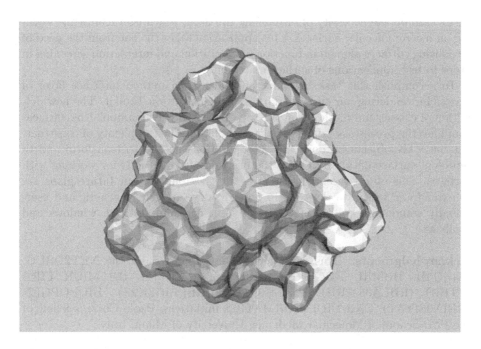

Fig. 6. In this figure are presented the aligned surface of the inhibitor (in blue) and the crystal surface of the inhibitor (in white)

The inhibitor was aligned on the enzyme's surface by similarities found on the surfaces. The results of the alignment were compared by measuring the root mean square deviation between the aligned surfaces and the surface computed from the crystallographic complex structure. The results of the best alignment are presented in Fig. 5. In Fig. 6 we can see how the inhibitor is well aligned to the crystal form, and how this kind of presentation is useful to evaluate the results of the alignment.

In both cases, we can evaluate the results of the search of similarity by inspecting the aligned surfaces with this extension of Visualization Toolkit. This kind of visualization has allowed the graphical rendering of complex data, such as surface data and correlated vertices. The possibility to see the surfaces aligned allows the user to evaluate the quality of the similarity found by inspecting the alignment, and to see if vertices tagged as correlated belong to complementary or similar surfaces in order to discriminate bad solutions from the good ones. The user can interact with the scene represented by varying the opacity of surfaces in order to make them transparent to improve the visibility of the surfaces and obtaining a more informative representation of the scene.

6   Discussion and Conclusions

Searching for surface similarities can help to define a protein function by homology, or to identify complementarities in protein-protein interactions. In this work we focused on the necessity of visualizing the data, which is crucial while working on macromolecular surfaces. This application takes the cue from the need of visualizing surfaces aligned in functional annotation and interaction screening in order to test applications of surface matching.

To accomplish this task, we have implemented a python interface layer of classes for rendering our surfaces data with Visualization Toolkit. The new definition of classes and methods on the top of the VTK for manipulating surfaces simplifies the processes of data visualization, hiding the complexity of constructing the VTK objects by doing all the operations needed to transform and to represent surface data, and giving the user only few methods for working with surfaces. This object oriented system is very flexible and our future plans are to extend and integrate the code in order to develop a more generic and user-friendly standalone application, where a user can interact with windows and toolbars.

Acknowledgments. This work have been supported by the NET2DRUG, EGEE-III, BBMRI, EDGE European project and by the MIUR FIRB LITBIO (RBLA0332RH), ITALBIONET (RBPR05ZK2Z), BIOPOPGEN (RBIN064YAT), CNR-BIOINFORMATICS initiatives. Paolo Cozzi is fellow of the PhD School of Molecular Medicine, University of Milan, Italy.

References

1. Watson, J.D., Laskowski, R.A., Thornton, J.M.: Predicting protein function from sequence and structural data. Curr. Opin. Struct. Biol. 15, 275–284 (2005)
2. Camacho, C.J., Vajda, S.: Protein-protein association kinetics and protein docking. Curr. Opin. Struct. Biol. 12, 36–40 (2002)
3. Merelli, I., Cozzi, P., Milanesi, L., D'Agostino, D., Clematis, A.: Images based system for surface matching in macromulecular screening. In: 2008 IEEE International Conference on Bioinformatics and Biomedicine (in press, 2008)
4. Schroeder, W.J., Avila, L.S., Hoffman, W.: Visualizing with VTK: a tutorial. Computer Graphics and Application 20, 20–27 (2000)
5. Schroeder, W.J., Martin, K., Lorensen, B.: The Visualization Toolkit - An object Oriented Approach to 3D Graphics. Kitware (2004)
6. Quevillon, E., Silventoinen, V., Pillai, S., Harte, N., Mulder, N., Apweiler, R., Lopez, R.: InterProScan: protein domains identifier. Nucleic Acids Res. 33(Web Server issue), W116–W120 (2005)
7. Berman, H.M., Westbrook, J., Feng, Z., Gilliland, G., Bhat, T.N., Weissig, H., Shindyalov, I.N., Bourne, P.E.: The Protein Data Bank. Nucleic Acids Res. 28(1), 235–242 (2000)

Author Index